Statistical Physics of Particles

Statistical physics has its origins in attempts to describe the thermal properties of matter in terms of its constituent particles, and has played a fundamental role in the development of quantum mechanics. It describes how new behavior emerges from interactions of many degrees of freedom, and as such has found applications outside physics in engineering, social sciences, and, increasingly, in biological sciences. This textbook introduces the central concepts and tools of statistical physics. It includes a chapter on probability and related issues such as the central limit theorem and information theory, not usually covered in existing texts. The book also covers interacting particles, and includes an extensive description of the van der Waals equation and its derivation by mean-field approximation. A companion volume, *Statistical Physics of Fields*, discusses non-mean field aspects of scaling and critical phenomena, through the perspective of renormalization group.

Based on lectures for a course in statistical mechanics taught by Professor Kardar at Massachusetts Institute of Technology (MIT), this textbook contains an integrated set of problems, with solutions to selected problems at the end of the book. It will be invaluable for graduate and advanced undergraduate courses in statistical physics. Additional solutions are available to lecturers on a password protected website at www.cambridge.org/9780521873420.

MEHRAN KARDAR is Professor of Physics at MIT, where he has taught and researched in the field of statistical physics for the past 20 years. He received his B.A. at Cambridge, and gained his Ph.D. at MIT. Professor Kardar has held research and visiting positions as a junior Fellow at Harvard, a Guggenheim Fellow at Oxford, UCSB, and at Berkeley as a Miller Fellow.

In this much-needed modern text, Kardar presents a remarkably clear view of statistical mechanics as a whole, revealing the relationships between different parts of this diverse subject. In two volumes, the classical beginnings of thermodynamics are connected smoothly to a thoroughly modern view of fluctuation effects, stochastic dynamics, and renormalization and scaling theory. Students will appreciate the precision and clarity in which difficult concepts are presented in generality and by example. I particularly like the wealth of interesting and instructive problems inspired by diverse phenomena throughout physics (and beyond!), which illustrate the power and broad applicability of statistical mechanics.

Statistical Physics of Particles includes a concise introduction to the mathematics of probability for physicists, an essential prerequisite to a true understanding of statistical mechanics, but which is unfortunately missing from most statistical mechanics texts. The old subject of kinetic theory of gases is given an updated treatment which emphasizes the connections to hydrodynamics.

As a graduate student at Harvard, I was one of many students making the trip to MIT from across the Boston area to attend Kardar's advanced statistical mechanics class. Finally, in *Statistical Physics of Fields* Kardar makes his fantastic course available to the physics community as a whole! The book provides an intuitive yet rigorous introduction to field-theoretic and related methods in statistical physics. The treatment of renormalization group is the best and most physical I've seen, and is extended to cover the often-neglected (or not properly explained!) but beautiful problems involving topological defects in two dimensions. The diversity of lattice models and techniques are also well-illustrated and complement these continuum approaches. The final two chapters provide revealing demonstrations of the applicability of renormalization and fluctuation concepts beyond equilibrium, one of the frontier areas of statistical mechanics.
Leon Balents, Department of Physics, University of California, Santa Barbara

Statistical Physics of Particles is the welcome result of an innovative and popular graduate course Kardar has been teaching at MIT for almost twenty years. It is a masterful account of the essentials of a subject which played a vital role in the development of twentieth century physics, not only surviving, but enriching the development of quantum mechanics. Its importance to science in the future can only increase with the rise of subjects such as quantitative biology.

Statistical Physics of Fields builds on the foundation laid by the Statistical Physics of Particles, with an account of the revolutionary developments of the past 35 years, many of which were facilitated by renormalization group ideas. Much of the subject matter is inspired by problems in condensed matter physics, with a number of pioneering contributions originally due to Kardar himself. This lucid exposition should be of particular interest to theorists with backgrounds in field theory and statistical mechanics.
David R Nelson, Arthur K Solomon Professor of Biophysics, Harvard University

If Landau and Lifshitz were to prepare a new edition of their classic Statistical Physics text they might produce a book not unlike this gem by Mehran Kardar. Indeed, Kardar is an extremely rare scientist, being both brilliant in formalism and an astoundingly careful and thorough teacher. He demonstrates both aspects of his range of talents in this pair of books, which belong on the bookshelf of every serious student of theoretical statistical physics.

Kardar does a particularly thorough job of explaining the subtleties of theoretical topics too new to have been included even in Landau and Lifshitz's most recent Third Edition (1980), such as directed paths in random media and the dynamics of growing surfaces, which are not in any text to my knowledge. He also provides careful discussion of topics that do appear in most modern texts on theoretical statistical physics, such as scaling and renormalization group.
H Eugene Stanley, Director, Center for Polymer Studies, Boston University

This is one of the most valuable textbooks I have seen in a long time. Written by a leader in the field, it provides a crystal clear, elegant and comprehensive coverage of the field of statistical physics. I'm sure this book will become "the" reference for the next generation of researchers, students and practitioners in statistical physics. I wish I had this book when I was a student but I will have the privilege to rely on it for my teaching.
Alessandro Vespignani, Center for Biocomplexity, Indiana University

Statistical Physics of Particles

Mehran Kardar

Department of Physics
Massachusetts Institute of Technology

CAMBRIDGE
UNIVERSITY PRESS

CAMBRIDGE
UNIVERSITY PRESS

Shaftesbury Road, Cambridge CB2 8EA, United Kingdom

One Liberty Plaza, 20th Floor, New York, NY 10006, USA

477 Williamstown Road, Port Melbourne, VIC 3207, Australia

314–321, 3rd Floor, Plot 3, Splendor Forum, Jasola District Centre, New Delhi – 110025, India

103 Penang Road, #05–06/07, Visioncrest Commercial, Singapore 238467

Cambridge University Press is part of Cambridge University Press & Assessment, a department of the University of Cambridge.

We share the University's mission to contribute to society through the pursuit of education, learning and research at the highest international levels of excellence.

www.cambridge.org
Information on this title: www.cambridge.org/9780521873420

First published 2007 (version 17, August 2022)

A catalogue record for this publication is available from the British Library

ISBN   978-0-521-87342-0   Hardback

Contents

Preface

Historically, the discipline of *statistical physics* originated in attempts to describe thermal properties of matter in terms of its constituent particles, but also played a fundamental role in the development of quantum mechanics. More generally, the formalism describes how new behavior emerges from interactions of many degrees of freedom, and as such has found applications in engineering, social sciences, and increasingly in biological sciences. This book introduces the central concepts and tools of this subject, and guides the reader to their applications through an integrated set of problems and solutions.

The material covered is directly based on my lectures for the first semester of an MIT graduate course on statistical mechanics, which I have been teaching on and off since 1988. (The material pertaining to the second semester is presented in a companion volume.) While the primary audience is physics graduate students in their first semester, the course has typically also attracted enterprising undergraduates. as well as students from a range of science and engineering departments. While the material is reasonably standard for books on statistical physics, students taking the course have found my exposition more useful, and have strongly encouraged me to publish this material. Aspects that make this book somewhat distinct are the chapters on probability and interacting particles. Probability is an integral part of statistical physics, which is not sufficiently emphasized in most textbooks. Devoting an entire chapter to this topic (and related issues such as the central limit theorem and information theory) provides valuable tools to the reader. In the context of interacting particles, I provide an extensive description of the van der Waals equation, including its derivation by mean-field approximation.

An essential part of learning the material is doing problems; an interesting selection of problems (and solutions) has been designed and integrated into the text. Following each chapter there are two sets of problems: solutions to the first set are included at the end of the book, and are intended to introduce additional topics and to reinforce technical tools. Pursuing these problems should also prove useful for students studying for qualifying exams. There are no solutions provided for a second set of problems, which can be used in assignments.

I am most grateful to my many former students for their help in formulating the material, problems, and solutions, typesetting the text and figures, and pointing out various typos and errors. The support of the National Science Foundation through research grants is also acknowledged.

1
Thermodynamics

1.1 Introduction

Thermodynamics *is a phenomenological description of properties of macroscopic systems in thermal equilibrium.*

Imagine yourself as a post-Newtonian physicist intent on understanding the behavior of such a simple system as a container of gas. How would you proceed? The prototype of a successful physical theory is classical mechanics, which describes the intricate motions of particles starting from simple basic laws and employing the mathematical machinery of calculus. By analogy, you may proceed as follows:

- Idealize the *system* under study as much as possible (as is the case of a point particle). The concept of mechanical work on the system is certainly familiar, yet there appear to be complications due to exchange of heat. The solution is first to examine *closed systems*, insulated by *adiabatic walls* that don't allow any exchange of heat with the surroundings. Of course, it is ultimately also necessary to study *open systems*, which may exchange heat with the outside world through *diathermic walls*.
- As the state of a point particle is quantified by its coordinates (and momenta), properties of the macroscopic system can also be described by a number of *thermodynamic coordinates* or *state functions*. The most familiar coordinates are those that relate to mechanical work, such as pressure and volume (for a fluid), surface tension and area (for a film), tension and length (for a wire), electric field and polarization (for a dielectric), etc. As we shall see, there are additional state functions not related to mechanical work. The state functions are well defined only when the system is in *equilibrium*, that is, when its properties do not change appreciably with time over the intervals of interest (observation times). The dependence on the observation time makes the concept of equilibrium subjective. For example, window glass is in equilibrium as a solid over many decades, but flows like a fluid over time scales of millennia. At the other extreme, it is perfectly legitimate to consider the equilibrium between matter and radiation in the early universe during the first minutes of the Big Bang.

- Finally, the relationship between the state functions is described by the laws of thermodynamics. As a *phenomenological* description, these laws are based on a number of empirical observations. A coherent logical and mathematical structure is then constructed on the basis of these observations, which leads to a variety of useful concepts, and to testable relationships among various quantities. The laws of thermodynamics can only be justified by a more fundamental (microscopic) theory of nature. For example, statistical mechanics attempts to obtain these laws starting from classical or quantum mechanical equations for the evolution of collections of particles.

1.2 The zeroth law

The zeroth law of thermodynamics describes the transitive nature of thermal equilibrium. It states:

If two systems, A and B, are separately in equilibrium with a third system, C, then they are also in equilibrium with one another.

Despite its apparent simplicity, the zeroth law has the consequence of implying the existence of an important state function, the *empirical temperature* Θ, such that systems in equilibrium are at the same temperature.

Fig. 1.1 Illustration of the zeroth law: systems *A* and *B*, which are initially separately in equilibrium with *C*, are placed in contact with each other.

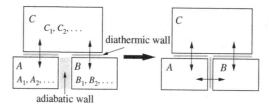

Let the equilibrium state of systems A, B, and C be described by the coordinates $\{A_1, A_2, \cdots\}$, $\{B_1, B_2, \cdots\}$, and $\{C_1, C_2, \cdots\}$, respectively. The assumption that A and C are in equilibrium implies a constraint between the coordinates of A and C, that is, a change in A_1 must be accompanied by some changes in $\{A_2, \cdots ; C_1, C_2, \cdots\}$ to maintain equilibrium of A and C. Denote this constraint by

$$f_{AC}(A_1, A_2, \cdots ; C_1, C_2, \cdots) = 0. \tag{1.1}$$

The equilibrium of B and C implies a similar constraint

$$f_{BC}(B_1, B_2, \cdots ; C_1, C_2, \cdots) = 0. \tag{1.2}$$

Note that each system is assumed to be separately in mechanical equilibrium. If they are allowed also to do work on each other, additional conditions (e.g., constant pressure) are required to describe their joint mechanical equilibrium.

Clearly we can state the above constraint in many different ways. For example, we can study the variations of C_1 as all of the other parameters are changed. This is equivalent to solving each of the above equations for C_1 to yield [1]

$$C_1 = F_{AC}(A_1, A_2, \cdots; C_2, \cdots),$$
$$C_1 = F_{BC}(B_1, B_2, \cdots; C_2, \cdots). \tag{1.3}$$

Thus if C is separately in equilibrium with A and B, we must have

$$F_{AC}(A_1, A_2, \cdots; C_2, \cdots) = F_{BC}(B_1, B_2, \cdots; C_2, \cdots). \tag{1.4}$$

However, according to the zeroth law there is also equilibrium between A and B, implying the constraint

$$f_{AB}(A_1, A_2, \cdots; B_1, B_2, \cdots) = 0. \tag{1.5}$$

We can select any set of parameters $\{A, B\}$ that satisfy the above equation, and substitute them in Eq. (1.4). The resulting equality must hold quite independently of any set of variables $\{C\}$ in this equation. We can then change these parameters, moving along the manifold constrained by Eq. (1.5), and Eq. (1.4) will remain valid irrespective of the state of C. Therefore, it must be possible to simplify Eq. (1.4) by canceling the coordinates of C. Alternatively, we can select any fixed set of parameters C, and ignore them henceforth, reducing the condition (1.5) for equilibrium of A and B to

$$\Theta_A(A_1, A_2, \cdots) = \Theta_B(B_1, B_2, \cdots), \tag{1.6}$$

that is, equilibrium is characterized by a function Θ of thermodynamic coordinates. This function specifies the *equation of state*, and *isotherms* of A are described by the condition $\Theta_A(A_1, A_2, \cdots) = \Theta$. While at this point there are many potential choices of Θ, the key point is the existence of a function that constrains the parameters of each system in thermal equilibrium.

There is a similarity between Θ and the force in a mechanical system. Consider two one-dimensional systems that can do work on each other as in the case of two springs connected together. Equilibrium is achieved when the forces exerted by each body on the other are equal. (Of course, unlike the scalar temperature, the vectorial force has a direction; a complication that we have ignored. The pressure of a gas piston provides a more apt analogy.) The mechanical equilibrium between several such bodies is also transitive, and the latter could have been used as a starting point for deducing the existence of a mechanical force.

---

[1] From a purely mathematical perspective, it is not necessarily possible to solve an arbitrary constraint condition for C_1. However, the requirement that the constraint describes real physical parameters clearly implies that we can obtain C_1 as a function of the remaining parameters.

As an example, let us consider the following three systems: (A) a wire of length L with tension F, (B) a paramagnet of magnetization M in a magnetic field B, and (C) a gas of volume V at pressure P.

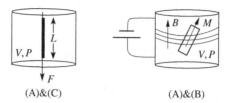

Observations indicate that when these systems are in equilibrium, the following constraints are satisfied between their coordinates:

$$\left(P+\frac{a}{V^2}\right)(V-b)(L-L_0)-c[F-K(L-L_0)]=0,$$
$$\left(P+\frac{a}{V^2}\right)(V-b)M-dB=0. \tag{1.7}$$

The two conditions can be organized into three empirical temperature functions as

$$\Theta \propto \left(P+\frac{a}{V^2}\right)(V-b)=c\left(\frac{F}{L-L_0}-K\right)=d\frac{B}{M}. \tag{1.8}$$

Note that the zeroth law severely constrains the form of the constraint equation describing the equilibrium between two bodies. Any arbitrary function cannot necessarily be organized into an equality of two empirical temperature functions.

The constraints used in the above example were in fact chosen to reproduce three well-known equations of state that will be encountered and discussed later in this book. In their more familiar forms they are written as

$$\begin{cases} (P+a/V^2)(V-b)=Nk_BT & \text{(van der Waals gas)} \\[4pt] M=(N\ \mu_B^2 B)/(3k_BT) & \text{(Curie paramagnet)} \\[4pt] F=(K+DT)(L-L_0) & \text{(Hooke's law for rubber)} \end{cases} \tag{1.9}$$

Note that we have employed the symbol for Kelvin temperature T, in place of the more general empirical temperature Θ. This concrete temperature scale can be constructed using the properties of the ideal gas.

The ideal gas temperature scale: while the zeroth law merely states the presence of isotherms, to set up a practical temperature scale at this stage, a reference system is necessary. The *ideal gas* occupies an important place in thermodynamics and provides the necessary reference. Empirical observations indicate that the product of pressure and volume is constant along the isotherms of any gas that is sufficiently dilute. The ideal gas refers to this *dilute* limit of

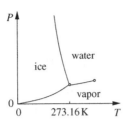

Fig. 1.3 The triple point of ice, water, and steam occurs at a unique point in the (P, T) phase diagram.

real gases, and the ideal gas temperature is proportional to the product. The constant of proportionality is determined by reference to the temperature of the triple point of the ice–water–gas system, which was set to 273.16 (degrees) kelvin (K) by the 10th General Conference on Weights and Measures in 1954. Using a dilute gas (i.e., as $P \to 0$) as thermometer, the temperature of a system can be obtained from

$$T(\text{K}) \equiv 273.16 \times \left(\lim_{P \to 0} (PV)_{\text{system}} / \lim_{P \to 0} (PV)_{\text{ice–water–gas}} \right). \tag{1.10}$$

1.3 The first law

In dealing with simple mechanical systems, conservation of energy is an important principle. For example, the location of a particle can be changed in a potential by external work, which is then related to a change in its potential energy. Observations indicate that a similar principle operates at the level of macroscopic bodies *provided that the system is properly insulated*, that is, when the only sources of energy are of mechanical origin. We shall use the following formulation of these observations:

The amount of work required to change the state of an otherwise adiabatically isolated system depends only on the initial and final states, and not on the means by which the work is performed, or on the intermediate stages through which the system passes.

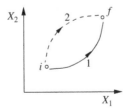

Fig. 1.4 The two adiabatic paths for changing macroscopic coordinates between the initial and final point result in the same change in internal energy.

For a particle moving in a potential, the required work can be used to construct a potential energy landscape. Similarly, for the thermodynamic system we can construct another state function, the internal energy $E(\mathbf{X})$. Up to a constant,

$E(\mathbf{X})$ can be obtained from the amount of work ΔW needed for an *adiabatic* transformation from an initial state $\mathbf{X_i}$ to a final state $\mathbf{X_f}$, using

$$\Delta W = E(\mathbf{X_f}) - E(\mathbf{X_i}). \tag{1.11}$$

Another set of observations indicate that once the adiabatic constraint is removed, the amount of work is no longer equal to the change in the internal energy. The difference $\Delta Q = \Delta E - \Delta W$ is defined as the *heat* intake of the system from its surroundings. Clearly, in such transformations, ΔQ and ΔW are not separately functions of state in that they depend on external factors such as the means of applying work, and not only on the final states. To emphasize this, for a differential transformation we write

$$đQ = dE - đW, \tag{1.12}$$

where $dE = \sum_i \partial_i E dX_i$ can be obtained by differentiation, while $đQ$ and $đW$ generally cannot. Also note that our convention is such that the signs of work and heat indicate the energy *added* to the system, and not vice versa. The first law of thermodynamics thus states that to change the state of a system we need a fixed amount of energy, which can be in the form of mechanical work or heat. This can also be regarded as a way of defining and quantifying the exchanged heat.

A *quasi-static* transformation is one that is performed sufficiently slowly so that the system is always in equilibrium. Thus, at any stage of the process, the thermodynamic coordinates of the system exist and can in principle be computed. For such transformations, the work done on the system (equal in magnitude but opposite in sign to the work done by the system) can be related to changes in these coordinates. As a mechanical example, consider the stretching of a spring or rubber band. To construct the potential energy of the system as a function of its length L, we can pull the spring sufficiently slowly so that at each stage the external force is matched by the internal force F from the spring. For such a quasi-static process, the change in the potential energy of the spring is $\int F dL$. If the spring is pulled abruptly, some of the external work is converted into kinetic energy and eventually lost as the spring comes to rest.

Generalizing from the above example, one can typically divide the state functions $\{\mathbf{X}\}$ into a set of *generalized displacements* $\{\mathbf{x}\}$, and their conjugate *generalized forces* $\{\mathbf{J}\}$, such that for an infinitesimal quasi-static transformation[2]

$$đW = \sum_i J_i dx_i. \tag{1.13}$$

[2] I denote force by the symbol $\mathbf{J}$ rather than $\mathbf{F}$, to reserve the latter for the free energy. I hope the reader is not confused with currents (sometimes also denoted by $\mathbf{J}$), which rarely appear in this book.

Table 1.1 *Generalized forces and displacements*

System	Force		Displacement	
Wire	tension	F	length	L
Film	surface tension	$\mathcal{S}$	area	A
Fluid	pressure	$-P$	volume	V
Magnet	magnetic field	H	magnetization	M
Dielectric	electric field	E	polarization	P
Chemical reaction	chemical potential	μ	particle number	N

Table 1.1 provides some common examples of such conjugate coordinates. Note that pressure P is by convention calculated from the force exerted by the system on the walls, as opposed to the force on a spring, which is exerted in the opposite direction. This is the origin of the negative sign that accompanies hydrostatic work.

The displacements are usually *extensive* quantities, that is, proportional to system size, while the forces are *intensive* and independent of size. The latter are indicators of equilibrium; for example, the pressure is uniform in a gas in equilibrium (in the absence of external potentials) and equal for two equilibrated gases in contact. As discussed in connection with the zeroth law, temperature plays a similar role when heat exchanges are involved. Is there a corresponding displacement, and if so what is it? This question will be answered in the following sections.

The ideal gas: we noted in connection with the zeroth law that the equation of state of the ideal gas takes a particularly simple form, $PV \propto T$. The internal energy of the ideal gas also takes a very simple form, as observed for example by *Joule's free expansion experiment*: measurements indicate that if an ideal gas expands adiabatically (but not necessarily quasi-statically), from a volume V_i to V_f, the initial and final temperatures are the same. As the transformation is adiabatic ($\Delta Q = 0$) and there is no external work done on the system ($\Delta W = 0$), the internal energy of the gas is unchanged. Since the pressure and volume of the gas change in the process, but its temperature does not, we conclude that the internal energy depends only on temperature, that is, $E(V, T) = E(T)$. This property of the ideal gas is in fact a consequence of the form of its equation of state, as will be proved in one of the problems at the end of this chapter.

Fig. 1.5 A gas initially confined in the left chamber is allowed to expand rapidly to both chambers.

Response functions are the usual method for characterizing the macroscopic behavior of a system. They are experimentally measured from the changes of thermodynamic coordinates with external probes. Some common response functions are as follows.

Heat capacities are obtained from the change in temperature upon addition of heat to the system. Since heat is not a function of state, the path by which it is supplied must also be specified. For example, for a gas we can calculate the heat capacities at constant volume or pressure, denoted by $C_V = \mathleft. dQ/dT \right|_V$ and $C_P = \left. dQ/dT \right|_P$, respectively. The latter is larger since some of the heat is used up in the work done in changes of volume:

$$C_V = \left. \frac{dQ}{dT} \right|_V = \left. \frac{dE - dW}{dT} \right|_V = \left. \frac{dE + PdV}{dT} \right|_V = \left. \frac{\partial E}{\partial T} \right|_V,$$

$$C_P = \left. \frac{dQ}{dT} \right|_P = \left. \frac{dE - dW}{dT} \right|_P = \left. \frac{dE + PdV}{dT} \right|_P = \left. \frac{\partial E}{\partial T} \right|_P + P \left. \frac{\partial V}{\partial T} \right|_P. \qquad (1.14)$$

Force constants measure the (infinitesimal) ratio of displacement to force and are generalizations of the spring constant. Examples include the isothermal compressibility of a gas $\kappa_T = - \left. \partial V/\partial P \right|_T /V$, and the susceptibility of a magnet $\chi_T = \left. \partial M/\partial B \right|_T /V$. From the equation of state of an ideal gas $PV \propto T$, we obtain $\kappa_T = 1/P$.

Thermal responses probe the change in the thermodynamic coordinates with temperature. For example, the expansivity of a gas is given by $\alpha_P = \left. \partial V/\partial T \right|_P /V$, which equals $1/T$ for the ideal gas.

Since the internal energy of an ideal gas depends only on its temperature, $\left. \partial E/\partial T \right|_V = \left. \partial E/\partial T \right|_P = dE/dT$, and Eq. (1.14) simplifies to

$$C_P - C_V = P \left. \frac{\partial V}{\partial T} \right|_P = PV\alpha_P = \frac{PV}{T} \equiv Nk_B. \qquad (1.15)$$

The last equality follows from extensivity: for a given amount of ideal gas, the constant PV/T is proportional to N, the number of particles in the gas; the ratio is Boltzmann's constant with a value of $k_B \approx 1.4 \times 10^{-23}\,\text{JK}^{-1}$.

1.4 The second law

The practical impetus for development of the science of thermodynamics in the nineteenth century was the advent of heat engines. The increasing reliance on machines to do work during the industrial revolution required better understanding of the principles underlying conversion of heat to work. It is quite interesting to see how such practical considerations as the efficiency of engines can lead to abstract ideas like entropy.

An idealized *heat engine* works by taking in a certain amount of heat Q_H, from a heat source (for example a coal fire), converting a portion of it to work

W, and dumping the remaining heat Q_C into a heat sink (e.g., atmosphere). The efficiency of the engine is calculated from

$$\eta = \frac{W}{Q_H} = \frac{Q_H - Q_C}{Q_H} \le 1. \tag{1.16}$$

An idealized *refrigerator* is like an engine running backward, that is, using work W to extract heat Q_C from a cold system, and dumping heat Q_H at a higher temperature. We can similarly define a figure of merit for the performance of a refrigerator as

$$\omega = \frac{Q_C}{W} = \frac{Q_C}{Q_H - Q_C}. \tag{1.17}$$

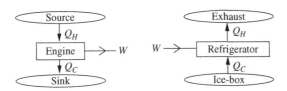

Fig. 1.6 The idealized engine and refrigerator.

The first law rules out so-called "perpetual motion machines of the first kind," that is, engines that produce work without consuming any energy. However, the conservation of energy is not violated by an engine that produces work by converting water to ice. Such a "perpetual motion machine of the second kind" would certainly solve the world's energy problems, but is ruled out by the second law of thermodynamics. The observation that the natural direction for the flow of heat is from hotter to colder bodies is the essence of the second law of thermodynamics. There is a number of different formulations of the second law, such as the following two statements:

Kelvin's statement. *No process is possible whose sole result is the complete conversion of heat into work.*

Clausius's statement. *No process is possible whose sole result is the transfer of heat from a colder to a hotter body.*

A perfect engine is ruled out by the first statement, a perfect refrigerator by the second. Since we shall use both statements, we first show that they are equivalent. Proof of the equivalence of the Kelvin and Clausius statements proceeds by showing that if one is violated, so is the other.

(a) Let us assume that there is a machine that violates Clausius's statement by taking heat Q from a cooler region to a hotter one. Now consider an engine operating between these two regions, taking heat Q_H from the hotter one and dumping Q_C at the colder sink. The combined system takes $Q_H - Q$ from the hot source, produces work equal to $Q_H - Q_C$, and dumps $Q_C - Q$ at the cold sink. If we adjust the engine output such that $Q_C = Q$, the net result is a 100% efficient engine, in violation of Kelvin's statement.

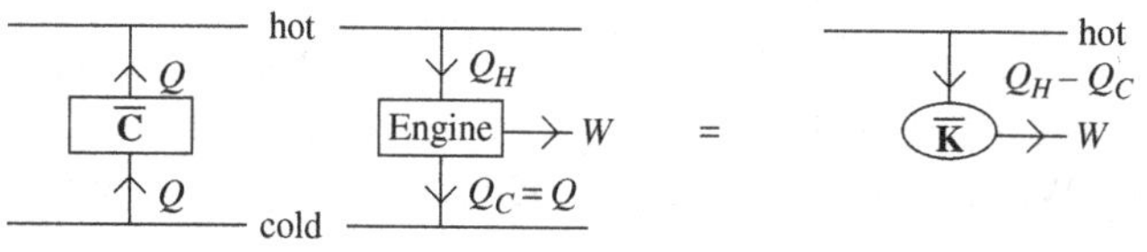

Fig. 1.7 A machine violating Clausius's statement ($\overline{C}$) can be connected to an engine, resulting in a combined device ($\overline{K}$) that violates Kelvin's statement.

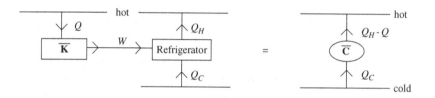

(b) Alternatively, assume a machine that violates Kelvin's law by taking heat Q and converting it completely to work. The work output of this machine can be used to run a refrigerator, with the net outcome of transferring heat from a colder to a hotter body, in violation of Clausius's statement.

Fig. 1.8 A machine violating Kelvin's statement can be connected to a refrigerator, resulting in violation of Clausius's statement.

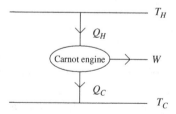

1.5 Carnot engines

A Carnot engine is any engine that is reversible, runs in a cycle, with all of its heat exchanges taking place at a source temperature T_H, and a sink temperature T_C.

Fig. 1.9 A Carnot engine operates between temperatures T_H and T_C, with no other heat exchanges.

A *reversible* process is one that can be run backward in time by simply reversing its inputs and outputs. It is the thermodynamic equivalent of frictionless motion in mechanics. Since time reversibility implies equilibrium, a reversible transformation must be quasi-static, but the reverse is not necessarily true (e.g., if there is energy dissipation due to friction). An engine that runs in a *cycle* returns to its original internal state at the end of the process. The distinguishing characteristic of the Carnot engine is that heat exchanges with the surroundings are carried out only at two temperatures.

The zeroth law allows us to select two isotherms at temperatures T_H and T_C for these heat exchanges. To complete the Carnot cycle we have to connect these isotherms by reversible *adiabatic* paths in the coordinate space. Since heat is not a function of state, we don't know how to construct such paths in general. Fortunately, we have sufficient information at this point to construct a Carnot engine using an ideal gas as its internal working substance. For the purpose of demonstration, let us compute the adiabatic curves for a monatomic ideal gas with an internal energy

$$E = \frac{3}{2}Nk_BT = \frac{3}{2}PV.$$

Along a quasi-static path

$$đQ = dE - đW = d\left(\frac{3}{2}PV\right) + PdV = \frac{5}{2}PdV + \frac{3}{2}VdP. \tag{1.18}$$

The adiabatic condition $đQ = 0$, then implies a path

$$\frac{dP}{P} + \frac{5}{3}\frac{dV}{V} = 0, \quad \Longrightarrow \quad PV^\gamma = \text{constant}, \tag{1.19}$$

with $\gamma = 5/3$.

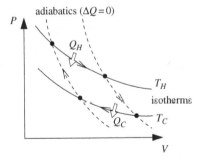

Fig. **1.10** The Carnot cycle for an ideal gas, with isothermal and adiabatic paths indicated by solid and dashed lines, respectively.

The adiabatic curves are clearly distinct from the isotherms, and we can select two such curves to intersect our isotherms, thereby completing a Carnot cycle. The assumption of $E \propto T$ is not necessary, and in one of the problems provided at the end of this chapter, you will construct adiabatics for any $E(T)$. In fact, a similar construction is possible for any two-parameter system with $E(J, x)$.

Carnot's theorem   *No engine operating between two reservoirs (at temperatures T_H and T_C) is more efficient than a Carnot engine operating between them.*

Since a Carnot engine is reversible, it can be run backward as a refrigerator. Use the non-Carnot engine to run the Carnot engine backward. Let us denote the heat exchanges of the non-Carnot and Carnot engines by Q_H, Q_C, and Q'_H, Q'_C, respectively. The net effect of the two engines is to transfer heat equal to $Q_H - Q'_H = Q_C - Q'_C$ from T_H to T_C. According to Clausius's statement, the quantity of transferred heat cannot be negative, that is, $Q_H \geq Q'_H$. Since the same quantity of work W is involved in this process, we conclude that

$$\frac{W}{Q_H} \leq \frac{W}{Q'_H}, \implies \eta_{\text{Carnot}} \geq \eta_{\text{non-Carnot}}. \tag{1.20}$$

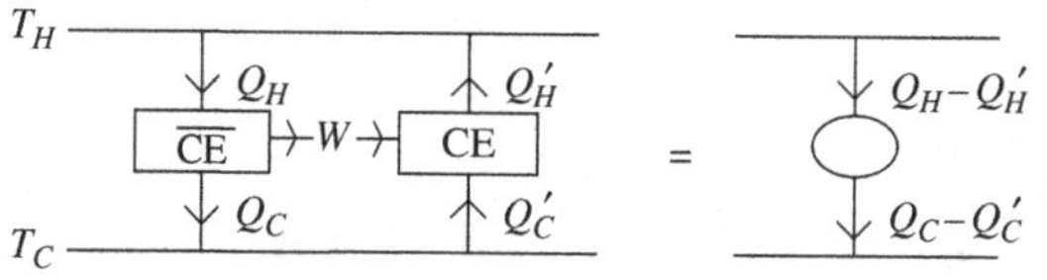

Fig. 1.11 A generic engine is used to run a Carnot engine in reverse.

Corollary   All reversible (Carnot) engines have the same *universal* efficiency $\eta(T_H, T_C)$, since each can be used to run any other one backward.

The thermodynamic temperature scale: as shown earlier, it is at least theoretically possible to construct a Carnot engine using an ideal gas (or any other two-parameter system) as working substance. We now find that independent of the material used, and design and construction, all such cyclic and reversible engines have the same maximum theoretical efficiency. Since this maximum efficiency is only dependent on the two temperatures, it can be used to construct a temperature scale. Such a temperature scale has the attractive property

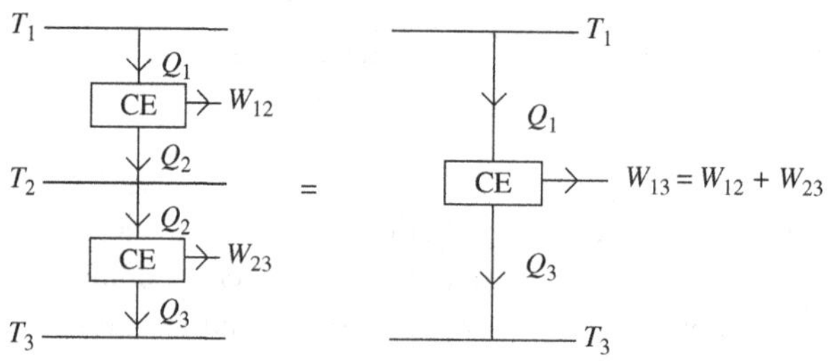

Fig. 1.12 Two Carnot engines connected in series are equivalent to a third.

of being independent of the properties of any material (e.g., the ideal gas). To construct such a scale we first obtain a constraint on the form of $\eta(T_H, T_C)$. Consider two Carnot engines running in series, one between temperatures T_1 and T_2, and the other between T_2 and T_3 ($T_1 > T_2 > T_3$). Denote the heat exchanges, and work outputs, of the two engines by Q_1, Q_2, W_{12}, and Q_2, Q_3, W_{23}, respectively. Note that the heat dumped by the first engine is taken in by the second, so that the combined effect is another Carnot engine (since each component is reversible) with heat exchanges Q_1, Q_3, and work output $W_{13} = W_{12} + W_{23}$.

Using the universal efficiency, the three heat exchanges are related by

$$\begin{cases} Q_2 = Q_1 - W_{12} = Q_1[1 - \eta(T_1, T_2)], \\ Q_3 = Q_2 - W_{23} = Q_2[1 - \eta(T_2, T_3)] = Q_1[1 - \eta(T_1, T_2)][1 - \eta(T_2, T_3)], \\ Q_3 = Q_1 - W_{13} = Q_1[1 - \eta(T_1, T_3)]. \end{cases}$$

Comparison of the final two expressions yields

$$[1 - \eta(T_1, T_3)] = [1 - \eta(T_1, T_2)][1 - \eta(T_2, T_3)]. \tag{1.21}$$

This property implies that $1 - \eta(T_1, T_2)$ can be written as a ratio of the form $f(T_2)/f(T_1)$, which by convention is set to T_2/T_1, that is,

$$1 - \eta(T_1, T_2) = \frac{Q_2}{Q_1} \equiv \frac{T_2}{T_1},$$

$$\implies \eta(T_H, T_C) = \frac{T_H - T_C}{T_H}. \tag{1.22}$$

Equation (1.22) defines temperature up to a constant of proportionality, which is again set by choosing the triple point of water, ice, and steam to 273.16 K. So far we have used the symbols Θ and T interchangeably. In fact, by running a Carnot cycle for a perfect gas, it can be proved (see problems) that the ideal gas and thermodynamic temperature scales are equivalent. Clearly, the thermodynamic scale is not useful from a practical standpoint; its advantage is conceptual, in that it is independent of the properties of any substance. All thermodynamic temperatures are positive, since according to Eq. (1.22) the heat extracted from a temperature T is proportional to it. If a negative temperature existed, an engine operating between it and a positive temperature would extract heat from both reservoirs and convert the sum total to work, in violation of Kelvin's statement of the second law.

1.6 Entropy

To construct finally the state function that is conjugate to temperature, we appeal to the following theorem:

Clausius's theorem   *For any cyclic transformation (reversible or not), $\oint \dbar Q/T \le 0$, where $\dbar Q$ is the heat increment supplied to the system at temperature T.*

Subdivide the cycle into a series of infinitesimal transformations in which the system receives energy in the form of heat $\dbar Q$ and work $\dbar W$. The system need not be in equilibrium at each interval. Direct all the heat exchanges

Fig. 1.13 The heat
exchanges of the system
are directed to a Carnot
engine with a
reservoir at T_0.

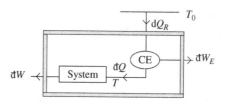

of the system to one port of a Carnot engine, which has another reservoir at a fixed temperature T_0. (There can be more than one Carnot engine as long as they all have one end connected to T_0.) Since the sign of đQ is not specified, the Carnot engine must operate a series of infinitesimal cycles in either direction. To deliver heat đQ to the system at some stage, the engine has to extract heat đQ_R from the fixed reservoir. If the heat is delivered to a part of the system that is locally at a temperature T, then according to Eq. (1.22),

$$\text{đ}Q_R = T_0 \frac{\text{đ}Q}{T}. \tag{1.23}$$

After the cycle is completed, the system and the Carnot engine return to their original states. The net effect of the combined process is extracting heat $Q_R = \oint \text{đ}Q_R$ from the reservoir and converting it to external work W. The work $W = Q_R$ is the sum total of the work elements done by the Carnot engine, *and* the work performed by the system in the complete cycle. By Kelvin's statement of the second law, $Q_R = W \leq 0$, that is,

$$T_0 \oint \frac{\text{đ}Q}{T} \leq 0, \quad \Longrightarrow \quad \oint \frac{\text{đ}Q}{T} \leq 0, \tag{1.24}$$

since $T_0 > 0$. Note that T in Eq. (1.24) refers to the temperature of the whole system only for quasi-static processes in which it can be uniquely defined throughout the cycle. Otherwise, it is just a local temperature (say at a boundary of the system) at which the Carnot engine deposits the element of heat.

Consequences of Clausius's theorem:

1. For a *reversible cycle* $\oint \text{đ}Q_{\text{rev}}/T = 0$, since by running the cycle in the opposite direction đ$Q_{\text{rev}} \to -\text{đ}Q_{\text{rev}}$, and by the above theorem đQ_{rev}/T is both non-negative and non-positive, hence zero. This result implies that the integral of đQ_{rev}/T between any two points A and B is independent of path, since for two paths (1) and (2)

$$\int_A^B \frac{\text{đ}Q_{\text{rev}}^{(1)}}{T_1} + \int_B^A \frac{\text{đ}Q_{\text{rev}}^{(2)}}{T_2} = 0, \quad \Longrightarrow \int_A^B \frac{\text{đ}Q_{\text{rev}}^{(1)}}{T_1} = \int_A^B \frac{\text{đ}Q_{\text{rev}}^{(2)}}{T_2}. \tag{1.25}$$

2. Using Eq. (1.25) we can construct yet another function of state, the *entropy S*. Since the integral is independent of path, and only depends on the two end-points, we can set

$$S(B) - S(A) \equiv \int_A^B \frac{\text{đ}Q_{\text{rev}}}{T}. \tag{1.26}$$

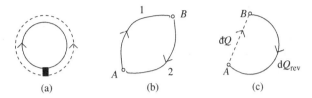

Fig. 1.14 (a) A reversible cycle. (b) Two reversible paths between A and B. (c) The cycle formed from a generic path between A and B, and a reversible one.

For reversible processes, we can now compute the heat from $đQ_{\text{rev}} = TdS$. This allows us to construct adiabatic curves for a general (multivariable) system from the condition of constant S. Note that Eq. (1.26) only defines the entropy up to an overall constant.

3. For a reversible (hence quasi-static) transformation, $đQ = TdS$ and $đW = \sum_i J_i dx_i$, and the first law implies

$$dE = đQ + đW = TdS + \sum_i J_i dx_i.\qquad(1.27)$$

We can see that in this equation S and T appear as conjugate variables, with S playing the role of a displacement, and T as the corresponding force. This identification allows us to make the correspondence between mechanical and thermal exchanges more precise, although we should keep in mind that unlike its mechanical analog, temperature is always positive. While to obtain Eq. (1.27) we had to appeal to reversible transformations, it is important to emphasize that it is a relation between functions of state. Equation (1.27) is likely the most fundamental and useful identity in thermodynamics.

4. The number of *independent variables* necessary to describe a thermodynamic system also follows from Eq. (1.27). If there are n methods of doing work on a system, represented by n conjugate pairs (J_i, x_i), then $n + 1$ independent coordinates are necessary to describe the system. (We shall ignore possible constraints between the mechanical coordinates.) For example, choosing $(E, \{x_i\})$ as coordinates, it follows from Eq. (1.27) that

$$\left.\frac{\partial S}{\partial E}\right|_{\mathbf{x}} = \frac{1}{T} \quad \text{and} \quad \left.\frac{\partial S}{\partial x_i}\right|_{E, x_{j \neq i}} = -\frac{J_i}{T}.\qquad(1.28)$$

($\mathbf{x}$ and $\mathbf{J}$ are shorthand notations for the parameter sets $\{x_i\}$ and $\{J_i\}$.)

Fig. 1.15 The initially isolated subsystems are allowed to interact, resulting in an increase of entropy.

5. Consider an irreversible change from A to B. Make a complete cycle by returning from B to A along a reversible path. Then

$$\int_A^B \frac{đQ}{T} + \int_B^A \frac{đQ_{\text{rev}}}{T} \leq 0, \quad \Longrightarrow \quad \int_A^B \frac{đQ}{T} \leq S(B) - S(A).\qquad(1.29)$$

In differential form, this implies that $dS \geq đQ/T$ for any transformation. In particular, consider adiabatically isolating a number of subsystems, each initially separately in equilibrium. As they come to a state of joint equilibrium, since the net $dQ = 0$, we must have $\delta S \geq 0$. Thus an adiabatic system attains a maximum value of entropy in equilibrium since spontaneous internal changes can only increase S. The direction of increasing entropy thus points out the arrow of time, and the path to equilibrium. The mechanical analog is a point mass placed on a surface, with gravity providing a downward force. As various constraints are removed, the particle will settle down at locations of decreasing height. The statement about the increase of entropy is thus no more mysterious than the observation that objects tend to fall down under gravity!

1.7 Approach to equilibrium and thermodynamic potentials

The time evolution of systems toward equilibrium is governed by the second law of thermodynamics. For example, in the previous section we showed that for an adiabatically isolated system entropy must increase in any spontaneous change and reaches a maximum in equilibrium. What about out-of-equilibrium systems that are not adiabatically isolated, and may also be subject to external mechanical work? It is usually possible to define other thermodynamic potentials that are extremized when the system is in equilibrium.

Fig. 1.16 Mechanical equilibrium of an extended spring.

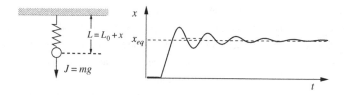

Enthalpy is the appropriate function when there is no heat exchange ($dQ = 0$), and the system comes to mechanical equilibrium with a *constant external force*. The minimum enthalpy principle merely formulates the observation that stable mechanical equilibrium is obtained by minimizing the net potential energy of the system *plus the external agent*.

For example, consider a spring of natural extension L_0 and spring constant K, subject to the force $J = mg$ exerted by a particle of mass m. For an extension $x = L - L_0$, the internal potential energy of the spring is $U(x) = Kx^2/2$, while there is a change of $-mgx$ in the potential energy of the particle. Mechanical equilibrium is obtained by minimizing $Kx^2/2 - mgx$ at an extension $x_{eq} = mg/K$. The spring at any other value of the displacement initially oscillates before coming to rest at x_{eq} due to friction. For a more general potential energy $U(x)$, the internally generated force $J_i = -dU/dx$ has to be balanced with the external force J at the equilibrium point.

For any set of displacements **x**, at constant (externally applied) generalized forces **J**, the work input to the system is $dW \leq \mathbf{J} \cdot \delta\mathbf{x}$. (Equality is achieved for a quasi-static change with $\mathbf{J} = \mathbf{J}_i$, but there is generally some loss of the external work to dissipation.) Since $dQ = 0$, using the first law, $\delta E \leq \mathbf{J} \cdot \delta \mathbf{x}$, and

$$\delta H \leq 0, \quad \text{where} \quad H = E - \mathbf{J} \cdot \mathbf{x} \qquad (1.30)$$

is the enthalpy. The variations of H *in equilibrium* are given by

$$dH = dE - d(\mathbf{J} \cdot \mathbf{x}) = T dS + \mathbf{J} \cdot d\mathbf{x} - \mathbf{x} \cdot d\mathbf{J} - \mathbf{J} \cdot d\mathbf{x} = T dS - \mathbf{x} \cdot d\mathbf{J}. \qquad (1.31)$$

The equality in Eq. (1.31), and the inequality in Eq. (1.30), are a possible source of confusion. Equation (1.30) refers to variations of H on approaching equilibrium as some parameter that is not a function of state is varied (e.g., the velocity of the particle joined to the spring in the above example). By contrast, Eq. (1.31) describes a relation between equilibrium coordinates. To differentiate the two cases, I will denote the former (non-equilibrium) variations by δ.

The coordinate set $(S, \mathbf{J})$ is the natural choice for describing the enthalpy, and it follows from Eq. (1.31) that

$$x_i = - \left. \frac{\partial H}{\partial J_i} \right|_{S, J_{j \neq i}}. \qquad (1.32)$$

Variations of the enthalpy with temperature are related to heat capacities at constant force, for example

$$C_P = \left. \frac{dQ}{dT} \right|_P = \left. \frac{dE + P dV}{dT} \right|_P = \left. \frac{d(E + PV)}{dT} \right|_P = \left. \frac{dH}{dT} \right|_P. \qquad (1.33)$$

Note, however, that a change of variables is necessary to express H in terms of T, rather than the more natural variable S.

Helmholtz free energy is useful for *isothermal* transformations in the absence of mechanical work ($dW = 0$). It is rather similar to enthalpy, with T taking the place of J. From Clausius's theorem, the heat intake of a system at a constant temperature T satisfies $dQ \leq T \delta S$. Hence $\delta E = dQ + dW \leq T \delta S$, and

$$\delta F \leq 0, \quad \text{where} \quad F = E - TS \qquad (1.34)$$

is the Helmholtz free energy. Since

$$dF = dE - d(TS) = T dS + \mathbf{J} \cdot d\mathbf{x} - S dT - T dS = -S dT + \mathbf{J} \cdot d\mathbf{x}, \qquad (1.35)$$

the coordinate set $(T, \mathbf{x})$ (the quantities kept constant during an isothermal transformation with no work) is most suitable for describing the free energy. The equilibrium forces and entropy can be obtained from

$$J_i = \left. \frac{\partial F}{\partial x_i} \right|_{T, x_{j \neq i}} \quad , \quad S = - \left. \frac{\partial F}{\partial T} \right|_{\mathbf{x}}. \qquad (1.36)$$

The internal energy can also be calculated from F using

$$E = F + TS = F - T \left. \frac{\partial F}{\partial T} \right|_{\mathbf{x}} = -T^2 \left. \frac{\partial (F/T)}{\partial T} \right|_{\mathbf{x}}. \qquad (1.37)$$

Table 1.2 *Inequalities satisfied by thermodynamic potentials*

	đ$Q = 0$	constant T
đ$W = 0$	$\delta S \geq 0$	$\delta F \leq 0$
constant $\mathbf{J}$	$\delta H \leq 0$	$\delta G \leq 0$

Gibbs free energy applies to isothermal transformations involving mechanical work at constant external force. The natural inequalities for work and heat input into the system are given by đ$W \leq \mathbf{J} \cdot \delta\mathbf{x}$ and đ$Q \leq T\delta S$. Hence $\delta E \leq T\delta S + \mathbf{J} \cdot \delta\mathbf{x}$, leading to

$$\delta G \leq 0, \quad \text{where} \quad G = E - TS - \mathbf{J} \cdot \mathbf{x} \tag{1.38}$$

is the Gibbs free energy. Variations of G are given by

$$dG = dE - d(TS) - d(\mathbf{J} \cdot \mathbf{x}) = TdS + \mathbf{J} \cdot d\mathbf{x} - SdT - TdS - \mathbf{x} \cdot d\mathbf{J} - \mathbf{J} \cdot d\mathbf{x}$$
$$= -SdT - \mathbf{x} \cdot d\mathbf{J}, \tag{1.39}$$

and most easily expressed in terms of $(T, \mathbf{J})$.

Table 1.2 summarizes the above results on thermodynamic functions. Equations (1.30), (1.34), and (1.38) are examples of Legendre transformations, used to change variables to the most natural set of coordinates for describing a particular situation.

So far, we have implicitly assumed a constant number of particles in the system. In chemical reactions, and in equilibrium between two phases, the number of particles in a given constituent may change. The change in the number of particles necessarily involves changes in the internal energy, which is expressed in terms of a *chemical work* đ$W = \mu \cdot d\mathbf{N}$. Here $\mathbf{N} = \{N_1, N_2, \cdots\}$ lists the number of particles of each species, and $\mu = \{\mu_1, \mu_2, \cdots\}$ the associated *chemical potentials* that measure the work necessary to add additional particles to the system. Traditionally, chemical work is treated differently from mechanical work and is not subtracted from E in the Gibbs free energy of Eq. (1.38). For chemical equilibrium in circumstances that involve no mechanical work, the appropriate state function is the **grand potential** given by

$$\mathcal{G} = E - TS - \mu \cdot \mathbf{N}. \tag{1.40}$$

$\mathcal{G}(T, \mu, \mathbf{x})$ is minimized in chemical equilibrium, and its variations in general satisfy

$$d\mathcal{G} = -SdT + \mathbf{J} \cdot d\mathbf{x} - \mathbf{N} \cdot d\mu. \tag{1.41}$$

Example. To illustrate the concepts of this section, consider N particles of supersaturated steam in a container of volume V at a temperature T. How can we describe the approach of steam to an equilibrium mixture with N_w particles in the liquid and N_s particles in the gas phase? The fixed coordinates describing this system are V, T, and N. The appropriate thermodynamic function from Table 1.2 is the Helmholtz free energy $F(V, T, N)$, whose variations satisfy

$$dF = d(E - TS) = -SdT - PdV + \mu dN. \tag{1.42}$$

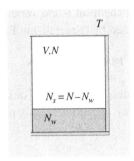

Fig. 1.17 Condensation of water from supersaturated steam.

Before the system reaches equilibrium at a particular value of N_w, it goes through a series of non-equilibrium states with smaller amounts of water. If the process is sufficiently slow, we can construct an out-of-equilibrium value for F as

$$F(V, T, N|N_w) = F_w(T, N_w) + F_s(V, T, N - N_w), \tag{1.43}$$

which depends on an additional variable N_w. (It is assumed that the volume occupied by water is small and irrelevant.) According to Eq. (1.34), the equilibrium point is obtained by minimizing F with respect to this variable. Since

$$\delta F = \left. \frac{\partial F_w}{\partial N_w} \right|_{T,V} \delta N_w - \left. \frac{\partial F_s}{\partial N_s} \right|_{T,V} \delta N_w, \tag{1.44}$$

and $\partial F/\partial N|_{T,V} = \mu$ from Eq. (1.42), the equilibrium condition can be obtained by equating the chemical potentials, that is, from $\mu_w(V, T) = \mu_s(V, T)$.

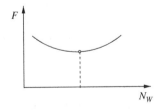

Fig. 1.18 The net free energy has a minimum as a function of the amount of water.

The identity of chemical potentials is the condition for chemical equilibrium. Naturally, to proceed further we need expressions for μ_w and μ_s.

1.8 Useful mathematical results

In the preceding sections we introduced a number of state functions. However, if there are n ways of doing mechanical work, $n+1$ independent parameters suffice to characterize an equilibrium state. There must thus be various constraints and relations between the thermodynamic parameters, some of which are explored in this section.

(1) Extensivity: including chemical work, variations of the *extensive* coordinates of the system are related by (generalizing Eq. (1.27))

$$dE = T dS + \mathbf{J} \cdot d\mathbf{x} + \mu \cdot d\mathbf{N}. \tag{1.45}$$

For fixed intensive coordinates, the extensive quantities are simply proportional to size or to the number of particles. This proportionality is expressed mathematically by

$$E(\lambda S, \lambda \mathbf{x}, \lambda \mathbf{N}) = \lambda E(S, \mathbf{x}, \mathbf{N}). \tag{1.46}$$

Evaluating the derivative of the above equation with respect to λ at $\lambda = 1$ results in

$$\left. \frac{\partial E}{\partial S} \right|_{\mathbf{x},\mathbf{N}} S + \sum_i \left. \frac{\partial E}{\partial x_i} \right|_{S,x_{j\neq i},\mathbf{N}} x_i + \sum_\alpha \left. \frac{\partial E}{\partial N_\alpha} \right|_{S,\mathbf{x},N_{\beta\neq\alpha}} N_\alpha = E(S, \mathbf{x}, \mathbf{N}). \tag{1.47}$$

The partial derivatives in the above equation can be identified from Eq. (1.45) as T, J_i, and μ_α, respectively. Substituting these values into Eq. (1.47) leads to

$$E = TS + \mathbf{J} \cdot \mathbf{x} + \mu \cdot \mathbf{N}. \tag{1.48}$$

Combining the variations of Eq. (1.48) with Eq. (1.45) leads to a constraint between the variations of intensive coordinates

$$S dT + \mathbf{x} \cdot d\mathbf{J} + \mathbf{N} \cdot d\mu = 0, \tag{1.49}$$

known as the *Gibbs–Duhem relation*. While Eq. (1.48) is sometimes referred to as the "fundamental equation of thermodynamics," I regard Eq. (1.45) as the more fundamental. The reason is that extensivity is an additional assumption, and in fact relies on short-range interactions between constituents of the system. It is violated for a large system controlled by gravity, such as a galaxy, while Eq. (1.45) remains valid.

Example. For a fixed amount of gas, variations of the chemical potential along an isotherm can be calculated as follows. Since $dT = 0$, the Gibbs–Duhem relation gives $-V dP + N d\mu = 0$, and

$$d\mu = \frac{V}{N} dP = k_B T \frac{dP}{P}, \tag{1.50}$$

where we have used the ideal gas equation of state $PV = Nk_B T$. Integrating the above equation gives

$$\mu = \mu_0 + k_B T \ln \frac{P}{P_0} = \mu_0 - k_B T \ln \frac{V}{V_0}, \qquad (1.51)$$

where (P_0, V_0, μ_0) refer to the coordinates of a reference point.

(2) Maxwell relations. Combining the mathematical rules of differentiation with thermodynamic relationships leads to several useful results. The most important of these are Maxwell's relations, which follow from the commutative property $[\partial_x \partial_y f(x, y) = \partial_y \partial_x f(x, y)]$ of derivatives. For example, it follows from Eq. (1.45) that

$$\left. \frac{\partial E}{\partial S} \right|_{\mathbf{x}, \mathbf{N}} = T, \quad \text{and} \quad \left. \frac{\partial E}{\partial x_i} \right|_{S, x_{j \neq i}, \mathbf{N}} = J_i. \qquad (1.52)$$

The joint second derivative of E is then given by

$$\frac{\partial^2 E}{\partial S \partial x_i} = \frac{\partial^2 E}{\partial x_i \partial S} = \left. \frac{\partial T}{\partial x_i} \right|_S = \left. \frac{\partial J_i}{\partial S} \right|_{x_i}. \qquad (1.53)$$

Since $(\partial y / \partial x) = (\partial x / \partial y)^{-1}$, the above equation can be inverted to give

$$\left. \frac{\partial S}{\partial J_i} \right|_{x_i} = \left. \frac{\partial x_i}{\partial T} \right|_S. \qquad (1.54)$$

Similar identities can be obtained from the variations of other state functions. Supposing that we are interested in finding an identity involving $\partial S / \partial x |_T$. We would like to find a state function whose variations include $S dT$ and $J dx$. The correct choice is $dF = d(E - TS) = -S dT + J dx$. Looking at the second derivative of F yields the Maxwell relation

$$- \left. \frac{\partial S}{\partial x} \right|_T = \left. \frac{\partial J}{\partial T} \right|_x. \qquad (1.55)$$

To calculate $\partial S / \partial J |_T$, consider $d(E - TS - Jx) = -S dT - x dJ$, which leads to the identity

$$\left. \frac{\partial S}{\partial J} \right|_T = \left. \frac{\partial x}{\partial T} \right|_J. \qquad (1.56)$$

There is a variety of mnemonics that are supposed to help you remember and construct Maxwell relations, such as magic squares, Jacobians, etc. I personally don't find any of these methods worth learning. The most logical approach is to remember the laws of thermodynamics and hence Eq. (1.27), and then to manipulate it so as to find the appropriate derivative using the rules of differentiation.

Example. To obtain $\partial \mu / \partial P |_{N,T}$ for an ideal gas, start with $d(E - TS + PV) = -S dT + V dP + \mu dN$. Clearly

$$\left. \frac{\partial \mu}{\partial P} \right|_{N,T} = \left. \frac{\partial V}{\partial N} \right|_{T,P} = \frac{V}{N} = \frac{k_B T}{P}, \qquad (1.57)$$

as in Eq. (1.50). From Eq. (1.27) it also follows that

$$\left.\frac{\partial S}{\partial V}\right|_{E,N} = \frac{P}{T} = -\frac{\partial E/\partial V|_{S,N}}{\partial E/\partial S|_{V,N}}, \tag{1.58}$$

where we have used Eq. (1.45) for the final identity. The above equation can be rearranged into

$$\left.\frac{\partial S}{\partial V}\right|_{E,N} \left.\frac{\partial E}{\partial S}\right|_{V,N} \left.\frac{\partial V}{\partial E}\right|_{S,N} = -1, \tag{1.59}$$

which is an illustration of the *chain rule* of differentiation.

(3) **The Gibbs phase rule**. In constructing a scale for temperature, we used the triple point of steam–water–ice in Fig. 1.3 as a reference standard. How can we be sure that there is indeed a unique coexistence point, and how robust is it? The phase diagram in Fig. 1.3 depicts the states of the system in terms of the two intensive parameters P and T. Quite generally, if there are n ways of performing work on a system that can also change its internal energy by transformations between c chemical constituents, the number of independent intensive parameters is $n + c$. Of course, including thermal exchanges there are $n + c + 1$ displacement-like variables in Eq. (1.45), but the intensive variables are constrained by Eq. (1.49); at least one of the parameters characterizing the system must be extensive. The system depicted in Fig. 1.3 corresponds to a one-component system (water) with only one means of doing work (hydrostatic), and is thus described by two independent intensive coordinates, for example, (P, T) or (μ, T). In a situation such as depicted in Fig. 1.17, where two phases (liquid and gas) coexist, there is an additional constraint on the intensive parameters, as the chemical potentials must be equal in the two phases. This constraint reduces the number of independent parameters by 1, and the corresponding manifold for coexisting phases in Fig. 1.3 is one-dimensional. At the triple point, where three phases coexist, we have to impose another constraint so that all three chemical potentials are equal. The Gibbs phase rule states that quite generally, if there are p phases in coexistence, the dimensionality (number of degrees of freedom) of the corresponding loci in the space of intensive parameters is

$$f = n + c + 1 - p. \tag{1.60}$$

The triple point of pure water thus occurs at a single point ($f = 1 + 1 + 1 - 3 = 0$) in the space of intensive parameters. If there are additional constituents, for example, a concentration of salt in the solution, the number of intensive quantities increases and the triple point can drift along a corresponding manifold.

1.9 Stability conditions

As noted earlier, the conditions derived in Section 1.7 are similar to the well-known requirements for mechanical stability: a particle moving freely in an

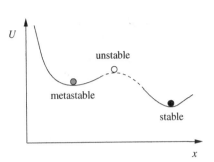

Fig. 1.19 Possible types of mechanical equilibrium for a particle in a potential. The convex portions (solid line) of the potential can be explored with a finite force J, while the concave (dashed line) portion around the unstable point is not accessible.

external potential $U(x)$ dissipates energy and settles to equilibrium at a minimum value of U. The vanishing of the force $J_i = -dU/dx$ is not by itself sufficient to ensure stability, as we must check that it occurs at a *minimum* of the potential energy, such that $d^2U/dx^2 > 0$. In the presence of an external force J, we must minimize the "enthalpy" $H = U - Jx$, which amounts to tilting the potential. At the new equilibrium point $x_{eq}(J)$, we must require $d^2H/dx^2 = d^2U/dx^2 > 0$. Thus only the *convex* portions of the potential $U(x)$ are physically accessible.

With more than one mechanical coordinate, the requirement that any change $\delta \mathbf{x}$ results in an increase in energy (or enthalpy) can be written as

$$\sum_{i,j} \frac{\partial^2 U}{\partial x_i \partial x_j} \delta x_i \delta x_j > 0. \tag{1.61}$$

We can express the above equation in more symmetric form, by noting that the corresponding change in forces is given by

$$\delta J_i = \delta \left(\frac{\partial U}{\partial x_i} \right) = \sum_j \frac{\partial^2 U}{\partial x_i \partial x_j} \delta x_j. \tag{1.62}$$

Thus Eq. (1.61) is equivalent to

$$\sum_i \delta J_i \delta x_i > 0. \tag{1.63}$$

When dealing with a thermodynamic system, we should allow for thermal and chemical inputs to the internal energy of the system. Including the corresponding pairs of conjugate coordinates, the condition for mechanical stability should generalize to

$$\delta T \delta S + \sum_i \delta J_i \delta x_i + \sum_\alpha \delta \mu_\alpha \delta N_\alpha > 0. \tag{1.64}$$

Before examining the consequences of the above condition, I shall provide a more standard derivation that is based on the uniformity of an extended thermodynamic body. Consider a homogeneous system at equilibrium, characterized by intensive state functions $(T, \mathbf{J}, \mu)$, and extensive variables $(E, \mathbf{x}, \mathbf{N})$. Now imagine that the system is arbitrarily divided into two *equal* parts, and that one part spontaneously transfers some energy to the other part in the form of work or heat. The two subsystems, A and B, initially have the same values for the intensive variables, while their extensive coordinates satisfy $E_A + E_B = E$, $\mathbf{x}_A + \mathbf{x}_B = \mathbf{x}$, and $\mathbf{N}_A + \mathbf{N}_B = \mathbf{N}$. After the exchange of energy between the two subsystems, the coordinates of A change to

$$(E_A + \delta E, \ \mathbf{x}_A + \delta\mathbf{x}, \ \mathbf{N}_A + \delta\mathbf{N}) \text{ and } (T_A + \delta T_A, \ \mathbf{J}_A + \delta\mathbf{J}_A, \ \mu_A + \delta\mu_A), \qquad (1.65)$$

and those of B to

$$(E_B - \delta E, \ \mathbf{x}_B - \delta\mathbf{x}, \ \mathbf{N}_B - \delta\mathbf{N}) \text{ and } (T_B + \delta T_B, \ \mathbf{J}_B + \delta\mathbf{J}_B, \ \mu_B + \delta\mu_B). \qquad (1.66)$$

Fig. 1.20 Spontaneous change between two halves of a homogeneous system.

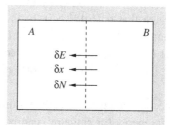

Note that the overall system is maintained at constant E, $\mathbf{x}$, and $\mathbf{N}$. Since the intensive variables are themselves functions of the extensive coordinates, to *first order* in the variations of $(E, \mathbf{x}, \mathbf{N})$, we have

$$\delta T_A = -\delta T_B \equiv \delta T, \quad \delta\mathbf{J}_A = -\delta\mathbf{J}_B \equiv \delta\mathbf{J}, \quad \delta\mu_A = -\delta\mu_B \equiv \delta\mu. \qquad (1.67)$$

Using Eq. (1.48), the entropy of the system can be written as

$$S = S_A + S_B = \left(\frac{E_A}{T_A} - \frac{\mathbf{J}_A}{T_A} \cdot \mathbf{x}_A - \frac{\mu_A}{T_A} \cdot \mathbf{N}_A \right) + \left(\frac{E_B}{T_B} - \frac{\mathbf{J}_B}{T_B} \cdot \mathbf{x}_B - \frac{\mu_B}{T_B} \cdot \mathbf{N}_B \right). \qquad (1.68)$$

Since, by assumption, we are expanding about the equilibrium point, the first-order changes vanish, and to second order

$$\delta S = \delta S_A + \delta S_B = 2 \left[\delta\left(\frac{1}{T_A} \right) \delta E_A - \delta\left(\frac{\mathbf{J}_A}{T_A} \right) \cdot \delta\mathbf{x}_A - \delta\left(\frac{\mu_A}{T_A} \right) \cdot \delta\mathbf{N}_A \right]. \qquad (1.69)$$

(We have used Eq. (1.67) to note that the second-order contribution of B is the same as A.) Equation (1.69) can be rearranged to

$$\delta S = -\frac{2}{T_A} \left[\delta T_A \left(\frac{\delta E_A - \mathbf{J}_A \cdot \delta\mathbf{x}_A - \mu_A \cdot \delta\mathbf{N}_A}{T_A} \right) + \delta\mathbf{J}_A \cdot \delta\mathbf{x}_A + \delta\mu_A \cdot \delta\mathbf{N}_A \right]$$
$$= -\frac{2}{T_A} \left[\delta T_A \delta S_A + \delta\mathbf{J}_A \cdot \delta\mathbf{x}_A + \delta\mu_A \cdot \delta\mathbf{N}_A \right]. \qquad (1.70)$$

The condition for stable equilibrium is that any change should lead to a decrease in entropy, and hence we must have

$$\delta T \delta S + \delta \mathbf{J} \cdot \delta \mathbf{x} + \delta \boldsymbol{\mu} \cdot \delta \mathbf{N} \geq 0. \tag{1.71}$$

We have now removed the subscript A, as Eq. (1.71) must apply to the whole system as well as to any part of it.

The above condition was obtained assuming that the overall system is kept at constant E, $\mathbf{x}$, and $\mathbf{N}$. In fact, since all coordinates appear symmetrically in this expression, the same result is obtained for any other set of constraints. For example, variations in δT and $\delta \mathbf{x}$ with $\delta \mathbf{N} = 0$, lead to

$$\begin{cases} \delta S = \left. \dfrac{\partial S}{\partial T} \right|_{\mathbf{x}} \delta T + \left. \dfrac{\partial S}{\partial x_i} \right|_{T} \delta x_i \\[2mm] \delta J_i = \left. \dfrac{\partial J_i}{\partial T} \right|_{\mathbf{x}} \delta T + \left. \dfrac{\partial J_i}{\partial x_j} \right|_{T} \delta x_j \end{cases}. \tag{1.72}$$

Substituting these variations into Eq. (1.71) leads to

$$\left. \frac{\partial S}{\partial T} \right|_{\mathbf{x}} (\delta T)^2 + \left. \frac{\partial J_i}{\partial x_j} \right|_{T} \delta x_i \delta x_j \geq 0. \tag{1.73}$$

Note that the cross terms proportional to $\delta T \delta x_i$ cancel due to the Maxwell relation in Eq. (1.56). Equation (1.73) is a *quadratic form*, and must be positive for *all choices* of δT and $\delta \mathbf{x}$. The resulting constraints on the coefficients are independent of how the system was initially partitioned into subsystems A and B, and represent the conditions for stable equilibrium. If only δT is non-zero, Eq. (1.71) requires $\partial S / \partial T |_{\mathbf{x}} \geq 0$, implying a positive heat capacity, since

$$C_{\mathbf{x}} = \left. \frac{\mathrm{d} Q}{\mathrm{d} T} \right|_{\mathbf{x}} = T \left. \frac{\partial S}{\partial T} \right|_{\mathbf{x}} \geq 0. \tag{1.74}$$

If only one of the δx_i in Eq. (1.71) is non-zero, the corresponding response function $\partial x_i / \partial J_i |_{T, x_{j \neq i}}$ must be positive. However, a more general requirement exists since all $\delta \mathbf{x}$ values may be chosen non-zero. The general requirement is that the matrix of coefficients $\partial J_i / \partial x_j |_{T}$ must be *positive definite*. A matrix is positive definite if all of its eigenvalues are positive. It is necessary, but not sufficient, that all the diagonal elements of such a matrix (the inverse response functions) be positive, leading to further constraints between the response functions. Including chemical work for a gas, the appropriate matrix is

$$\begin{bmatrix} -\left. \dfrac{\partial P}{\partial V} \right|_{T,N} & -\left. \dfrac{\partial P}{\partial N} \right|_{T,V} \\[3mm] \left. \dfrac{\partial \mu}{\partial V} \right|_{T,N} & \left. \dfrac{\partial \mu}{\partial N} \right|_{T,V} \end{bmatrix}. \tag{1.75}$$

In addition to the positivity of the response functions $\kappa_{T,N} = -V^{-1} \partial V / \partial P |_{T,N}$ and $\partial N / \partial \mu |_{T,V}$, the determinant of the matrix must be positive, requiring

$$\left. \frac{\partial P}{\partial N} \right|_{T,V} \left. \frac{\partial \mu}{\partial V} \right|_{T,N} - \left. \frac{\partial P}{\partial V} \right|_{T,N} \left. \frac{\partial \mu}{\partial N} \right|_{T,V} \geq 0. \tag{1.76}$$

Another interesting consequence of Eq. (1.71) pertains to the critical point of a gas where $\partial P/\partial V|_{T_c,N} = 0$. Assuming that the critical isotherm can be analytically expanded as

$$\delta P(T = T_c) = \left.\frac{\partial P}{\partial V}\right|_{T_c,N} \delta V + \frac{1}{2}\left.\frac{\partial^2 P}{\partial V^2}\right|_{T_c,N}\delta V^2 + \frac{1}{6}\left.\frac{\partial^3 P}{\partial V^3}\right|_{T_c,N}\delta V^3 + \cdots, \quad (1.77)$$

the stability condition $-\delta P\delta V \geq 0$ implies that $\partial^2 P/\partial V^2|_{T_c,N}$ must be zero, and the third derivative negative, if the first derivative vanishes. This condition is used to obtain the critical point of the gas from approximations to the isotherms (as we shall do in a later chapter for the van der Waals equation). In fact, it is usually not justified to make a Taylor expansion around the critical point as in Eq. (1.77), although the constraint $-\delta P\delta V \geq 0$ remains applicable.

Fig. 1.21 Stability condition at criticality, illustrated for van der Waals isotherms.

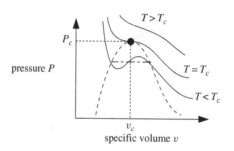

1.10 The third law

Differences in entropy between two states can be computed using the second law, from $\Delta S = \int dQ_{rev}/T$. Low-temperature experiments indicate that $\Delta S(\mathbf{X}, T)$ vanishes as T goes to zero for any set of coordinates $\mathbf{X}$. This observation is independent of the other laws of thermodynamics, leading to the formulation of a third law by Nernst, which states:

The entropy of all systems at zero absolute temperature is a universal constant that can be taken to be zero.

The above statement actually implies that

$$\lim_{T\to 0} S(\mathbf{X}, T) = 0, \quad (1.78)$$

which is a stronger requirement than the vanishing of the differences $\Delta S(\mathbf{X}, T)$. This extended condition has been tested for metastable phases of a substance. Certain materials, such as sulfur or phosphine, can exist in a number of rather similar crystalline structures (*allotropes*). Of course, at a given temperature only one of these structures is truly stable. Let us imagine that, as the high-temperature equilibrium phase A is cooled slowly, it makes a transition at a temperature T^* to phase B, releasing *latent heat L*. Under more rapid cooling conditions the transition is avoided, and phase A persists in metastable equilibrium. The entropies in the two phases can be calculated by measuring the heat capacities $C_A(T)$ and $C_B(T)$. Starting from $T = 0$, the entropy at a temperature slightly above T^* can be computed along the two possible paths as

$$S(T^* + \epsilon) = S_A(0) + \int_0^{T^*} dT' \frac{C_A(T')}{T'} = S_B(0) + \int_0^{T^*} dT' \frac{C_B(T')}{T'} + \frac{L}{T^*}. \qquad (1.79)$$

By such measurements we can indeed verify that $S_A(0) = S_B(0) \equiv 0$.

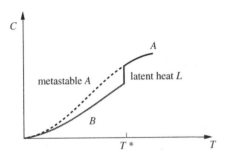

Fig. 1.22 Heat capacity measurements on allotropes of the same material.

Consequences of the third law:

1. Since $S(T = 0, \mathbf{X}) = 0$ for all coordinates $\mathbf{X}$,

$$\lim_{T \to 0} \frac{\partial S}{\partial \mathbf{X}} \bigg|_T = 0. \qquad (1.80)$$

2. Heat capacities must vanish as $T \to 0$ since

$$S(T, \mathbf{X}) - S(0, \mathbf{X}) = \int_0^T dT' \frac{C_\mathbf{X}(T')}{T'}, \qquad (1.81)$$

and the integral diverges as $T \to 0$ unless

$$\lim_{T \to 0} C_\mathbf{X}(T) = 0. \qquad (1.82)$$

3. Thermal expansivities also vanish as $T \to 0$ since

$$\alpha_J = \frac{1}{x} \frac{\partial x}{\partial T} \bigg|_J = \frac{1}{x} \frac{\partial S}{\partial J} \bigg|_T. \qquad (1.83)$$

The second equality follows from the Maxwell relation in Eq. (1.56). The vanishing of the latter is guaranteed by Eq. (1.80).

4. It is impossible to cool any system to absolute zero temperature in a finite number of steps. For example, we can cool a gas by an adiabatic reduction in pressure. Since the curves of S versus T for different pressures must join at $T = 0$, successive steps involve progressively smaller changes, in S and in T, on approaching zero temperature. Alternatively, the unattainability of zero temperatures implies that $S(T = 0, P)$ is independent of P. This is a weaker statement of the third law, which also implies the equality of zero temperature entropy for different substances.

Fig. 1.23 An infinite number of steps is required to cool a gas to $T = 0$ by a series of isothermal decompressions.

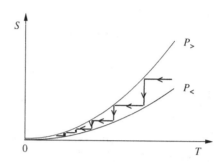

In the following chapters, we shall attempt to justify the laws of thermodynamics from a microscopic point of view. The first law is clearly a reflection of the conservation of energy, which also operates at the microscopic level. The zeroth and second laws suggest an irreversible approach to equilibrium, a concept that has no analog at the particulate level. It is justified as reflecting the collective tendencies of large numbers of degrees of freedom. In statistical mechanics the entropy per particle is calculated as $S/N = k_B \ln(g_N)/N$, where g_N is the degeneracy of the states (number of configurations with the same energy) of a system of N particles. The third law of thermodynamics thus requires that $\lim_{N \to \infty} \ln(g_N)/N \to 0$ at $T = 0$, limiting the possible number of ground states for a many-body system.

The above condition *does not hold* within the framework of classical statistical mechanics, as there are examples of both non-interacting (such as an ideal gas) and interacting (the frustrated spins in a triangular antiferromagnet) systems with a large number of (degenerate) ground states, and a finite zero-temperature entropy. However, classical mechanics is inapplicable at very low temperatures and energies where quantum effects become important. The third law is then equivalent to a restriction on the degeneracy of ground states of a quantum mechanical system.[3] While this can be proved for a non-interacting

---

[3] For any spin system with rotational symmetry, such as a ferromagnet, there are of course many possible ground states related by rotations. However, the number of such states does not grow with the number of spins N, thus such degeneracy does not affect the absence of a thermodynamic entropy at zero temperature.

system of identical particles (as we shall demonstrate in the final chapter), there is no general proof of its validity with interactions. Unfortunately, the onset of quantum effects (and other possible origins of the breaking of classical degeneracy) is system-specific. Hence it is not a priori clear how low the temperature must be before the predictions of the third law can be observed. Another deficiency of the law is its inapplicability to glassy phases. Glasses result from the freezing of supercooled liquids into configurations with extremely slow dynamics. While not truly equilibrium phases (and hence subject to all the laws of thermodynamics), they are effectively so due to the slowness of the dynamics. A possible test of the applicability of the third law to glasses is discussed in the problems.

Problems for chapter 1

1. *Surface tension*: thermodynamic properties of the interface between two phases are described by a state function called the surface tension $\mathcal{S}$. It is defined in terms of the work required to increase the surface area by an amount dA through $đW = \mathcal{S}dA$.

   (a) By considering the work done against surface tension in an infinitesimal change in radius, show that the pressure inside a spherical drop of water of radius R is larger than outside pressure by $2\mathcal{S}/R$. What is the air pressure inside a soap bubble of radius R?

   (b) A water droplet condenses on a solid surface. There are three surface tensions involved, $\mathcal{S}_{aw}$, $\mathcal{S}_{sw}$, and $\mathcal{S}_{sa}$, where a, s, and w refer to air, solid, and water, respectively. Calculate the angle of contact, and find the condition for the appearance of a water film (complete wetting).

   (c) In the realm of "large" bodies gravity is the dominant force, while at "small" distances surface tension effects are all important. At room temperature, the surface tension of water is $\mathcal{S}_o \approx 7 \times 10^{-2}\,\mathrm{N\,m^{-1}}$. Estimate the typical length scale that separates "large" and "small" behaviors. Give a couple of examples for where this length scale is important.

****

2. *Surfactants*: surfactant molecules such as those in soap or shampoo prefer to spread on the air–water surface rather than dissolve in water. To see this, float a hair on the surface of water and gently touch the water in its vicinity with a piece of soap. (This is also why a piece of soap can power a toy paper boat.)

   (a) The air–water surface tension $\mathcal{S}_o$ (assumed to be temperature-independent) is reduced roughly by $N\,k_B T/A$, where N is the number of surfactant particles, and A is the area. Explain this result qualitatively.

   (b) Place a drop of water on a clean surface. Observe what happens to the air–water surface contact angle as you gently touch the droplet surface with a small piece of soap, and explain the observation.

(c) More careful observations show that at higher surfactant densities

$$\left.\frac{\partial S}{\partial A}\right|_T = \frac{Nk_BT}{(A-Nb)^2} - \frac{2a}{A}\left(\frac{N}{A}\right)^2 \quad , \quad \text{and} \quad \left.\frac{\partial T}{\partial S}\right|_A = -\frac{A-Nb}{Nk_B},$$

where a and b are constants. Obtain the expression for $S(A, T)$ and explain qualitatively the origin of the corrections described by a and b.

(d) Find an expression for $C_S - C_A$ in terms of $\left.\frac{\partial E}{\partial A}\right|_T = \left.\frac{\partial E}{\partial A}\right|_S$, S, $\left.\frac{\partial S}{\partial A}\right|_T$, and $\left.\frac{\partial T}{\partial S}\right|_A$.

****

3. *Temperature scales*: prove the equivalence of the ideal gas temperature scale Θ, and the thermodynamic scale T, by performing a Carnot cycle on an ideal gas. The ideal gas satisfies $PV = N k_B \Theta$, and its internal energy E is a function of Θ only. However, *you may not assume that $E \propto \Theta$*. You may wish to proceed as follows:

(a) Calculate the heat exchanges Q_H and Q_C as a function of Θ_H, Θ_C, and the volume expansion factors.

(b) Calculate the volume expansion factor in an adiabatic process as a function of Θ.

(c) Show that $Q_H/Q_C = \Theta_H/\Theta_C$.

****

4. *Equations of state*: the equation of state constrains the form of internal energy as in the following examples.

(a) Starting from $dE = TdS - PdV$, show that the equation of state $PV = Nk_BT$ in fact implies that E can only depend on T.

(b) What is the most general equation of state consistent with an internal energy that depends only on temperature?

(c) Show that for a van der Waals gas C_V is a function of temperature alone.

****

5. The *Clausius–Clapeyron equation* describes the variation of boiling point with pressure. It is usually derived from the condition that the chemical potentials of the gas and liquid phases are the same at coexistence. For an alternative derivation, consider a Carnot engine using one mole of water. At the source (P, T) the latent heat L is supplied converting water to steam. There is a volume increase V associated with this process. The pressure is adiabatically decreased to $P - dP$. At the sink $(P - dP, T - dT)$ steam is condensed back to water.

(a) Show that the work output of the engine is $W = VdP + \mathcal{O}(dP^2)$. Hence obtain the Clausius–Clapeyron equation

$$\left.\frac{dP}{dT}\right|_{\text{boiling}} = \frac{L}{TV}. \tag{1}$$

(b) What is wrong with the following argument: "The heat Q_H supplied at the source to convert one mole of water to steam is $L(T)$. At the sink $L(T - dT)$ is supplied to condense one mole of steam to water. The difference $dT dL/dT$ must equal the work $W = V dP$, equal to $L dT/T$ from Eq. (1). Hence $dL/dT = L/T$, implying that L is proportional to T!"

(c) Assume that L is approximately temperature-independent, and that the volume change is dominated by the volume of steam treated as an ideal gas, that is, $V = N k_B T/P$. Integrate Eq. (1) to obtain $P(T)$.

(d) A hurricane works somewhat like the engine described above. Water evaporates at the warm surface of the ocean, steam rises up in the atmosphere, and condenses to water at the higher and cooler altitudes. The Coriolis force converts the upward suction of the air to spiral motion. (Using ice and boiling water, you can create a little storm in a tea cup.) Typical values of warm ocean surface and high altitude temperatures are $80°$ F and $-120°$ F, respectively. The warm water surface layer must be at least 200 feet thick to provide sufficient water vapor, as the hurricane needs to condense about 90 million tons of water vapor per hour to maintain itself. Estimate the maximum possible efficiency, and power output, of such a hurricane. (The latent heat of vaporization of water is about 2.3×10^6 J kg^{-1}.)

(e) Due to gravity, atmospheric pressure $P(h)$ drops with the height h. By balancing the forces acting on a slab of air (behaving like a perfect gas) of thickness dh, show that $P(h) = P_0 \exp(-mgh/kT)$, where m is the average mass of a molecule in air.

(f) Use the above results to estimate the boiling temperature of water on top of Mount Everest ($h \approx 9$ km). The latent heat of vaporization of water is about 2.3×10^6 J kg^{-1}.

****

6. *Glass*: liquid quartz, if cooled slowly, crystallizes at a temperature T_m, and releases latent heat L. Under more rapid cooling conditions, the liquid is supercooled and becomes glassy

(a) As both phases of quartz are almost incompressible, there is no work input, and changes in internal energy satisfy $dE = T dS + \mu dN$. Use the extensivity condition to obtain the expression for μ in terms of E, T, S, and N.

(b) The heat capacity of crystalline quartz is approximately $C_X = \alpha T^3$, while that of glassy quartz is roughly $C_G = \beta T$, where α and β are constants.
Assuming that the third law of thermodynamics applies to both crystalline and glass phases, calculate the entropies of the two phases at temperatures $T \leq T_m$.

(c) At zero temperature the local bonding structure is similar in glass and crystalline quartz, so that they have approximately the same internal energy E_0. Calculate the internal energies of both phases at temperatures $T \leq T_m$.

(d) Use the condition of thermal equilibrium between two phases to compute the equilibrium melting temperature T_m in terms of α and β.

(e) Compute the latent heat L in terms of α and β.

(f) Is the result in the previous part correct? If not, which of the steps leading to it is most likely to be incorrect?

****

---

7. *Filament*: for an elastic filament it is found that, at a finite range in temperature, a displacement x requires a force

$$J = ax - bT + cTx,$$

where a, b, and c are constants. Furthermore, its heat capacity at constant displacement is proportional to temperature, that is, $C_x = A(x)T$.

(a) Use an appropriate Maxwell relation to calculate $\partial S/\partial x|_T$.
(b) Show that A has to in fact be independent of x, that is, $dA/dx = 0$.
(c) Give the expression for $S(T, x)$ assuming $S(0, 0) = S_0$.
(d) Calculate the heat capacity at constant tension, that is, $C_J = T \, \partial S/\partial T|_J$ as a function of T and J.

****

8. *Hard core gas*: a gas obeys the equation of state $P(V - Nb) = Nk_BT$, and has a heat capacity C_V independent of temperature. (N is kept fixed in the following.)

(a) Find the Maxwell relation involving $\partial S/\partial V|_{T,N}$.
(b) By calculating $dE(T, V)$, show that E is a function of T (and N) only.
(c) Show that $\gamma \equiv C_P/C_V = 1 + Nk_B/C_V$ (independent of T and V).
(d) By writing an expression for $E(P, V)$, or otherwise, show that an adiabatic change satisfies the equation $P(V - Nb)^\gamma = $ constant.

****

9. *Superconducting transition*: many metals become superconductors at low temperatures T, and magnetic fields B. The heat capacities of the two phases at zero magnetic field are approximately given by

$$\begin{cases} C_s(T) = V\alpha T^3 & \text{in the superconducting phase} \\ C_n(T) = V\left[\beta T^3 + \gamma T\right] & \text{in the normal phase} \end{cases},$$

where V is the volume, and $\{\alpha, \beta, \gamma\}$ are constants. (There is no appreciable change in volume at this transition, and mechanical work can be ignored throughout this problem.)

(a) Calculate the entropies $S_s(T)$ and $S_n(T)$ of the two phases at zero field, using the third law of thermodynamics.
(b) Experiments indicate that there is no latent heat ($L = 0$) for the transition between the normal and superconducting phases at zero field. Use this information to obtain the transition temperature T_c, as a function of α, β, and γ.

(c) At zero temperature, the electrons in the superconductor form bound Cooper pairs. As a result, the internal energy of the superconductor is reduced by an amount $V\Delta$, that is, $E_n(T = 0) = E_0$ and $E_s(T = 0) = E_0 - V\Delta$ for the metal and superconductor, respectively. Calculate the internal energies of both phases at finite temperatures.

(d) By comparing the Gibbs free energies (or chemical potentials) in the two phases, obtain an expression for the energy gap Δ in terms of α, β, and γ.

(e) In the presence of a magnetic field B, inclusion of magnetic work results in $dE = TdS + BdM + \mu dN$, where M is the magnetization. The superconducting phase is a perfect diamagnet, expelling the magnetic field from its interior, such that $M_s = -VB/(4\pi)$ in appropriate units. The normal metal can be regarded as approximately non-magnetic, with $M_n = 0$. Use this information, in conjunction with previous results, to show that the superconducting phase becomes normal for magnetic fields larger than

$$B_c(T) = B_0 \left(1 - \frac{T^2}{T_c^2} \right),$$

giving an expression for B_0.

****

10. *Photon gas Carnot cycle*: the aim of this problem is to obtain the black-body radiation relation, $E(T, V) \propto VT^4$, starting from the equation of state, by performing an infinitesimal Carnot cycle on the photon gas.

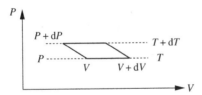

(a) Express the work done, W, in the above cycle, in terms of dV and dP.

(b) Express the heat absorbed, Q, in expanding the gas *along an isotherm*, in terms of P, dV, and an appropriate derivative of $E(T, V)$.

(c) Using the efficiency of the Carnot cycle, relate the above expressions for W and Q to T and dT.

(d) Observations indicate that the pressure of the photon gas is given by $P = AT^4$, where $A = \pi^2 k_B^4 / 45 (\hbar c)^3$ is a constant. Use this information to obtain $E(T, V)$, assuming $E(0, V) = 0$.

(e) Find the relation describing the *adiabatic paths* in the above cycle.

****

11. *Irreversible processes*

(a) Consider two substances, initially at temperatures T_1^0 and T_2^0, coming to equilibrium at a final temperature T_f through heat exchange. By relating the direction of heat flow to the temperature difference, show that the change in the total entropy, which can be written as

$$\Delta S = \Delta S_1 + \Delta S_2 \geq \int_{T_1^0}^{T_f} \frac{dQ_1}{T_1} + \int_{T_2^0}^{T_f} \frac{dQ_2}{T_2} = \int \frac{T_1 - T_2}{T_1 T_2} dQ,$$

must be positive. This is an example of the more general condition that "*in a closed system, equilibrium is characterized by the maximum value of entropy S.*"

(b) Now consider a gas with adjustable volume V, and diathermal walls, embedded in a heat bath of constant temperature T, *and* fixed pressure P. The change in the entropy of the bath is given by

$$\Delta S_{\text{bath}} = \frac{\Delta Q_{\text{bath}}}{T} = -\frac{\Delta Q_{\text{gas}}}{T} = -\frac{1}{T} \left(\Delta E_{\text{gas}} + P \Delta V_{\text{gas}} \right).$$

By considering the change in entropy of the combined system establish that "*the equilibrium of a gas at fixed T and P is characterized by the minimum of the Gibbs free energy G = E + PV − TS.*"

****

12. *The Solar System* originated from a dilute gas of particles, sufficiently separated from other such clouds to be regarded as an isolated system. Under the action of gravity the particles coalesced to form the Sun and planets.

(a) The motion and organization of planets is much more ordered than the original dust cloud. Why does this not violate the second law of thermodynamics?

(b) The nuclear processes of the Sun convert protons to heavier elements such as carbon. Does this further organization lead to a reduction in entropy?

(c) The evolution of life and intelligence requires even further levels of organization. How is this achieved on Earth without violating the second law?

****

2
Probability

2.1 General definitions

The laws of thermodynamics are based on observations of *macroscopic bodies*, and encapsulate their thermal properties. On the other hand, matter is composed of atoms and molecules whose motions are governed by more fundamental laws (classical or quantum mechanics). It should be possible, in principle, to derive the behavior of a macroscopic body from the knowledge of its components. This is the problem addressed by kinetic theory in the following chapter. Actually, describing the full dynamics of the enormous number of particles involved is quite a daunting task. As we shall demonstrate, for discussing equilibrium properties of a macroscopic system, full knowledge of the behavior of its constituent particles is not necessary. All that is required is the *likelihood* that the particles are in a particular microscopic state. Statistical mechanics is thus an inherently *probabilistic* description of the system, and familiarity with manipulations of probabilities is an important prerequisite. The purpose of this chapter is to review some important results in the theory of probability, and to introduce the notations that will be used in the following chapters.

The entity under investigation is a *random variable x*, which has a set of possible *outcomes* $S \equiv \{x_1, x_2, \cdots\}$. The outcomes may be *discrete* as in the case of a coin toss, $S_{coin} = \{head, tail\}$, or a dice throw, $S_{dice} = \{1, 2, 3, 4, 5, 6\}$, or *continuous* as for the velocity of a particle in a gas, $S_{\vec{v}} = \{-\infty < v_x, v_y, v_z < \infty\}$, or the energy of an electron in a metal at zero temperature, $S_\epsilon = \{0 \le \epsilon \le \epsilon_F\}$. An *event* is any subset of outcomes $E \subset S$, and is assigned a *probability $p(E)$*, for example, $p_{dice}(\{1\}) = 1/6$, or $p_{dice}(\{1, 3\}) = 1/3$. From an axiomatic point of view, the probabilities must satisfy the following conditions:

(i) *Positivity.* $p(E) \ge 0$, that is, all probabilities must be real and non-negative.
(ii) *Additivity.* $p(A \text{ or } B) = p(A) + p(B)$, if A and B are disconnected events.
(iii) *Normalization.* $p(S) = 1$, that is, the random variable must have some outcome in S.

From a practical point of view, we would like to know how to assign probability values to various outcomes. There are two possible approaches:

1. *Objective* probabilities are obtained *experimentally* from the relative frequency of the occurrence of an outcome in many tests of the random variable. If the random process is repeated N times, and the event A occurs N_A times, then

$$p(A) = \lim_{N \to \infty} \frac{N_A}{N}.$$

   For example, a series of $N = 100, 200, 300$ throws of a dice may result in $N_1 = 19, 30, 48$ occurrences of 1. The ratios $0.19, 0.15, 0.16$ provide an increasingly more reliable estimate of the probability $p_{\text{dice}}(\{1\})$.

2. *Subjective* probabilities provide a *theoretical* estimate based on the uncertainties related to lack of precise knowledge of outcomes. For example, the assessment $p_{\text{dice}}(\{1\}) = 1/6$ is based on the knowledge that there are six possible outcomes to a dice throw, and that in the absence of any prior reason to believe that the dice is biased, all six are equally likely. All assignments of probability in statistical mechanics are subjectively based. The consequences of such subjective assignments of probability have to be checked against measurements, and they may need to be modified as more information about the outcomes becomes available.

2.2 One random variable

As the properties of a discrete random variable are rather well known, here we focus on continuous random variables, which are more relevant to our purposes. Consider a random variable x, whose outcomes are real numbers, that is, $\mathcal{S}_x = \{-\infty < x < \infty\}$.

- *The cumulative probability function* (CPF) $P(x)$ is the probability of an outcome with *any value* less than x, that is, $P(x) = \text{prob}(E \subset [-\infty, x])$. $P(x)$ must be a monotonically increasing function of x, with $P(-\infty) = 0$ and $P(+\infty) = 1$.

Fig. 2.1 A typical cumulative probability function.

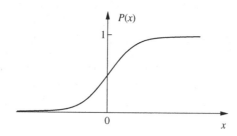

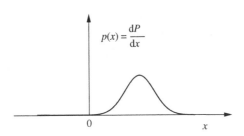

Fig. 2.2 A typical probability density function.

- *The probability density function* (PDF) is defined by $p(x) \equiv \mathrm{d}P(x)/\mathrm{d}x$. Hence, $p(x)\mathrm{d}x = \mathrm{prob}(E \in [x, x+\mathrm{d}x])$. As a probability density, it is *positive*, and normalized such that

$$\mathrm{prob}(\mathcal{S}) = \int_{-\infty}^{\infty} \mathrm{d}x \, p(x) = 1. \qquad (2.1)$$

Note that since $p(x)$ is a *probability density*, it has dimensions of $[x]^{-1}$, and changes its value if the units measuring x are modified. Unlike $P(x)$, the PDF has no upper bound, that is, $0 < p(x) < \infty$, and may contain divergences as long as they are integrable.

- *The expectation value* of any function, $F(x)$, of the random variable is

$$\langle F(x) \rangle = \int_{-\infty}^{\infty} \mathrm{d}x \, p(x) F(x). \qquad (2.2)$$

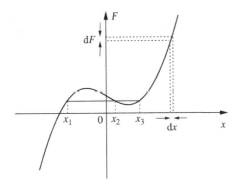

Fig. 2.3 Obtaining the PDF for the function $F(x)$.

The function $F(x)$ is itself a random variable, with an associated PDF of $p_F(f)\mathrm{d}f = \mathrm{prob}(F(x) \in [f, f+\mathrm{d}f])$. There may be multiple solutions x_i to the equation $F(x) = f$, and

$$p_F(f)\mathrm{d}f = \sum_i p(x_i)\mathrm{d}x_i, \quad \Longrightarrow \quad p_F(f) = \sum_i p(x_i) \left| \frac{\mathrm{d}x}{\mathrm{d}F} \right|_{x=x_i}. \qquad (2.3)$$

The factors of $|\mathrm{d}x/\mathrm{d}F|$ are the *Jacobians* associated with the change of variables from x to F. For example, consider $p(x) = \lambda \exp(-\lambda|x|)/2$, and the function

$F(x) = x^2$. There are two solutions to $F(x) = f$, located at $x_\pm = \pm\sqrt{f}$, with corresponding Jacobians $|\pm f^{-1/2}/2|$. Hence,

$$P_F(f) = \frac{\lambda}{2}\exp\left(-\lambda\sqrt{f}\right)\left(\left|\frac{1}{2\sqrt{f}}\right| + \left|\frac{-1}{2\sqrt{f}}\right|\right) = \frac{\lambda\exp\left(-\lambda\sqrt{f}\right)}{2\sqrt{f}},$$

for $f > 0$, and $p_F(f) = 0$ for $f < 0$. Note that $p_F(f)$ has an (integrable) divergence at $f = 0$.

Fig. 2.4 Probability density functions for x, and $F = x^2$.

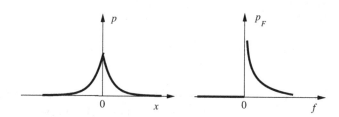

- *Moments* of the PDF are expectation values for powers of the random variable. The nth moment is

$$m_n \equiv \langle x^n \rangle = \int \mathrm{d}x p(x) x^n. \tag{2.4}$$

- *The characteristic function* is the generator of moments of the distribution. It is simply the Fourier transform of the PDF, defined by

$$\tilde{p}(k) = \left\langle \mathrm{e}^{-\mathrm{i}kx} \right\rangle = \int \mathrm{d}x p(x)\,\mathrm{e}^{-\mathrm{i}kx}. \tag{2.5}$$

The PDF can be recovered from the characteristic function through the inverse Fourier transform

$$p(x) = \frac{1}{2\pi}\int \mathrm{d}k \tilde{p}(k)\,\mathrm{e}^{+\mathrm{i}kx}. \tag{2.6}$$

Moments of the distribution are obtained by expanding $\tilde{p}(k)$ in powers of k,

$$\tilde{p}(k) = \left\langle \sum_{n=0}^{\infty} \frac{(-\mathrm{i}k)^n}{n!} x^n \right\rangle = \sum_{n=0}^{\infty} \frac{(-\mathrm{i}k)^n}{n!}\langle x^n \rangle. \tag{2.7}$$

Moments of the PDF around any point x_0 can also be generated by expanding

$$\mathrm{e}^{\mathrm{i}kx_0}\tilde{p}(k) = \left\langle \mathrm{e}^{-\mathrm{i}k(x-x_0)} \right\rangle = \sum_{n=0}^{\infty} \frac{(-\mathrm{i}k)^n}{n!}\langle (x-x_0)^n \rangle. \tag{2.8}$$

• *The cumulant generating function* is the logarithm of the characteristic function. Its expansion generates the *cumulants* of the distribution defined through

$$\ln \tilde{p}(k) = \sum_{n=1}^{\infty} \frac{(-ik)^n}{n!} \langle x^n \rangle_c . \qquad (2.9)$$

Relations between moments and cumulants can be obtained by expanding the logarithm of $\tilde{p}(k)$ in Eq. (2.7), and using

$$\ln(1+\epsilon) = \sum_{n=1}^{\infty} (-1)^{n+1} \frac{\epsilon^n}{n} . \qquad (2.10)$$

The first four cumulants are called the *mean, variance, skewness,* and *curtosis* (or kurtosis) of the distribution, respectively, and are obtained from the moments as

$$\begin{aligned}
\langle x \rangle_c &= \langle x \rangle , \\
\langle x^2 \rangle_c &= \langle x^2 \rangle - \langle x \rangle^2 , \\
\langle x^3 \rangle_c &= \langle x^3 \rangle - 3 \langle x^2 \rangle \langle x \rangle + 2 \langle x \rangle^3 , \\
\langle x^4 \rangle_c &= \langle x^4 \rangle - 4 \langle x^3 \rangle \langle x \rangle - 3 \langle x^2 \rangle^2 + 12 \langle x^2 \rangle \langle x \rangle^2 - 6 \langle x \rangle^4 .
\end{aligned} \qquad (2.11)$$

The cumulants provide a useful and compact way of describing a PDF.

An important theorem allows easy computation of moments in terms of the cumulants: represent the nth cumulant graphically as a *connected cluster* of n points. The mth moment is then obtained by summing all possible subdivisions of m points into groupings of smaller (connected or disconnected) clusters. The contribution of each subdivision to the sum is the product of the connected cumulants that it represents. Using this result, the first four moments are computed graphically.

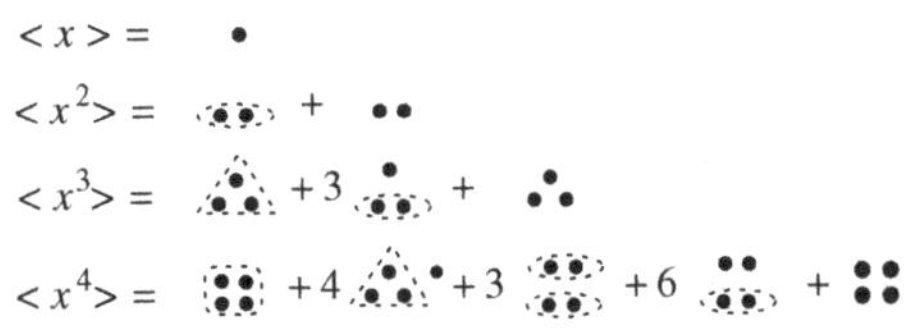

Fig. 2.5 Graphical computation of the first four moments.

The corresponding algebraic expressions are

$$\begin{aligned}
\langle x \rangle &= \langle x \rangle_c , \\
\langle x^2 \rangle &= \langle x^2 \rangle_c + \langle x \rangle_c^2 , \\
\langle x^3 \rangle &= \langle x^3 \rangle_c + 3 \langle x^2 \rangle_c \langle x \rangle_c + \langle x \rangle_c^3 , \\
\langle x^4 \rangle &= \langle x^4 \rangle_c + 4 \langle x^3 \rangle_c \langle x \rangle_c + 3 \langle x^2 \rangle_c^2 + 6 \langle x^2 \rangle_c \langle x \rangle_c^2 + \langle x \rangle_c^4 .
\end{aligned} \qquad (2.12)$$

This theorem, which is the starting point for various diagrammatic computations in statistical mechanics and field theory, is easily proved by equating the expressions in Eqs (2.7) and (2.9) for $\tilde{p}(k)$:

$$\sum_{m=0}^{\infty} \frac{(-ik)^m}{m!} \langle x^m \rangle = \exp\left[\sum_{n=1}^{\infty} \frac{(-ik)^n}{n!} \langle x^n \rangle_c\right] = \prod_n \sum_{p_n} \left[\frac{(-ik)^{np_n}}{p_n!} \left(\frac{\langle x^n \rangle_c}{n!}\right)^{p_n}\right]. \quad (2.13)$$

Matching the powers of $(-ik)^m$ on the two sides of the above expression leads to

$$\langle x^m \rangle = \sum_{\{p_n\}}' m! \prod_n \frac{1}{p_n!(n!)^{p_n}} \langle x^n \rangle_c^{p_n}. \quad (2.14)$$

The sum is restricted such that $\sum np_n = m$, and leads to the graphical interpretation given above, as the numerical factor is simply the number of ways of breaking m points into $\{p_n\}$ clusters of n points.

2.3 Some important probability distributions

The properties of three commonly encountered probability distributions are examined in this section.

(1) *The normal (Gaussian) distribution* describes a continuous real random variable x, with

$$p(x) = \frac{1}{\sqrt{2\pi\sigma^2}} \exp\left[-\frac{(x-\lambda)^2}{2\sigma^2}\right]. \quad (2.15)$$

The corresponding characteristic function also has a Gaussian form,

$$\tilde{p}(k) = \int_{-\infty}^{\infty} dx \frac{1}{\sqrt{2\pi\sigma^2}} \exp\left[-\frac{(x-\lambda)^2}{2\sigma^2} - ikx\right] = \exp\left[-ik\lambda - \frac{k^2\sigma^2}{2}\right]. \quad (2.16)$$

Cumulants of the distribution can be identified from $\ln \tilde{p}(k) = -ik\lambda - k^2\sigma^2/2$, using Eq. (2.9), as

$$\langle x \rangle_c = \lambda, \quad \langle x^2 \rangle_c = \sigma^2, \quad \langle x^3 \rangle_c = \langle x^4 \rangle_c = \cdots = 0. \quad (2.17)$$

The normal distribution is thus completely specified by its two first cumulants. This makes the computation of moments using the cluster expansion (Eqs. (2.12)) particularly simple, and

$$\begin{aligned}
\langle x \rangle &= \lambda, \\
\langle x^2 \rangle &= \sigma^2 + \lambda^2, \\
\langle x^3 \rangle &= 3\sigma^2\lambda + \lambda^3, \\
\langle x^4 \rangle &= 3\sigma^4 + 6\sigma^2\lambda^2 + \lambda^4,
\end{aligned} \quad (2.18)$$

$$\cdots$$

The normal distribution serves as the starting point for most perturbative computations in field theory. The vanishing of higher cumulants implies that all graphical computations involve only products of one-point, and two-point (known as propagators), clusters.

(2) *The binomial distribution*: consider a random variable with two outcomes A and B (e.g., a coin toss) of relative probabilities p_A and $p_B = 1 - p_A$. The probability that in N trials the event A occurs exactly N_A times (e.g., 5 heads in 12 coin tosses) is given by the binomial distribution

$$p_N(N_A) = \binom{N}{N_A} p_A^{N_A} p_B^{N - N_A}. \tag{2.19}$$

The prefactor,

$$\binom{N}{N_A} = \frac{N!}{N_A!(N - N_A)!}, \tag{2.20}$$

is just the coefficient obtained in the binomial expansion of $(p_A + p_B)^N$, and gives the number of possible orderings of N_A events A and $N_B = N - N_A$ events B. The characteristic function for this discrete distribution is given by

$$\tilde{p}_N(k) = \left\langle e^{-ikN_A} \right\rangle = \sum_{N_A=0}^{N} \frac{N!}{N_A!(N - N_A)!} p_A^{N_A} p_B^{N - N_A} e^{-ikN_A} = \left(p_A e^{-ik} + p_B \right)^N. \tag{2.21}$$

The resulting cumulant generating function is

$$\ln \tilde{p}_N(k) = N \ln \left(p_A e^{-ik} + p_B \right) = N \ln \tilde{p}_1(k), \tag{2.22}$$

where $\ln \tilde{p}_1(k)$ is the cumulant generating function for a single trial. Hence, the cumulants after N trials are simply N times the cumulants in a single trial. In each trial, the allowed values of N_A are 0 and 1 with respective probabilities p_B and p_A, leading to $\left\langle N_A^\ell \right\rangle = p_A$, for all ℓ. After N trials the first two cumulants are

$$\langle N_A \rangle_c = Np_A, \quad \left\langle N_A^2 \right\rangle_c = N \left(p_A - p_A^2 \right) = Np_Ap_B. \tag{2.23}$$

A measure of fluctuations around the mean is provided by the *standard deviation*, which is the square root of the variance. While the mean of the binomial distribution scales as N, its standard deviation only grows as $\sqrt{N}$. Hence, the *relative* uncertainty becomes smaller for large N.

The binomial distribution is straightforwardly generalized to a *multinomial* distribution, when the several outcomes $\{A, B, \cdots, M\}$ occur with probabilities $\{p_A, p_B, \cdots, p_M\}$. The probability of finding outcomes $\{N_A, N_B, \cdots, N_M\}$ in a total of $N = N_A + N_B + \cdots + N_M$ trials is

$$p_N(\{N_A, N_B, \cdots, N_M\}) = \frac{N!}{N_A!N_B!\cdots N_M!} p_A^{N_A} p_B^{N_B} \cdots p_M^{N_M}. \tag{2.24}$$

(3) *The Poisson distribution*: the classical example of a Poisson process is radioactive decay. Observing a piece of radioactive material over a time interval T shows that:

(a) The probability of one and only one event (decay) in the interval $[t, t+dt]$ is proportional to dt as $dt \to 0$.
(b) The probabilities of events at different intervals are independent of each other.

Fig. 2.6 Subdividing the time interval into small segments of size dt.

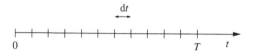

The probability of observing exactly M decays in the interval T is given by the Poisson distribution. It is obtained as a limit of the binomial distribution by subdividing the interval into $N = T/dt \gg 1$ segments of size dt. In each segment, an event occurs with probability $p = \alpha dt$, and there is no event with probability $q = 1 - \alpha dt$. As the probability of more than one event in dt is too small to consider, the process is equivalent to a binomial one. Using Eq. (2.21), the characteristic function is given by

$$\tilde{p}(k) = \left(pe^{-ik} + q\right)^n = \lim_{dt \to 0}\left[1 + \alpha dt\left(e^{-ik} - 1\right)\right]^{T/dt} = \exp\left[\alpha(e^{-ik} - 1)T\right]. \quad (2.25)$$

The Poisson PDF is obtained from the inverse Fourier transform in Eq. (2.6) as

$$p(x) = \int_{-\infty}^{\infty} \frac{dk}{2\pi} \exp\left[\alpha\left(e^{-ik} - 1\right)T + ikx\right] = e^{-\alpha T}\int_{-\infty}^{\infty}\frac{dk}{2\pi}e^{ikx}\sum_{M=0}^{\infty}\frac{(\alpha T)^M}{M!}e^{-ikM}, \quad (2.26)$$

using the power series for the exponential. The integral over k is

$$\int_{-\infty}^{\infty}\frac{dk}{2\pi}e^{ik(x-M)} = \delta(x - M), \quad (2.27)$$

leading to

$$p_{\alpha T}(x) = \sum_{M=0}^{\infty}e^{-\alpha T}\frac{(\alpha T)^M}{M!}\delta(x - M). \quad (2.28)$$

This answer clearly realizes that the only possible values of x are integers M. The probability of M events is thus $p_{\alpha T}(M) = e^{-\alpha T}(\alpha T)^M/M!$. The cumulants of the distribution are obtained from the expansion

$$\ln\tilde{p}_{\alpha T}(k) = \alpha T(e^{-ik} - 1) = \alpha T\sum_{n=1}^{\infty}\frac{(-ik)^n}{n!}, \quad \implies \quad \langle M^n\rangle_c = \alpha T. \quad (2.29)$$

All cumulants have the same value, and the moments are obtained from Eqs. (2.12) as

$$\langle M \rangle = (\alpha T), \quad \langle M^2 \rangle = (\alpha T)^2 + (\alpha T), \quad \langle M^3 \rangle = (\alpha T)^3 + 3(\alpha T)^2 + (\alpha T). \quad (2.30)$$

Example. Assuming that stars are randomly distributed in the Galaxy (clearly unjustified) with a density n, what is the probability that the nearest star is at a distance R? Since the probability of finding a star in a small volume dV is ndV, and they are assumed to be independent, the number of stars in a volume V is described by a Poisson process as in Eq. (2.28), with $\alpha = n$. The probability $p(R)$ of encountering the first star at a distance R is the product of the probabilities $p_{nV}(0)$ of finding zero stars in the volume $V = 4\pi R^3/3$ around the origin, and $p_{ndV}(1)$ of finding one star in the shell of volume $dV = 4\pi R^2 dR$ at a distance R. Both $p_{nV}(0)$ and $p_{ndV}(1)$ can be calculated from Eq. (2.28), and

$$p(R)dR = p_{nV}(0)\, p_{ndV}(1) = e^{-4\pi R^3 n/3}\, e^{-4\pi R^2 ndR}\, 4\pi R^2 ndR,$$

$$\implies \quad p(R) = 4\pi R^2 n \exp\left(-\frac{4\pi}{3} R^3 n\right). \quad (2.31)$$

2.4 Many random variables

With more than one random variable, the set of outcomes is an N-dimensional space, $\mathcal{S}_\mathbf{x} = \{-\infty < x_1, x_2, \cdots, x_N < \infty\}$. For example, describing the location and velocity of a gas particle requires six coordinates.

- *The joint PDF* $p(\mathbf{x})$ is the probability density of an outcome in a volume element $d^N\mathbf{x} = \prod_{i=1}^N dx_i$ around the point $\mathbf{x} = \{x_1, x_2, \cdots, x_N\}$. The joint PDF is normalized such that

$$p_\mathbf{x}(\mathcal{S}) = \int d^N\mathbf{x}\, p(\mathbf{x}) = 1. \quad (2.32)$$

If, and only if, the N random variables are *independent*, the joint PDF is the product of individual PDFs,

$$p(\mathbf{x}) = \prod_{i=1}^N p_i(x_i). \quad (2.33)$$

- *The unconditional PDF* describes the behavior of a subset of random variables, independent of the values of the others. For example, if we are interested only in the location of a gas particle, an unconditional PDF can be constructed by integrating over all velocities at a given location, $p(\vec{x}) = \int d^3\vec{v}\, p(\vec{x}, \vec{v})$; more generally,

$$p(x_1, \cdots, x_m) = \int \prod_{i=m+1}^N dx_i\, p(x_1, \cdots, x_N). \quad (2.34)$$

- *The conditional PDF* describes the behavior of a subset of random variables, for specified values of the others. For example, the PDF for the velocity of a particle at a particular location $\vec{x}$, denoted by $p(\vec{v}|\vec{x})$, is proportional to the joint PDF $p(\vec{v}|\vec{x}) = p(\vec{x}, \vec{v})/\mathcal{N}$. The constant of proportionality, obtained by normalizing $p(\vec{v}|\vec{x})$, is

$$\mathcal{N} = \int \mathrm{d}^3\vec{v}\, p(\vec{x}, \vec{v}) = p(\vec{x}), \qquad (2.35)$$

the unconditional PDF for a particle at $\vec{x}$. In general, the unconditional PDFs are obtained from *Bayes' theorem* as

$$p(x_1, \cdots, x_m | x_{m+1}, \cdots, x_N) = \frac{p(x_1, \cdots, x_N)}{p(x_{m+1}, \cdots, x_N)}. \qquad (2.36)$$

Note that if the random variables are independent, the unconditional PDF is equal to the conditional PDF.

- *The expectation value* of a function $F(\mathbf{x})$ is obtained as before from

$$\langle F(\mathbf{x}) \rangle = \int \mathrm{d}^N\mathbf{x}\, p(\mathbf{x}) F(\mathbf{x}). \qquad (2.37)$$

- *The joint characteristic function* is obtained from the N-dimensional Fourier transformation of the joint PDF,

$$\tilde{p}(\mathbf{k}) = \left\langle \exp\left(-i\sum_{j=1}^{N} k_j x_j\right) \right\rangle. \qquad (2.38)$$

The *joint moments* and *joint cumulants* are generated by $\tilde{p}(\mathbf{k})$ and $\ln\tilde{p}(\mathbf{k})$, respectively, as

$$\left\langle x_1^{n_1} x_2^{n_2} \cdots x_N^{n_N} \right\rangle = \left[\frac{\partial}{\partial(-ik_1)}\right]^{n_1} \left[\frac{\partial}{\partial(-ik_2)}\right]^{n_2} \cdots \left[\frac{\partial}{\partial(-ik_N)}\right]^{n_N} \tilde{p}(\mathbf{k} = \mathbf{0}),$$

$$\left\langle x_1^{n_1} * x_2^{n_2} * \cdots x_N^{n_N} \right\rangle_c = \left[\frac{\partial}{\partial(-ik_1)}\right]^{n_1} \left[\frac{\partial}{\partial(-ik_2)}\right]^{n_2} \cdots \left[\frac{\partial}{\partial(-ik_N)}\right]^{n_N} \ln\tilde{p}(\mathbf{k} = \mathbf{0}).$$

$$(2.39)$$

The previously described graphical relation between joint moments (all clusters of labeled points) and joint cumulant (connected clusters) is still applicable. For example, from

we obtain

$$\langle x_1 x_2 \rangle = \langle x_1 \rangle_c \langle x_2 \rangle_c + \langle x_1 * x_2 \rangle_c, \quad \text{and}$$

$$\left\langle x_1^2 x_2 \right\rangle = \langle x_1 \rangle_c^2 \langle x_2 \rangle_c + \left\langle x_1^2 \right\rangle_c \langle x_2 \rangle_c + 2\langle x_1 * x_2 \rangle_c \langle x_1 \rangle_c + \left\langle x_1^2 * x_2 \right\rangle_c.$$

$$(2.40)$$

The *connected correlation* $\langle x_\alpha * x_\beta \rangle_c$ is zero if x_α and x_β are independent random variables.

- *The joint Gaussian distribution* is the generalization of Eq. (2.15) to N random variables, as

$$p(\mathbf{x}) = \frac{1}{\sqrt{(2\pi)^N \det[C]}} \exp\left[-\frac{1}{2}\sum_{mn} \left(C^{-1}\right)_{mn}(x_m - \lambda_m)(x_n - \lambda_n)\right], \quad (2.41)$$

where C is a symmetric matrix, and C^{-1} is its inverse. The simplest way to get the normalization factor is to make a linear transformation from the variables $y_j = x_j - \lambda_j$, using the unitary matrix that diagonalizes C. This reduces the normalization to that of the product of N Gaussians whose variances are determined by the eigenvalues of C. The product of the eigenvalues is the determinant $\det[C]$. (This also indicates that the matrix C must be positive definite.) The corresponding joint characteristic function is obtained by similar manipulations, and is given by

$$\tilde{p}(\mathbf{k}) = \exp\left[-ik_m\lambda_m - \frac{1}{2}C_{mn}k_m k_n\right], \quad (2.42)$$

where the summation convention (implicit summation over a repeated index) is used. The joint cumulants of the Gaussian are then obtained from $\ln \tilde{p}(\mathbf{k})$ as

$$\langle x_m \rangle_c = \lambda_m, \quad \langle x_m * x_n \rangle_c = C_{mn}, \quad (2.43)$$

with all higher cumulants equal to zero. In the special case of $\{\lambda_m\} = 0$, all odd *moments* of the distribution are zero, while the general rules for relating moments to cumulants indicate that any even moment is obtained by summing over all ways of grouping the involved random variables into pairs, for example,

$$\langle x_a x_b x_c x_d \rangle = C_{ab}C_{cd} + C_{ac}C_{bd} + C_{ad}C_{bc}. \quad (2.44)$$

In the context of field theories, this result is referred to as *Wick's theorem*.

2.5 Sums of random variables and the central limit theorem

Consider the sum $X = \sum_{i=1}^{N} x_i$, where x_i are random variables with a joint PDF of $p(\mathbf{x})$. The PDF for X is

$$p_X(x) = \int d^N\mathbf{x}\, p(\mathbf{x})\delta\left(x - \sum x_i\right) = \int \prod_{i=1}^{N-1} dx_i\, p(x_1, \cdots, x_{N-1}, x - x_1 \cdots - x_{N-1}),$$

$$(2.45)$$

and the corresponding characteristic function (using Eq. (2.38)) is given by

$$\tilde{p}_X(k) = \left\langle \exp\left(-ik\sum_{j=1}^{N} x_j\right)\right\rangle = \tilde{p}\left(k_1 = k_2 = \cdots = k_N = k\right). \quad (2.46)$$

Cumulants of the sum are obtained by expanding $\ln \tilde{p}_X(k)$,

$$\ln \tilde{p}\,(k_1 = k_2 = \cdots = k_N = k) = -ik \sum_{i_1=1}^{N} \langle x_{i_1} \rangle_c + \frac{(-ik)^2}{2} \sum_{i_1, i_2}^{N} \langle x_{i_1} x_{i_2} \rangle_c + \cdots, \qquad (2.47)$$

as

$$\langle X \rangle_c = \sum_{i=1}^{N} \langle x_i \rangle_c, \quad \langle X^2 \rangle_c = \sum_{i,j}^{N} \langle x_i x_j \rangle_c, \cdots. \qquad (2.48)$$

If the random variables are independent, $p(\mathbf{x}) = \prod p_i(x_i)$, and $\tilde{p}_X(k) = \prod \tilde{p}_i(k)$, the cross cumulants in Eq. (2.48) vanish, and the nth cumulant of X is simply the sum of the individual cumulants, $\langle X^n \rangle_c = \sum_{i=1}^{N} \langle x_i^n \rangle_c$. When all the N random variables are independently taken from the same distribution $p(x)$, this implies $\langle X^n \rangle_c = N \langle x^n \rangle_c$, generalizing the result obtained previously for the binomial distribution. For large values of N, the average value of the sum is proportional to N, while fluctuations around the mean, as measured by the standard deviation, grow only as $\sqrt{N}$. The random variable $y = (X - N\langle x \rangle_c)/\sqrt{N}$ has zero mean, and cumulants that scale as $\langle y^n \rangle_c \propto N^{1-n/2}$. As $N \to \infty$, only the second cumulant survives, and the PDF for y converges to the normal distribution,

$$\lim_{N \to \infty} p\left(y = \frac{\sum_{i=1}^{N} x_i - N\langle x \rangle_c}{\sqrt{N}} \right) = \frac{1}{\sqrt{2\pi \langle x^2 \rangle_c}} \exp\left(-\frac{y^2}{2 \langle x^2 \rangle_c} \right). \qquad (2.49)$$

(Note that the Gaussian distribution is the only distribution with only first and second cumulants.)

The convergence of the PDF for the sum of many random variables to a normal distribution is an essential result in the context of statistical mechanics where such sums are frequently encountered. The *central limit theorem* states a more general form of this result: it is not necessary for the random variables to be independent, as the condition $\sum_{i_1, \cdots, i_m}^{N} \langle x_{i_1} \cdots x_{i_m} \rangle_c \ll \mathcal{O}(N^{m/2})$ is sufficient for the validity of Eq. (2.49).

Note that the above discussion implicitly assumes that the cumulants of the individual random variables (appearing in Eq. (2.48)) are finite. What happens if this is not the case, that is, when the variables are taken from a very wide PDF? The sum may still converge to a so-called *Levy distribution*. Consider a sum of N *independent, identically distributed* random variables, with the mean set to zero for convenience. The variance does not exist if the individual PDF falls off slowly at large values as $p_i(x) = p_1(x) \propto 1/|x|^{1+\alpha}$, with $0 < \alpha \leq 2$. ($\alpha > 0$ is required to make sure that the distribution is normalizable; while for $\alpha > 2$ the variance is finite.) The behavior of $p_1(x)$ at large x determines the behavior of $\tilde{p}_1(k)$ at small k, and simple power counting indicates that the expansion of $\tilde{p}_1(k)$ is singular, starting with $|k|^\alpha$. Based on this argument we conclude that

$$\ln \tilde{p}_X(k) = N \ln \tilde{p}_1(k) = N[-a|k|^\alpha + \text{higher order terms}]. \qquad (2.50)$$

As before we can define a rescaled variable $y = X/N^{1/\alpha}$ to get rid of the N dependence of the leading term in the above equation, resulting in

$$\lim_{N \to \infty} \tilde{p}_y(k) = -a|k|^{\alpha}. \tag{2.51}$$

The higher-order terms appearing in Eq. (2.50) scale with negative powers of N and vanish as $N \to \infty$. The simplest example of a Levy distribution is obtained for $\alpha = 1$, and corresponds to $p_y = a/[\pi(y^2 + a^2)]$. (This is the Cauchy distribution discussed in the problems section.) For other values of α the distribution does not have a simple closed form, but can be written as the asymptotic series

$$p_{\alpha}(y) = \frac{1}{\pi} \sum_{n=1}^{\infty} (-1)^{n+1} \sin\left(\frac{n\pi}{2}\alpha\right) \frac{\Gamma(1+n\alpha)}{n!} \frac{a^n}{y^{1+n\alpha}}. \tag{2.52}$$

Such distributions describe phenomena with large rare events, characterized here by a tail that falls off slowly as $p_{\alpha}(y \to \infty) \sim y^{-1-\alpha}$.

2.6 Rules for large numbers

To describe equilibrium properties of macroscopic bodies, statistical mechanics has to deal with the very large number N of microscopic degrees of freedom. Actually, taking the *thermodynamic limit* of $N \to \infty$ leads to a number of simplifications, some of which are described in this section.

There are typically three types of N dependence encountered in the thermodynamic limit:

(a) *Intensive* quantities, such as temperature T, and generalized forces, for example, pressure P, and magnetic field $\vec{B}$, are independent of N, that is, $\mathcal{O}(N^0)$.

(b) *Extensive* quantities, such as energy E, entropy S, and generalized displacements, for example, volume V, and magnetization $\vec{M}$, are proportional to N, that is, $\mathcal{O}(N^1)$.

(c) *Exponential* dependence, that is, $\mathcal{O}(\exp(N\phi))$, is encountered in enumerating discrete micro-states, or computing available volumes in phase space.

Other asymptotic dependencies are certainly not ruled out a priori. For example, the Coulomb energy of N ions at fixed density scales as $Q^2/R \sim N^{5/3}$. Such dependencies are rarely encountered in everyday physics. The Coulomb interaction of ions is quickly screened by counter-ions, resulting in an extensive overall energy. (This is not the case in astrophysical problems since the gravitational energy is not screened. For example, the entropy of a black hole is proportional to the square of its mass.)

In statistical mechanics we frequently encounter sums or integrals of exponential variables. Performing such sums in the thermodynamic limit is considerably simplified due to the following results.

(1) *Summation of exponential quantities*: consider the sum

$$S = \sum_{i=1}^{\mathcal{N}} \mathcal{E}_i, \tag{2.53}$$

where each term is positive, with an exponential dependence on N, that is,

$$0 \leq \mathcal{E}_i \sim \mathcal{O}\big(\exp(N\phi_i)\big), \tag{2.54}$$

and the number of terms $\mathcal{N}$ is proportional to some power of N.

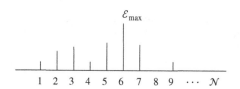

Such a sum can be approximated by its largest term $\mathcal{E}_{max}$, in the following sense. Since for each term in the sum, $0 \leq \mathcal{E}_i \leq \mathcal{E}_{max}$,

$$\mathcal{E}_{max} \leq S \leq \mathcal{N}\mathcal{E}_{max}. \tag{2.55}$$

An intensive quantity can be constructed from $\ln S/N$, which is bounded by

$$\frac{\ln \mathcal{E}_{max}}{N} \leq \frac{\ln S}{N} \leq \frac{\ln \mathcal{E}_{max}}{N} + \frac{\ln \mathcal{N}}{N}. \tag{2.56}$$

For $\mathcal{N} \propto N^p$, the ratio $\ln \mathcal{N}/N$ vanishes in the large N limit, and

$$\lim_{N \to \infty} \frac{\ln S}{N} = \frac{\ln \mathcal{E}_{max}}{N} = \phi_{max}. \tag{2.57}$$

(2) *Saddle point integration*: similarly, an integral of the form

$$\mathcal{I} = \int dx \exp\big(N\phi(x)\big) \tag{2.58}$$

can be approximated by the maximum value of the integrand, obtained at a point x_{max} that maximizes the exponent $\phi(x)$. Expanding the exponent around this point gives

$$\mathcal{I} = \int dx \exp\left\{N\left[\phi(x_{max}) - \frac{1}{2}|\phi''(x_{max})|(x - x_{max})^2 + \cdots\right]\right\}. \tag{2.59}$$

Note that at the maximum, the first derivative $\phi'(x_{\max})$ is zero, while the second derivative $\phi''(x_{\max})$ is negative. Terminating the series at the quadratic order results in

$$\mathcal{J} \approx e^{N\phi(x_{\max})} \int dx \exp\left[-\frac{N}{2}|\phi''(x_{\max})|(x-x_{\max})^2\right] \approx \sqrt{\frac{2\pi}{N|\phi''(x_{\max})|}} e^{N\phi(x_{\max})}, \quad (2.60)$$

where the range of integration has been extended to $[-\infty, \infty]$. The latter is justified since the integrand is negligibly small outside the neighborhood of $x_{\max}$.

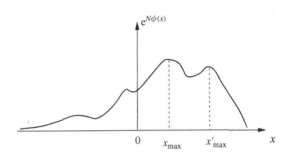

Fig. **2.8** Saddle point evaluation of an "exponential" integral.

There are two types of correction to the above result. Firstly, there are higher-order terms in the expansion of $\phi(x)$ around $x_{\max}$. These corrections can be looked at perturbatively, and lead to a series in powers of $1/N$. Secondly, there may be additional local maxima for the function. A maximum at $x'_{\max}$ leads to a similar Gaussian integral that can be added to Eq. (2.60). Clearly such contributions are smaller by $\mathcal{O}\left(\exp\{-N[\phi(x_{\max}) - \phi(x'_{\max})]\}\right)$. Since all these corrections vanish in the thermodynamic limit,

$$\lim_{N\to\infty} \frac{\ln \mathcal{J}}{N} = \lim_{N\to\infty} \left[\phi(x_{\max}) - \frac{1}{2N}\ln\left(\frac{N|\phi''(x_{\max})|}{2\pi}\right) + \mathcal{O}\left(\frac{1}{N^2}\right)\right] = \phi(x_{\max}). \quad (2.61)$$

The *saddle point* method for evaluating integrals is the extension of the above result to more general integrands, and integration paths in the complex plane. (The appropriate extremum in the complex plane is a saddle point.) The simplified version presented above is sufficient for our needs.

Stirling's approximation for $N!$ at large N can be obtained by saddle point integration. In order to get an integral representation of $N!$, start with the result

$$\int_0^\infty dx\, e^{-\alpha x} = \frac{1}{\alpha}. \quad (2.62)$$

Repeated differentiation of both sides of the above equation with respect to α leads to

$$\int_0^\infty dx\, x^N e^{-\alpha x} = \frac{N!}{\alpha^{N+1}}. \quad (2.63)$$

Although the above derivation only applies to integer N, it is possible to define by analytical continuation a function

$$\Gamma(N+1) \equiv N! = \int_0^\infty dx\, x^N e^{-x},\tag{2.64}$$

for all N. While the integral in Eq. (2.64) is not exactly in the form of Eq. (2.58), it can still be evaluated by a similar method. The integrand can be written as $\exp\left(N\phi(x)\right)$, with $\phi(x) = \ln x - x/N$. The exponent has a maximum at $x_{max} = N$, with $\phi(x_{max}) = \ln N - 1$, and $\phi''(x_{max}) = -1/N^2$. Expanding the integrand in Eq. (2.64) around this point yields

$$N! \approx \int dx\, \exp\left[N\ln N - N - \frac{1}{2N}(x-N)^2\right] \approx N^N e^{-N}\sqrt{2\pi N},\tag{2.65}$$

where the integral is evaluated by extending its limits to $[-\infty, \infty]$. Stirling's formula is obtained by taking the logarithm of Eq. (2.65) as

$$\ln N! = N\ln N - N + \frac{1}{2}\ln(2\pi N) + \mathcal{O}\left(\frac{1}{N}\right).\tag{2.66}$$

2.7 Information, entropy, and estimation

Information: consider a random variable with a discrete set of outcomes $\mathcal{S} = \{x_i\}$, occurring with probabilities $\{p(i)\}$, for $i = 1, \cdots, M$. In the context of information theory there is a precise meaning to the *information content* of a probability distribution. Let us construct a message from N independent outcomes of the random variable. Since there are M possibilities for each character in this message, it has an apparent information content of $N\ln_2 M$ bits; that is, this many binary bits of information have to be transmitted to convey the message precisely. On the other hand, the probabilities $\{p(i)\}$ limit the types of messages that are likely. For example, if $p_2 \gg p_1$, it is very unlikely to construct a message with more x_1 than x_2. In particular, in the limit of large N, we expect the message to contain "roughly" $\{N_i = Np_i\}$ occurrences of each symbol.[1] The number of typical messages thus corresponds to the number of ways of rearranging the $\{N_i\}$ occurrences of $\{x_i\}$, and is given by the multinomial coefficient

$$g = \frac{N!}{\prod_{i=1}^M N_i!}.\tag{2.67}$$

This is much smaller than the total number of messages M^n. To specify one out of g possible sequences requires

$$\ln_2 g \approx -N\sum_{i=1}^M p_i \ln_2 p_i \qquad (\text{for } N \to \infty)\tag{2.68}$$

---

[1] More precisely, the probability of finding any N_i that is different from Np_i by more than $\mathcal{O}(\sqrt{N})$ becomes exponentially small in N, as $N \to \infty$.

bits of information. The last result is obtained by applying Stirling's approximation for $\ln N!$. It can also be obtained by noting that

$$1 = \left(\sum_i p_i\right)^N = \sum_{\{N_i\}} N! \prod_{i=1}^{M} \frac{p_i^{N_i}}{N_i!} \approx g \prod_{i=1}^{M} p_i^{Np_i}, \qquad (2.69)$$

where the sum has been replaced by its largest term, as justified in the previous section.

Shannon's theorem proves more rigorously that the minimum number of bits necessary to ensure that the percentage of errors in N trials vanishes in the $N \to \infty$ limit is $\ln_2 g$. For any non-uniform distribution, this is less than the $N \ln_2 M$ bits needed in the absence of any information on relative probabilities. The difference per trial is thus attributed to the information content of the probability distribution, and is given by

$$I[\{p_i\}] = \ln_2 M + \sum_{i=1}^{M} p_i \ln_2 p_i. \qquad (2.70)$$

Entropy: Eq. (2.67) is encountered frequently in statistical mechanics in the context of mixing M distinct components; its natural logarithm is related to the *entropy of mixing*. More generally, we can define an *entropy* for *any probability distribution* as

$$S = -\sum_{i=1}^{M} p(i) \ln p(i) = -\langle \ln p(i) \rangle. \qquad (2.71)$$

The above entropy takes a minimum value of zero for the delta function distribution $p(i) = \delta_{i,j}$, and a maximum value of $\ln M$ for the uniform distribution $p(i) = 1/M$. S is thus a measure of dispersity (disorder) of the distribution, and does not depend on the values of the random variables $\{x_i\}$. A one-to-one mapping to $f_i = F(x_i)$ leaves the entropy unchanged, while a many-to-one mapping makes the distribution more ordered and decreases S. For example, if the two values, x_1 and x_2, are mapped onto the same f, the change in entropy is

$$\Delta S(x_1, x_2 \to f) = \left[p_1 \ln \frac{p_1}{p_1 + p_2} + p_2 \ln \frac{p_2}{p_1 + p_2}\right] < 0. \qquad (2.72)$$

Estimation: the entropy S can also be used to quantify subjective estimates of probabilities. In the absence of any information, the best *unbiased estimate* is that all M outcomes are equally likely. This is the distribution of maximum entropy. If additional information is available, the unbiased estimate is obtained by maximizing the entropy subject to the constraints imposed by this information. For example, if it is known that $\langle F(x) \rangle = f$, we can maximize

$$S(\alpha, \beta, \{p_i\}) = -\sum_i p(i) \ln p(i) - \alpha \left(\sum_i p(i) - 1\right) - \beta \left(\sum_i p(i) F(x_i) - f\right), \qquad (2.73)$$

where the Lagrange multipliers α and β are introduced to impose the constraints of normalization, and $\langle F(x) \rangle = f$, respectively. The result of the optimization is a distribution $p_i \propto \exp\left(-\beta F(x_i)\right)$, where the value of β is fixed by the constraint. This process can be generalized to an arbitrary number of conditions. It is easy to see that if the first $n = 2k$ moments (and hence n cumulants) of a distribution are specified, the unbiased estimate is the exponential of an nth-order polynomial.

In analogy with Eq. (2.71), we can define an entropy for a continuous random variable ($\mathcal{S}_x = \{-\infty < x < \infty\}$) as

$$S = -\int dx\, p(x) \ln p(x) = -\langle \ln p(x) \rangle. \tag{2.74}$$

There are, however, problems with this definition, as for example S is not invariant under a one-to-one mapping. (After a change of variable to $f = F(x)$, the entropy is changed by $\langle |F'(x)| \rangle$.) As discussed in the following chapter, canonically conjugate pairs offer a suitable choice of coordinates in classical statistical mechanics, since the Jacobian of a canonical transformation is unity. The ambiguities are also removed if the continuous variable is discretized. This happens quite naturally in quantum statistical mechanics where it is usually possible to work with a discrete ladder of states. The appropriate volume for discretization of phase space is then set by Planck's constant $\hbar$.

Problems for chapter 2

1. *Characteristic functions*: calculate the characteristic function, the mean, and the variance of the following probability density functions:

   (a) *Uniform* $p(x) = \frac{1}{2a}$ for $-a < x < a$, and $p(x) = 0$ otherwise.

   (b) *Laplace* $p(x) = \frac{1}{2a} \exp\left(-\frac{|x|}{a}\right)$.

   (c) *Cauchy* $p(x) = \frac{a}{\pi(x^2 + a^2)}$.

   The following two probability density functions are defined for $x \geq 0$. Compute only the mean and variance for each.

   (d) *Rayleigh* $p(x) = \frac{x}{a^2} \exp(-\frac{x^2}{2a^2})$.

   (e) *Maxwell* $p(x) = \sqrt{\frac{2}{\pi}} \frac{x^2}{a^3} \exp(-\frac{x^2}{2a^2})$.

   ****

2. *Directed random walk*: the motion of a particle in three dimensions is a series of independent steps of length ℓ. Each step makes an angle θ with the z axis, with a probability density $p(\theta) = 2\cos^2(\theta/2)/\pi$; while the angle ϕ is uniformly distributed between 0 and 2π. (Note that the solid angle factor of $\sin\theta$ is already included in the definition of $p(\theta)$, which is correctly normalized to unity.) The particle (walker) starts at the origin and makes a large number of steps N.

   (a) Calculate the expectation values $\langle z \rangle$, $\langle x \rangle$, $\langle y \rangle$, $\langle z^2 \rangle$, $\langle x^2 \rangle$, and $\langle y^2 \rangle$, and the covariances $\langle xy \rangle$, $\langle xz \rangle$, and $\langle yz \rangle$.

(b) Use the central limit theorem to estimate the probability density $p(x, y, z)$ for the particle to end up at the point (x, y, z).

****

3. *Tchebycheff inequality*: consider any probability density $p(x)$ for $(-\infty < x < \infty)$, with mean λ, and variance σ^2. Show that the total probability of outcomes that are more than $n\sigma$ away from λ is less than $1/n^2$, that is,

$$\int_{|x-\lambda| \geq n\sigma} dx\, p(x) \leq \frac{1}{n^2}.$$

Hint. Start with the integral defining σ^2, and break it up into parts corresponding to $|x - \lambda| > n\sigma$, and $|x - \lambda| < n\sigma$.

****

4. *Optimal selection*: in many specialized populations, there is little variability among the members. Is this a natural consequence of optimal selection?

(a) Let $\{r_\alpha\}$ be n random numbers, each independently chosen from a probability density $p(r)$, with $r \in [0, 1]$. Calculate the probability density $p_n(x)$ for the largest value of this set, that is, for $x = \max\{r_1, \cdots, r_n\}$.

(b) If each r_α is uniformly distributed between 0 and 1, calculate the mean and variance of x as a function of n, and comment on their behavior at large n.

****

5. *Benford's law* describes the observed probabilities of the *first digit* in a great variety of data sets, such as stock prices. Rather counter-intuitively, the digits 1 through 9 occur with probabilities $0.301, 0.176, 0.125, 0.097, 0.079, 0.067, 0.058, 0.051, 0.046$, respectively. The key observation is that this distribution is invariant under a change of scale, for example, if the stock prices were converted from dollars to Persian rials! Find a formula that fits the above probabilities on the basis of this observation.

****

6. *Information*: consider the velocity of a gas particle in one dimension $(-\infty < v < \infty)$.

(a) Find the unbiased probability density $p_1(v)$, subject only to the constraint that the average *speed* is c, that is, $\langle |v| \rangle = c$.

(b) Now find the probability density $p_2(v)$, given only the constraint of average kinetic energy, $\langle mv^2/2 \rangle = mc^2/2$.

(c) Which of the above statements provides more information on the velocity? Quantify the difference in information in terms of $I_2 - I_1 \equiv (\langle \ln p_2 \rangle - \langle \ln p_1 \rangle)/\ln 2$.

****

7. *Dice*: a dice is loaded such that 6 occurs twice as often as 1.

(a) Calculate the unbiased probabilities for the six faces of the dice.

(b) What is the information content (in bits) of the above statement regarding the dice?

****

8. *Random matrices*: as a model for energy levels of complex nuclei, Wigner considered $N \times N$ symmetric matrices whose elements are random. Let us assume that each element M_{ij} (for $i \geq j$) is an independent random variable taken from the probability density function

$$p(M_{ij}) = \frac{1}{2a} \quad \text{for} \quad -a < M_{ij} < a, \quad \text{and} \quad p(M_{ij}) = 0 \quad \text{otherwise.}$$

(a) Calculate the characteristic function for each element M_{ij}.

(b) Calculate the characteristic function for the trace of the matrix, $T \equiv \text{tr} M = \sum_i M_{ii}$.

(c) What does the central limit theorem imply about the probability density function of the trace at large N?

(d) For large N, each eigenvalue λ_α ($\alpha = 1, 2, \cdots, N$) of the matrix M is distributed according to a probability density function

$$p(\lambda) = \frac{2}{\pi \lambda_0} \sqrt{1 - \frac{\lambda^2}{\lambda_0^2}} \quad \text{for} \quad -\lambda_0 < \lambda < \lambda_0, \quad \text{and} \quad p(\lambda) = 0 \quad \text{otherwise}$$

(known as the Wigner semicircle rule). Find the variance of λ.

(*Hint.* Changing variables to $\lambda = \lambda_0 \sin \theta$ simplifies the integrals.)

(e) If, in the previous result, we have $\lambda_0^2 = 4Na^2/3$, can the eigenvalues be independent of each other?

****

9. *Random deposition*: a mirror is plated by evaporating a gold electrode in vaccum by passing an electric current. The gold atoms fly off in all directions, and a portion of them sticks to the glass (or to other gold atoms already on the glass plate). Assume that each column of deposited atoms is independent of neighboring columns, and that the average deposition rate is d layers per second.

(a) What is the probability of m atoms deposited at a site after a time t? What fraction of the glass is not covered by any gold atoms?

(b) What is the variance in the thickness?

****

10. *Diode*: the current I across a diode is related to the applied voltage V via $I = I_0 [\exp(eV/kT) - 1]$. The diode is subject to a random potential V of zero mean and variance σ^2, which is Gaussian distributed. Find the probability density $p(I)$ for the current I flowing through the diode. Find the most probable value for I, the mean value of I, and indicate them on a sketch of $p(I)$.

****

11. *Mutual information*: consider random variables x and y, distributed according to a joint probability $p(x, y)$. The mutual information between the two variables is defined by

$$M(x, y) \equiv \sum_{x,y} p(x, y) \ln \left(\frac{p(x, y)}{p_x(x) p_y(y)} \right),$$

where p_x and p_y denote the *unconditional* probabilities for x and y.

(a) Relate $M(x, y)$ to the entropies $S(x, y)$, $S(x)$, and $S(y)$ obtained from the corresponding probabilities.

(b) Calculate the mutual information for the joint Gaussian form

$$p(x, y) \propto \exp \left(-\frac{ax^2}{2} - \frac{by^2}{2} - cxy \right).$$

****

12. *Semi-flexible polymer in two dimensions*: configurations of a model polymer can be described by either a set of vectors $\{\mathbf{t}_i\}$ of length a in two dimensions (for $i = 1, \cdots, N$), or alternatively by the angles $\{\phi_i\}$ between successive vectors, as indicated in the figure below.

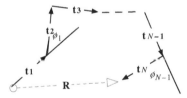

The polymer is at a temperature T, and subject to an energy

$$\mathcal{H} = -\kappa \sum_{i=1}^{N-1} \mathbf{t}_i \cdot \mathbf{t}_{i+1} = -\kappa a^2 \sum_{i=1}^{N-1} \cos \phi_i,$$

where κ is related to the bending rigidity, such that the probability of any configuration is proportional to $\exp(-\mathcal{H}/k_B T)$.

(a) Show that $\langle \mathbf{t}_m \cdot \mathbf{t}_n \rangle \propto \exp(-|n - m|/\xi)$, and obtain an expression for the *persistence length* $\ell_p = a\xi$. (You can leave the answer as the ratio of simple integrals.)

(b) Consider the end-to-end distance $\mathbf{R}$ as illustrated in the figure. Obtain an expression for $\langle R^2 \rangle$ in the limit of $N \gg 1$.

(c) Find the probability $p(\mathbf{R})$ in the limit of $N \gg 1$.

(d) If the ends of the polymer are pulled apart by a force **F**, the probabilities for polymer configurations are modified by the Boltzmann weight $\exp\left(\frac{\mathbf{F}\cdot\mathbf{R}}{k_B T}\right)$. By expanding this weight, or otherwise, show that

$$\langle \mathbf{R} \rangle = K^{-1}\mathbf{F} + \mathcal{O}(F^3),$$

and give an expression for the Hookian constant K in terms of quantities calculated before.

* * * * * * * *

3
Kinetic theory of gases

3.1 General definitions

Kinetic theory *studies the macroscopic properties of large numbers of particles, starting from their (classical) equations of motion.*

Thermodynamics describes the equilibrium behavior of *macroscopic objects* in terms of concepts such as work, heat, and entropy. The phenomenological laws of thermodynamics tell us how these quantities are constrained as a system approaches its equilibrium. At the *microscopic level*, we know that these systems are composed of particles (atoms, molecules), whose interactions and dynamics are reasonably well understood in terms of more fundamental theories. If these microscopic descriptions are complete, we should be able to account for the macroscopic behavior, that is, derive the laws governing the macroscopic state functions in equilibrium. Kinetic theory attempts to achieve this objective. In particular, we shall try to answer the following questions:

1. How can we define "equilibrium" for a system of moving particles?
2. Do all systems naturally evolve towards an equilibrium state?
3. What is the time evolution of a system that is not quite in equilibrium?

The simplest system to study, the veritable workhorse of thermodynamics, is the dilute (nearly ideal) gas. A typical volume of gas contains of the order of 10^{23} particles, and in kinetic theory we try to deduce the macroscopic properties of the gas from the time evolution of the set of atomic coordinates. At any time t, the *microstate* of a system of N particles is described by specifying the positions $\vec{q}_i(t)$, and momenta $\vec{p}_i(t)$, of all particles. The microstate thus corresponds to a point $\mu(t)$, in the $6N$-dimensional *phase space* $\Gamma = \prod_{i=1}^{N}\{\vec{q}_i, \vec{p}_i\}$. The time evolution of this point is governed by the canonical equations

$$\begin{cases} \dfrac{\mathrm{d}\vec{q}_i}{\mathrm{d}t} = \dfrac{\partial \mathcal{H}}{\partial \vec{p}_i} \\[2ex] \dfrac{\mathrm{d}\vec{p}_i}{\mathrm{d}t} = -\dfrac{\partial \mathcal{H}}{\partial \vec{q}_i} \end{cases}, \tag{3.1}$$

where the *Hamiltonian* $\mathcal{H}(\mathbf{p}, \mathbf{q})$ describes the total energy in terms of the set of coordinates $\mathbf{q} \equiv \{\vec{q}_1, \vec{q}_2, \cdots, \vec{q}_N\}$, and momenta $\mathbf{p} \equiv \{\vec{p}_1, \vec{p}_2, \cdots, \vec{p}_N\}$.

The microscopic equations of motion have *time reversal symmetry*, that is, if all the momenta are suddenly reversed, $\mathbf{p} \to -\mathbf{p}$, at $t = 0$, the particles retrace their previous trajectory, $\mathbf{q}(t) = \mathbf{q}(-t)$. This follows from the invariance of $\mathcal{H}$ under the transformation $T(\mathbf{p}, \mathbf{q}) \to (-\mathbf{p}, \mathbf{q})$.

Fig. 3.1 The phase space density is proportional to the number of representative points in an infinitesimal volume.

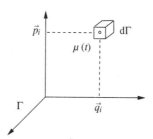

As formulated within thermodynamics, the *macrostate M* of an ideal gas in equilibrium is described by a small number of state functions such as E, T, P, and N. The space of macrostates is considerably smaller than the phase space spanned by microstates. Therefore, there must be a very large number of microstates μ corresponding to the same macrostate M.

This many-to-one correspondence suggests the introduction of a statistical *ensemble* of microstates. Consider $\mathcal{N}$ copies of a particular macrostate, each described by a different *representative point* $\mu(t)$ in the phase space Γ. Let $\mathrm{d}\mathcal{N}(\mathbf{p}, \mathbf{q}, t)$ equal the number of representative points in an infinitesimal volume $\mathrm{d}\Gamma = \prod_{i=1}^{N} \mathrm{d}^3\vec{p}_i \mathrm{d}^3\vec{q}_i$ around the point $(\mathbf{p}, \mathbf{q})$. A phase space density $\rho(\mathbf{p}, \mathbf{q}, t)$ is then defined by

$$\rho(\mathbf{p}, \mathbf{q}, t)\mathrm{d}\Gamma = \lim_{\mathcal{N} \to \infty} \frac{\mathrm{d}\mathcal{N}(\mathbf{p}, \mathbf{q}, t)}{\mathcal{N}}. \tag{3.2}$$

This quantity can be compared with the *objective* probability introduced in the previous section. Clearly, $\int \mathrm{d}\Gamma\rho = 1$ and ρ is a properly normalized *probability density function* in phase space. To compute macroscopic values for various functions $\mathcal{O}(\mathbf{p}, \mathbf{q})$, we shall use the *ensemble averages*

$$\langle \mathcal{O} \rangle = \int \mathrm{d}\Gamma\rho(\mathbf{p}, \mathbf{q}, t)\mathcal{O}(\mathbf{p}, \mathbf{q}). \tag{3.3}$$

When the exact microstate μ is specified, the system is said to be in a *pure state*. On the other hand, when our knowledge of the system is probabilistic, in the sense of its being taken from an ensemble with density $\rho(\Gamma)$, it is said to belong to a *mixed state*. It is difficult to describe equilibrium in the context of a pure state, since $\mu(t)$ is constantly changing in time according to Eqs. (3.1). Equilibrium is more conveniently described for mixed states by examining the time evolution of the phase space density

$\rho(t)$, which is governed by the Liouville equation introduced in the next section.

3.2 Liouville's theorem

Liouville's theorem *states that the phase space density $\rho(\Gamma, t)$ behaves like an incompressible fluid.*

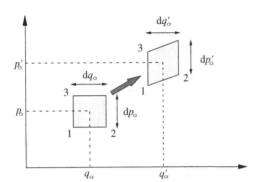

Fig. 3.2 Time evolution of a volume element in phase space.

Follow the evolution of $d\mathcal{N}$ pure states in an infinitesimal volume $d\Gamma = \prod_{i=1}^{N} d^3\vec{p}_i \, d^3\vec{q}_i$ around the point $(\mathbf{p}, \mathbf{q})$. According to Eqs. (3.1), after an interval δt these states have moved to the vicinity of another point $(\mathbf{p}', \mathbf{q}')$, where

$$q'_\alpha = q_\alpha + \dot{q}_\alpha \delta t + \mathcal{O}(\delta t^2), \quad p'_\alpha = p_\alpha + \dot{p}_\alpha \delta t + \mathcal{O}(\delta t^2). \tag{3.4}$$

In the above expression, the q_α and p_α refer to any of the $6N$ coordinates and momenta, and $\dot{q}_\alpha$ and $\dot{p}_\alpha$ are the corresponding velocities. The original volume element $d\Gamma$ is in the shape of a hypercube of sides dp_α and dq_α. In the time interval δt it gets distorted, and the projected sides of the new volume element are given by

$$\begin{cases} dq'_\alpha = dq_\alpha + \dfrac{\partial \dot{q}_\alpha}{\partial q_\alpha} dq_\alpha \delta t + \mathcal{O}(\delta t^2) \\[2mm] dp'_\alpha = dp_\alpha + \dfrac{\partial \dot{p}_\alpha}{\partial p_\alpha} dp_\alpha \delta t + \mathcal{O}(\delta t^2) \end{cases}. \tag{3.5}$$

To order of δt^2, the new volume element is $d\Gamma' = \prod_{i=1}^{N} d^3\vec{p}_i{}' \, d^3\vec{q}_i{}'$. From Eqs. (3.5) it follows that for each pair of conjugate coordinates

$$dq'_\alpha \cdot dp'_\alpha = dq_\alpha \cdot dp_\alpha \left[1 + \left(\frac{\partial \dot{q}_\alpha}{\partial q_\alpha} + \frac{\partial \dot{p}_\alpha}{\partial p_\alpha} \right) \delta t + \mathcal{O}(\delta t^2) \right]. \tag{3.6}$$

But since the time evolution of coordinates and momenta are governed by the canonical Eqs. (3.1), we have

$$\frac{\partial \dot{q}_\alpha}{\partial q_\alpha} = \frac{\partial}{\partial q_\alpha} \frac{\partial \mathcal{H}}{\partial p_\alpha} = \frac{\partial^2 \mathcal{H}}{\partial p_\alpha \partial q_\alpha}, \quad \text{and} \quad \frac{\partial \dot{p}_\alpha}{\partial p_\alpha} = \frac{\partial}{\partial p_\alpha} \left(-\frac{\partial \mathcal{H}}{\partial q_\alpha} \right) = -\frac{\partial^2 \mathcal{H}}{\partial q_\alpha \partial p_\alpha}. \tag{3.7}$$

Thus the projected area in Eq. (3.6) is unchanged for any pair of coordinates, and hence the volume element is unaffected, $d\Gamma' = d\Gamma$. All the pure states $d\mathcal{N}$ originally in the vicinity of $(\mathbf{p}, \mathbf{q})$ are transported to the neighborhood of $(\mathbf{p}', \mathbf{q}')$, but occupy exactly the same volume. The ratio $d\mathcal{N}/d\Gamma$ is left unchanged, and ρ behaves like the density of an incompressible fluid.

The incompressibility condition $\rho(\mathbf{p}', \mathbf{q}', t + \delta t) = \rho(\mathbf{p}, \mathbf{q}, t)$ can be written in differential form as

$$\frac{d\rho}{dt} = \frac{\partial \rho}{\partial t} + \sum_{\alpha=1}^{3N} \left(\frac{\partial \rho}{\partial p_\alpha} \cdot \frac{dp_\alpha}{dt} + \frac{\partial \rho}{\partial q_\alpha} \cdot \frac{dq_\alpha}{dt} \right) = 0. \tag{3.8}$$

Note the distinction between $\partial \rho / \partial t$ and $d\rho / dt$: the former *partial* derivative refers to the changes in ρ at a particular location in phase space, while the latter *total* derivative follows the evolution of a volume of fluid as it moves in phase space. Substituting from Eq. (3.1) into Eq. (3.8) leads to

$$\frac{\partial \rho}{\partial t} = \sum_{\alpha=1}^{3N} \left(\frac{\partial \rho}{\partial p_\alpha} \cdot \frac{\partial \mathcal{H}}{\partial q_\alpha} - \frac{\partial \rho}{\partial q_\alpha} \cdot \frac{\partial \mathcal{H}}{\partial p_\alpha} \right) = -\{\rho, \mathcal{H}\}, \tag{3.9}$$

where we have introduced the *Poisson bracket* of two functions in phase space as

$$\{A, B\} \equiv \sum_{\alpha=1}^{3N} \left(\frac{\partial A}{\partial q_\alpha} \cdot \frac{\partial B}{\partial p_\alpha} - \frac{\partial A}{\partial p_\alpha} \cdot \frac{\partial B}{\partial q_\alpha} \right) = -\{B, A\}. \tag{3.10}$$

Consequences of Liouville's theorem:

1. Under the action of time reversal, $(\mathbf{p}, \mathbf{q}, t) \to (-\mathbf{p}, \mathbf{q}, -t)$, the Poisson bracket $\{\rho, \mathcal{H}\}$ changes sign, and Eq. (3.9) implies that the density reverses its evolution, that is, $\rho(\mathbf{p}, \mathbf{q}, t) = \rho(-\mathbf{p}, \mathbf{q}, -t)$.

2. The time evolution of the ensemble average in Eq. (3.3) is given by (using Eq. (3.9))

$$\frac{d \langle \mathcal{O} \rangle}{dt} = \int d\Gamma \frac{\partial \rho(\mathbf{p}, \mathbf{q}, t)}{\partial t} \mathcal{O}(\mathbf{p}, \mathbf{q}) = \sum_{\alpha=1}^{3N} \int d\Gamma \mathcal{O}(\mathbf{p}, \mathbf{q}) \left(\frac{\partial \rho}{\partial p_\alpha} \cdot \frac{\partial \mathcal{H}}{\partial q_\alpha} - \frac{\partial \rho}{\partial q_\alpha} \cdot \frac{\partial \mathcal{H}}{\partial p_\alpha} \right). \tag{3.11}$$

The partial derivatives of ρ in the above equation can be removed by using the method of integration by parts, that is, $\int f \rho' = -\int \rho f'$ since ρ vanishes on the boundaries of the integrations, leading to

$$\frac{d \langle \mathcal{O} \rangle}{dt} = -\sum_{\alpha=1}^{3N} \int d\Gamma \rho \left[\left(\frac{\partial \mathcal{O}}{\partial p_\alpha} \cdot \frac{\partial \mathcal{H}}{\partial q_\alpha} - \frac{\partial \mathcal{O}}{\partial q_\alpha} \cdot \frac{\partial \mathcal{H}}{\partial p_\alpha} \right) + \mathcal{O} \left(\frac{\partial^2 \mathcal{H}}{\partial p_\alpha \partial q_\alpha} - \frac{\partial^2 \mathcal{H}}{\partial q_\alpha \partial p_\alpha} \right) \right]$$

$$= -\int d\Gamma \rho \{\mathcal{H}, \mathcal{O}\} = \langle \{\mathcal{O}, \mathcal{H}\} \rangle. \tag{3.12}$$

Note that the total time derivative *cannot* be taken inside the integral sign, that is,

$$\frac{\mathrm{d}\langle\mathcal{O}\rangle}{\mathrm{d}t} \neq \int \mathrm{d}\Gamma \frac{\mathrm{d}\rho(\mathbf{p}, \mathbf{q}, t)}{\mathrm{d}t} \mathcal{O}(\mathbf{p}, \mathbf{q}). \tag{3.13}$$

This common mistake yields $\mathrm{d}\langle\mathcal{O}\rangle/\mathrm{d}t = 0$.

3. If the members of the ensemble correspond to an equilibrium macroscopic state, the ensemble averages must be independent of time. This can be achieved by a stationary density $\partial\rho_{\mathrm{eq}}/\partial t = 0$, that is, by requiring

$$\{\rho_{\mathrm{eq}}, \mathcal{H}\} = 0. \tag{3.14}$$

A possible solution to the above equation is for ρ_{eq} to be a function of $\mathcal{H}$, that is, $\rho_{\mathrm{eq}}(\mathbf{p}, \mathbf{q}) = \rho(\mathcal{H}(\mathbf{p}, \mathbf{q}))$. It is then easy to verify that $\{\rho(\mathcal{H}), \mathcal{H}\} = \rho'(\mathcal{H})\{\mathcal{H}, \mathcal{H}\} = 0$. This solution implies that the value of ρ is constant on surfaces of constant energy $\mathcal{H}$, in phase space. This is indeed the *basic assumption of statistical mechanics*. For example, in the microcanonical ensemble, the total energy E of an isolated system is specified. All members of the ensemble must then be located on the surface $\mathcal{H}(\mathbf{p}, \mathbf{q}) = E$ in phase space. Equation (3.9) implies that a uniform density of points on this surface is stationary in time. The assumption of statistical mechanics is that the macrostate is indeed represented by such a uniform density of microstates. This is equivalent to replacing the objective measure of probability in Eq. (3.2) with a subjective one.

There may be additional *conserved quantities* associated with the Hamiltonian that satisfy $\{L_n, \mathcal{H}\} = 0$. In the presence of such quantities, a stationary density exists for any function of the form $\rho_{\mathrm{eq}}(\mathbf{p}, \mathbf{q}) = \rho(\mathcal{H}(\mathbf{p}, \mathbf{q}), L_1(\mathbf{p}, \mathbf{q}), L_2(\mathbf{p}, \mathbf{q}), \cdots)$. Clearly, the value of L_n is not changed during the evolution of the system, since

$$\begin{aligned}
\frac{\mathrm{d}L_n(\mathbf{p}, \mathbf{q})}{\mathrm{d}t} &\equiv \frac{L_n(\mathbf{p}(t+\mathrm{d}t), \mathbf{q}(t+\mathrm{d}t)) - L_n(\mathbf{p}(t), \mathbf{q}(t))}{\mathrm{d}t} \\
&= \sum_{\alpha=1}^{3N} \left(\frac{\partial L_n}{\partial p_\alpha} \cdot \frac{\partial p_\alpha}{\partial t} + \frac{\partial L_n}{\partial q_\alpha} \cdot \frac{\partial q_\alpha}{\partial t} \right) \\
&= -\sum_{\alpha=1}^{3N} \left(\frac{\partial L_n}{\partial p_\alpha} \cdot \frac{\partial \mathcal{H}}{\partial q_\alpha} - \frac{\partial L_n}{\partial q_\alpha} \cdot \frac{\partial \mathcal{H}}{\partial p_\alpha} \right) = \{L_n, \mathcal{H}\} = 0.
\end{aligned} \tag{3.15}$$

Hence, the functional dependence of ρ_{eq} on these quantities merely indicates that all *accessible states*, that is, those that can be connected without violating any conservation law, are equally likely.

4. The above postulate for ρ_{eq} answers the first question posed at the beginning of this chapter. However, in order to answer the second question, and to justify the basic assumption of statistical mechanics, we need to show that non-stationary densities converge onto the stationary solution ρ_{eq}. This contradicts the time reversal symmetry noted in point (1) above: for any solution $\rho(t)$ converging to ρ_{eq}, there is a time-reversed solution that diverges from it. The best that can be hoped for is to show that the solutions $\rho(t)$ are in the neighborhood of ρ_{eq} the majority of the time, so that *time averages* are dominated by the stationary solution. This brings us to the problem of *ergodicity*, which is whether it is justified to replace time averages with ensemble averages. In measuring the properties of any system, we deal with only one representative of the equilibrium ensemble. However, most macroscopic properties do not have instantaneous values and require some form

of averaging. For example, the pressure P exerted by a gas results from the impact of particles on the walls of the container. The number and momenta of these particles vary at different times and different locations. The measured pressure reflects an average over many characteristic microscopic times. If over this time scale the representative point of the system moves around and uniformly samples the accessible points in phase space, we may replace the time average with the ensemble average. For a few systems it is possible to prove an *ergodic theorem*, which states that the representative point comes arbitrarily close to all accessible points in phase space after a sufficiently long time. However, the proof usually works for time intervals that grow exponentially with the number of particles N, and thus exceed by far any reasonable time scale over which the pressure of a gas is typically measured. As such the proofs of the ergodic theorem have so far little to do with the reality of macroscopic equilibrium.

3.3 The Bogoliubov–Born–Green–Kirkwood–Yvon hierarchy

The full phase space density contains much more information than necessary for description of equilibrium properties. For example, knowledge of the one-particle distribution is sufficient for computing the pressure of a gas. A one-particle density refers to the expectation value of finding *any* of the N particles at location $\vec{q}$, with momentum $\vec{p}$, at time t, which is computed from the full density ρ as

$$f_1(\vec{p}, \vec{q}, t) = \left\langle \sum_{i=1}^{N} \delta^3(\vec{p} - \vec{p}_i)\delta^3(\vec{q} - \vec{q}_i) \right\rangle$$

$$= N \int \prod_{i=2}^{N} d^3\vec{p}_i \, d^3\vec{q}_i \rho(\vec{p}_1 = \vec{p}, \vec{q}_1 = \vec{q}, \vec{p}_2, \vec{q}_2, \cdots, \vec{p}_N, \vec{q}_N, t).$$

(3.16)

To obtain the second identity above, we used the first pair of delta functions to perform one set of integrals, and then assumed that the density is symmetric with respect to permuting the particles. Similarly, a two-particle density can be computed from

$$f_2(\vec{p}_1, \vec{q}_1, \vec{p}_2, \vec{q}_2, t) = N(N-1) \int \prod_{i=3}^{N} dV_i \, \rho(\vec{p}_1, \vec{q}_1, \vec{p}_2, \vec{q}_2, \cdots, \vec{p}_N, \vec{q}_N, t), \quad (3.17)$$

where $dV_i = d^3\vec{p}_i \, d^3\vec{q}_i$ is the contribution of particle i to phase space volume. The general s-particle density is defined by

$$f_s(\vec{p}_1, \cdots, \vec{q}_s, t) = \frac{N!}{(N-s)!} \int \prod_{i=s+1}^{N} dV_i \, \rho(\mathbf{p}, \mathbf{q}, t) = \frac{N!}{(N-s)!} \rho_s(\vec{p}_1, \cdots, \vec{q}_s, t), \quad (3.18)$$

where ρ_s is a standard *unconditional PDF* for the coordinates of s particles, and $\rho_N \equiv \rho$. While ρ_s is properly normalized to unity when integrated over all its variables, the s-particle density has a normalization of $N!/(N-s)!$. We shall use the two quantities interchangeably.

The evolution of the few-body densities is governed by the BBGKY hierarchy of equations attributed to Bogoliubov, Born, Green, Kirkwood, and Yvon. The simplest non-trivial Hamiltonian studied in kinetic theory is

$$\mathcal{H}(\mathbf{p}, \mathbf{q}) = \sum_{i=1}^{N} \left[\frac{\vec{p}_i^{\,2}}{2m} + U(\vec{q}_i) \right] + \frac{1}{2} \sum_{(i,j)=1}^{N} \mathcal{V}(\vec{q}_i - \vec{q}_j). \tag{3.19}$$

This Hamiltonian provides an adequate description of a weakly interacting gas. In addition to the classical kinetic energy of particles of mass m, it contains an *external potential U*, and a *two-body interaction $\mathcal{V}$*, between the particles. In principle, three- and higher-body interactions should also be included for a complete description, but they are not very important in the dilute gas (nearly ideal) limit.

For evaluating the time evolution of f_s, it is convenient to divide the Hamiltonian into

$$\mathcal{H} = \mathcal{H}_s + \mathcal{H}_{N-s} + \mathcal{H}', \tag{3.20}$$

where $\mathcal{H}_s$ and $\mathcal{H}_{N-s}$ include only interactions among each group of particles,

$$\begin{aligned}
\mathcal{H}_s &= \sum_{n=1}^{s} \left[\frac{\vec{p}_n^{\,2}}{2m} + U(\vec{q}_n) \right] + \frac{1}{2} \sum_{(n,m)=1}^{s} \mathcal{V}(\vec{q}_n - \vec{q}_m), \\
\mathcal{H}_{N-s} &= \sum_{i=s+1}^{N} \left[\frac{\vec{p}_i^{\,2}}{2m} + U(\vec{q}_i) \right] + \frac{1}{2} \sum_{(i,j)=s+1}^{N} \mathcal{V}(\vec{q}_i - \vec{q}_j),
\end{aligned} \tag{3.21}$$

while the interparticle interactions are contained in

$$\mathcal{H}' = \sum_{n=1}^{s} \sum_{i=s+1}^{N} \mathcal{V}(\vec{q}_n - \vec{q}_i). \tag{3.22}$$

From Eq. (3.18), the time evolution of f_s (or ρ_s) is obtained as

$$\frac{\partial \rho_s}{\partial t} = \int \prod_{i=s+1}^{N} \mathrm{d}V_i \, \frac{\partial \rho}{\partial t} = - \int \prod_{i=s+1}^{N} \mathrm{d}V_i \, \{\rho, \mathcal{H}_s + \mathcal{H}_{N-s} + \mathcal{H}'\}, \tag{3.23}$$

where Eq. (3.9) is used for the evolution of ρ. The three Poisson brackets in Eq. (3.23) will now be evaluated in turn. Since the first s coordinates are not integrated, the order of integrations and differentiations for the Poisson bracket may be reversed in the first term, and

$$\int \prod_{i=s+1}^{N} \mathrm{d}V_i \, \{\rho, \mathcal{H}_s\} = \left\{ \left(\int \prod_{i=s+1}^{N} \mathrm{d}V_i \, \rho \right), \mathcal{H}_s \right\} = \{\rho_s, \mathcal{H}_s\}. \tag{3.24}$$

Writing the Poisson brackets explicitly, the second term of Eq. (3.23) takes the form

$$-\int \prod_{i=s+1}^{N} \mathrm{d}V_i \, \{\rho, \mathcal{H}_{N-s}\} = \int \prod_{i=s+1}^{N} \mathrm{d}V_i \sum_{j=1}^{N} \left[\frac{\partial \rho}{\partial \vec{p}_j} \cdot \frac{\partial \mathcal{H}_{N-s}}{\partial \vec{q}_j} - \frac{\partial \rho}{\partial \vec{q}_j} \cdot \frac{\partial \mathcal{H}_{N-s}}{\partial \vec{p}_j} \right]$$

(using Eq. (3.21))

$$= \int \prod_{i=s+1}^{N} dV_i \sum_{j=s+1}^{N} \left[\frac{\partial \rho}{\partial \vec{p}_j} \cdot \left(\frac{\partial U}{\partial \vec{q}_j} + \frac{1}{2} \sum_{k=s+1}^{N} \frac{\partial \mathcal{V}(\vec{q}_j - \vec{q}_k)}{\partial \vec{q}_j} \right) - \frac{\partial \rho}{\partial \vec{q}_j} \cdot \frac{\vec{p}_j}{m} \right] = 0. \quad (3.25)$$

The last equality is obtained after performing the integrations by part: the term multiplying $\partial \rho / \partial \vec{p}_j$ has no dependence on $\vec{p}_j$, while $\vec{p}_j/m$ does not depend on $\vec{q}_j$. The final term in Eq. (3.23), involving the Poisson bracket with $\mathcal{H}'$, is

$$\int \prod_{i=s+1}^{N} dV_i \sum_{j=1}^{N} \left[\frac{\partial \rho}{\partial \vec{p}_j} \cdot \frac{\partial \mathcal{H}'}{\partial \vec{q}_j} - \frac{\partial \rho}{\partial \vec{q}_j} \cdot \frac{\partial \mathcal{H}'}{\partial \vec{p}_j} \right]$$

$$= \int \prod_{i=s+1}^{N} dV_i \left[\sum_{n=1}^{s} \frac{\partial \rho}{\partial \vec{p}_n} \cdot \sum_{j=s+1}^{N} \frac{\partial \mathcal{V}(\vec{q}_n - \vec{q}_j)}{\partial \vec{q}_n} + \sum_{j=s+1}^{N} \frac{\partial \rho}{\partial \vec{p}_j} \cdot \sum_{n=1}^{s} \frac{\partial \mathcal{V}(\vec{q}_j - \vec{q}_n)}{\partial \vec{q}_j} \right],$$

where the sum over all particles has been subdivided into the two groups. (Note that $\mathcal{H}'$ in Eq. (3.22) has no dependence on the momenta.) Integration by parts shows that the second term in the above expression is zero. The first term involves the sum of $(N-s)$ expressions that are equal by symmetry and simplifies to

$$(N-s) \int \prod_{i=s+1}^{N} dV_i \sum_{n=1}^{s} \frac{\partial \mathcal{V}(\vec{q}_n - \vec{q}_{s+1})}{\partial \vec{q}_n} \cdot \frac{\partial \rho}{\partial \vec{p}_n}$$

$$= (N-s) \sum_{n=1}^{s} \int dV_{s+1} \frac{\partial \mathcal{V}(\vec{q}_n - \vec{q}_{s+1})}{\partial \vec{q}_n} \cdot \frac{\partial}{\partial \vec{p}_n} \left[\int \prod_{i=s+2}^{N} dV_i \, \rho \right]. \quad (3.26)$$

Note that the quantity in the above square brackets is ρ_{s+1}. Thus, adding up Eqs. (3.24), (3.25), and (3.26),

$$\frac{\partial \rho_s}{\partial t} - \{\mathcal{H}_s, \rho_s\} = (N-s) \sum_{n=1}^{s} \int dV_{s+1} \frac{\partial \mathcal{V}(\vec{q}_n - \vec{q}_{s+1})}{\partial \vec{q}_n} \cdot \frac{\partial \rho_{s+1}}{\partial \vec{p}_n}, \quad (3.27)$$

or in terms of the densities f_s,

$$\frac{\partial f_s}{\partial t} - \{\mathcal{H}_s, f_s\} = \sum_{n=1}^{s} \int dV_{s+1} \frac{\partial \mathcal{V}(\vec{q}_n - \vec{q}_{s+1})}{\partial \vec{q}_n} \cdot \frac{\partial f_{s+1}}{\partial \vec{p}_n}. \quad (3.28)$$

In the absence of interactions with other particles, the density ρ_s for a group of s particles evolves as the density of an incompressible fluid (as required by Liouville's theorem), and is described by the *streaming terms* on the left-hand side of Eq. (3.27). However, because of interactions with the remaining $N-s$ particles, the flow is modified by the *collision terms* on the right-hand side. The collision integral is the sum of terms corresponding to a potential collision of any of the particles in the group of s, with any of the remaining $N-s$ particles. To describe the probability of finding the additional particle that collides with a member of this group, the result must depend on the joint PDF of $s+1$ particles described by ρ_{s+1}. This results in a hierarchy of equations in which $\partial \rho_1 / \partial t$ depends on ρ_2, $\partial \rho_2 / \partial t$ depends on ρ_3, etc., which is at least as complicated as the original equation for the full phase space density.

To proceed further, a physically motivated approximation for terminating the hierarchy is needed.

3.4 The Boltzmann equation

To estimate the relative importance of the different terms appearing in Eqs. (3.28), let us examine the first two equations in the hierarchy,

$$\left[\frac{\partial}{\partial t} - \frac{\partial U}{\partial \vec{q}_1} \cdot \frac{\partial}{\partial \vec{p}_1} + \frac{\vec{p}_1}{m} \cdot \frac{\partial}{\partial \vec{q}_1}\right] f_1 = \int \mathrm{d}V_2 \frac{\partial \mathcal{V}(\vec{q}_1 - \vec{q}_2)}{\partial \vec{q}_1} \cdot \frac{\partial f_2}{\partial \vec{p}_1}, \tag{3.29}$$

and

$$\left[\frac{\partial}{\partial t} - \frac{\partial U}{\partial \vec{q}_1} \cdot \frac{\partial}{\partial \vec{p}_1} - \frac{\partial U}{\partial \vec{q}_2} \cdot \frac{\partial}{\partial \vec{p}_2} + \frac{\vec{p}_1}{m} \cdot \frac{\partial}{\partial \vec{q}_1} + \frac{\vec{p}_2}{m} \cdot \frac{\partial}{\partial \vec{q}_2} - \frac{\partial \mathcal{V}(\vec{q}_1 - \vec{q}_2)}{\partial \vec{q}_1} \cdot \left(\frac{\partial}{\partial \vec{p}_1} - \frac{\partial}{\partial \vec{p}_2}\right)\right] f_2$$

$$= \int \mathrm{d}V_3 \left[\frac{\partial \mathcal{V}(\vec{q}_1 - \vec{q}_3)}{\partial \vec{q}_1} \cdot \frac{\partial}{\partial \vec{p}_1} + \frac{\partial \mathcal{V}(\vec{q}_2 - \vec{q}_3)}{\partial \vec{q}_2} \cdot \frac{\partial}{\partial \vec{p}_2}\right] f_3. \tag{3.30}$$

Note that two of the streaming terms in Eq. (3.30) have been combined by using $\partial \mathcal{V}(\vec{q}_1 - \vec{q}_2)/\partial \vec{q}_1 = -\partial \mathcal{V}(\vec{q}_2 - \vec{q}_1)/\partial \vec{q}_2$, which is valid for a *symmetric* potential such that $\mathcal{V}(\vec{q}_1 - \vec{q}_2) = \mathcal{V}(\vec{q}_2 - \vec{q}_1)$.

Time scales: all terms within square brackets in the above equations have dimensions of inverse time, and we estimate their relative magnitudes by dimensional analysis, using typical velocities and length scales. The typical speed of a gas particle at room temperature is $v \approx 10^2 \, \mathrm{m\,s^{-1}}$. For terms involving the external potential U, or the interatomic potential $\mathcal{V}$, an appropriate length scale can be extracted from the range of variations of the potential.

(a) The terms proportional to

$$\frac{1}{\tau_U} \sim \frac{\partial U}{\partial \vec{q}} \cdot \frac{\partial}{\partial \vec{p}}$$

involve spatial variations of the external potential $U(\vec{q})$, which take place over macroscopic distances L. We shall refer to the associated time τ_U as an *extrinsic* time scale, as it can be made arbitrarily long by increasing system size. For a typical value of $L \approx 10^{-3} \, \mathrm{m}$, we get $\tau_U \approx L/v \approx 10^{-5} \, \mathrm{s}$.

(b) From the terms involving the interatomic potential $\mathcal{V}$, we can extract two additional time scales, which are *intrinsic* to the gas under study. In particular, the *collision duration*

$$\frac{1}{\tau_c} \sim \frac{\partial \mathcal{V}}{\partial \vec{q}} \cdot \frac{\partial}{\partial \vec{p}}$$

is the typical time over which two particles are within the effective range d of their interaction. For *short-range* interactions (including van der Waals and Lennard-Jones, despite their power law decaying tails), $d \approx 10^{-10} \, \mathrm{m}$ is of the order of a typical atomic size, resulting in $\tau_c \approx 10^{-12} \, \mathrm{s}$. This is usually the shortest time scale in the problem. The analysis is somewhat more complicated for *long-range* interactions, such as the Coulomb gas in a plasma. For a neutral plasma, the Debye screening length λ replaces d in the above equation, as discussed in the problems.

(c) There are also collision terms on the right-hand side of Eqs. (3.28), which depend on f_{s+1}, and lead to an inverse time scale

$$\frac{1}{\tau_\times} \sim \int \, \mathrm{d}V \, \frac{\partial \mathcal{V}}{\partial \vec{q}} \cdot \frac{\partial}{\partial \vec{p}} \, \frac{f_{s+1}}{f_s} \sim \int \, \mathrm{d}V \, \frac{\partial \mathcal{V}}{\partial \vec{q}} \cdot \frac{\partial}{\partial \vec{p}} \, N \frac{\rho_{s+1}}{\rho_s}.$$

The integrals are only non-zero over the volume of the interparticle potential d^3. The term f_{s+1}/f_s is related to the probability of finding another particle per unit volume, which is roughly the particle density $n = N/V \approx 10^{26} \, \mathrm{m}^{-3}$. We thus obtain the *mean free time*

$$\tau_\times \approx \frac{\tau_c}{nd^3} \approx \frac{1}{nvd^2}, \tag{3.31}$$

which is the typical distance a particle travels between collisions. For short-range interactions, $\tau_\times \approx 10^{-8} \, \mathrm{s}$ is much longer than τ_c, and the collision terms on the right-hand side of Eqs. (3.28) are smaller by a factor of $nd^3 \approx (10^{26} \, \mathrm{m}^{-3})(10^{-10} \, \mathrm{m})^3 \approx 10^{-4}$.

Fig. 3.3 The mean free time between collisions is estimated by requiring that there is only one other particle in the volume swept in time $\tau_\times$.

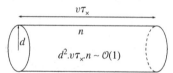

The *Boltzmann equation* is obtained for short-range interactions in the dilute regime by exploiting $\tau_c/\tau_\times \approx nd^3 \ll 1$. (By contrast, for long-range interactions such that $nd^3 \gg 1$, the *Vlasov equation* is obtained by dropping the collision terms on the left-hand side, as discussed in the problems.) From the above discussion, it is apparent that Eq. (3.29) is different from the rest of the hierarchy: it is the only one in which the collision terms are absent from the left-hand side. For all other equations, the right-hand side is smaller by a factor of nd^3, while in Eq. (3.29) it may indeed dominate the left-hand side. Thus a possible approximation scheme is to truncate the equations after the first two, by setting the right-hand side of Eq. (3.30) to zero.

Setting the right-hand side of the equation for f_2 to zero implies that the two-body density evolves as in an isolated two-particle system. The relatively simple mechanical processes that govern this evolution result in streaming terms for f_2 that are proportional to both τ_U^{-1} and τ_c^{-1}. The two sets of terms can be more or less treated independently: the former describe the evolution of the center of mass of the two particles, while the latter govern the dependence on relative coordinates.

The density f_2 is proportional to the joint PDF ρ_2 for finding one particle at $(\vec{p}_1, \vec{q}_1)$, and another at $(\vec{p}_2, \vec{q}_2)$, at the same time t. It is reasonable to expect that at distances much larger than the range of the potential $\mathcal{V}$, the particles are independent, that is,

$$\begin{cases} \rho_2(\vec{p}_1, \vec{q}_1, \vec{p}_2, \vec{q}_2, t) \longrightarrow \rho_1(\vec{p}_1, \vec{q}_1, t)\rho_1(\vec{p}_2, \vec{q}_2, t), \quad \text{or} \\[2mm] f_2(\vec{p}_1, \vec{q}_1, \vec{p}_2, \vec{q}_2, t) \longrightarrow f_1(\vec{p}_1, \vec{q}_1, t)f_1(\vec{p}_2, \vec{q}_2, t), \quad \text{for} \quad |\vec{q}_2 - \vec{q}_1| \gg d. \end{cases} \quad (3.32)$$

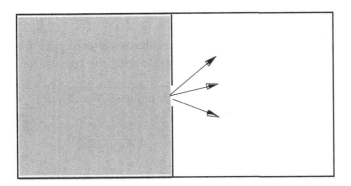

Fig. 3.4 Particles escaping into an initially empty container through a hole.

The above statement should be true even for situations out of equilibrium. For example, imagine that the gas particles in a chamber are suddenly allowed to invade an empty volume after the removal of a barrier. The density f_1 will undergo a complicated evolution, and its relaxation time will be at least comparable to τ_U. The two-particle density f_2 will also reach its final value at a comparable time interval. However, it is expected to relax to a form similar to Eq. (3.32) over a much shorter time of the order of τ_c.

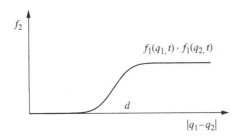

Fig. 3.5 The schematic behavior of a two-body density as a function of relative coordinates.

For the collision term on the right-hand side of Eq. (3.29), we actually need the precise dependence of f_2 on the relative coordinates and momenta at separations comparable to d. At time intervals longer than τ_c (but possibly shorter than τ_U), the "steady state" behavior of f_2 at small relative distances is obtained by equating the largest streaming terms in Eq. (3.30), that is,

$$\left[\frac{\vec{p}_1}{m} \cdot \frac{\partial}{\partial \vec{q}_1} + \frac{\vec{p}_2}{m} \cdot \frac{\partial}{\partial \vec{q}_2} - \frac{\partial \mathcal{V}(\vec{q}_1 - \vec{q}_2)}{\partial \vec{q}_1} \cdot \left(\frac{\partial}{\partial \vec{p}_1} - \frac{\partial}{\partial \vec{p}_2} \right) \right] f_2 = 0. \quad (3.33)$$

We expect $f_2(\vec{q}_1, \vec{q}_2)$ to have slow variations over the center of mass coordinate $\vec{Q} = (\vec{q}_1 + \vec{q}_2)/2$, and large variations over the relative coordinate $\vec{q} = \vec{q}_2 - \vec{q}_1$. Therefore, $\partial f_2/\partial \vec{q} \gg \partial f_2/\partial \vec{Q}$, and $\partial f_2/\vec{q}_2 \approx -\partial f_2/\partial \vec{q}_1 \approx \partial f_2/\partial \vec{q}$, leading to

$$\frac{\partial \mathcal{V}(\vec{q}_1 - \vec{q}_2)}{\partial \vec{q}_1} \cdot \left(\frac{\partial}{\partial \vec{p}_1} - \frac{\partial}{\partial \vec{p}_2} \right) f_2 = -\left(\frac{\vec{p}_1 - \vec{p}_2}{m} \right) \cdot \frac{\partial}{\partial \vec{q}} f_2. \tag{3.34}$$

The above equation provides a precise mathematical expression for how f_2 is constrained along the trajectories that describe the collision of the two particles.

The collision term on the right-hand side of Eq. (3.29) can now be written as

$$\begin{aligned}
\left. \frac{\mathrm{d}f_1}{\mathrm{d}t} \right|_{\text{coll.}} &= \int \mathrm{d}^3\vec{p}_2 \, \mathrm{d}^3\vec{q}_2 \, \frac{\partial \mathcal{V}(\vec{q}_1 - \vec{q}_2)}{\partial \vec{q}_1} \cdot \left(\frac{\partial}{\partial \vec{p}_1} - \frac{\partial}{\partial \vec{p}_2} \right) f_2 \\
&\approx \int \mathrm{d}^3\vec{p}_2 \, \mathrm{d}^3\vec{q} \, \left(\frac{\vec{p}_2 - \vec{p}_1}{m} \right) \cdot \frac{\partial}{\partial \vec{q}} f_2 \left(\vec{p}_1, \vec{q}_1, \vec{p}_2, \vec{q}; t \right).
\end{aligned} \tag{3.35}$$

The first identity is obtained from Eq. (3.29) by noting that the added term proportional to $\partial f_2/\partial \vec{p}_2$ is a complete derivative and integrates to zero, while the second equality follows from Eq. (3.34), after the change of variables to $\vec{q} = \vec{q}_2 - \vec{q}_1$. (Since it relies on establishing the "steady state" in the relative coordinates, this approximation is valid as long as we examine events in time with a resolution longer than τ_c.)

Kinematics of collision and scattering: the integrand in Eq. (3.35) is a derivative of f_2 with respect to $\vec{q}$ along the direction of relative motion $\vec{p} = \vec{p}_2 - \vec{p}_1$ of the colliding particles. To perform this integration we introduce a convenient coordinate system for $\vec{q}$, guided by the formalism used to describe the scattering of particles. Naturally, we choose one axis to be parallel to $\vec{p}_2 - \vec{p}_1$, with the corresponding coordinate a that is negative before a collision, and positive afterwards. The other two coordinates of $\vec{q}$ are represented by an

Fig. 3.6 The parameters used to describe the collision of two particles in their center of mass (at $a = 0$) frame, which moves with a conserved momentum $\vec{P} = \vec{p}_1 + \vec{p}_2 = \vec{p}_1' + \vec{p}_2'$. In an elastic collision, the relative momentum $\vec{p}$ is rotated by solid angle $\hat{\Omega}$.

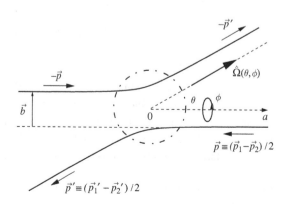

impact vector $\vec{b}$ that is $\vec{0}$ for a head-on collision ($[\vec{p}_1 - \vec{p}_2] \parallel [\vec{q}_1 - \vec{q}_2]$). We can now integrate over a to get

$$\left. \frac{\mathrm{d}f_1}{\mathrm{d}t} \right|_{\mathrm{coll.}} = \int \mathrm{d}^3\vec{p}_2 \, \mathrm{d}^2\vec{b} \, |\vec{v}_1 - \vec{v}_2| \left[f_2(\vec{p}_1, \vec{q}_1, \vec{p}_2, \vec{b}, +; t) - f_2(\vec{p}_1, \vec{q}_1, \vec{p}_2, \vec{b}, -; t) \right],$$

(3.36)

where $|\vec{v}_1 - \vec{v}_2| = |\vec{p}_1 - \vec{p}_2|/m$ is the relative speed of the two particles, with $(\vec{b}, -)$ and $(\vec{b}, +)$ referring to relative coordinates before and after the collision. Note that $\mathrm{d}^2\vec{b}\,|\vec{v}_1 - \vec{v}_2|$ is just the flux of particles impinging on the element of area $\mathrm{d}^2\vec{b}$.

In principle, the integration over a is from $-\infty$ to $+\infty$, but as the variations of f_2 are only significant over the interaction range d, we can evaluate the above quantities at separations of a few d from the collision point. This is a good compromise, allowing us to evaluate f_2 away from the collisions, but at small enough separations so that we can ignore the difference between $\vec{q}_1$ and $\vec{q}_2$. This amounts to a coarse graining in space that eliminates variations on scales finer than d. With these provisos, it is tempting to close the equation for f_1, by using the assumption of uncorrelated particles in Eq. (3.32). Clearly some care is necessary as a naive substitution gives zero! The key observation is that the densities f_2 for situations corresponding to before and after the collision have to be treated differently. For example, soon after opening of the slot separating empty and full gas containers, the momenta of the gas particles are likely to point away from the slot. Collisions will tend to randomize momenta, yielding a more isotropic distribution. However, the densities f_2 before and after the collision are related by streaming, implying that $f_2(\vec{p}_1, \vec{q}_1, \vec{p}_2, \vec{b}, +; t) = f_2(\vec{p}_1{}', \vec{q}_1, \vec{p}_2{}', \vec{b}, -; t)$, where $\vec{p}_1{}'$ and $\vec{p}_2{}'$ are momenta whose collision at an impact vector $\vec{b}$ results in production of outgoing particles with momenta $\vec{p}_1$ and $\vec{p}_2$. They can be obtained using time reversal symmetry, by integrating the equations of motion for incoming colliding particles of momenta $-\dot{p}_1$ and $-\vec{p}_2$. In terms of these momenta, we can write

$$\left. \frac{\mathrm{d}f_1}{\mathrm{d}t} \right|_{\mathrm{coll.}} = \int \mathrm{d}^3\vec{p}_2 \, \mathrm{d}^2\vec{b} \, |\vec{v}_1 - \vec{v}_2| \left[f_2(\vec{p}_1{}', \vec{q}_1, \vec{p}_2{}', \vec{b}, -; t) - f_2(\vec{p}_1, \vec{q}_1, \vec{p}_2, \vec{b}, -; t) \right].$$

(3.37)

It is sometimes more convenient to describe the scattering of two particles in terms of the relative momenta $\vec{p} = \vec{p}_1 - \vec{p}_2$ and $\vec{p}' = \vec{p}_1{}' - \vec{p}_2{}'$, before and after the collision. For a given $\vec{b}$, the initial momentum $\vec{p}$ is deterministically transformed to the final momentum $\vec{p}'$. To find the functional form $\vec{p}'(|\vec{p}|, \vec{b})$, one must integrate the equations of motion. However, it is possible to make some general statements based on conservation laws: in elastic collisions, the magnitude of $\vec{p}$ is preserved, and it merely rotates to a final direction indicated by the angles $(\theta, \phi) \equiv \hat{\Omega}(\vec{b})$ (a unit vector) in spherical coordinates. Since there is a one-to-one correspondence between the impact vector $\vec{b}$ and the *solid angle* Ω, we make a change of variables between the two, resulting in

$$\frac{df_1}{dt}\bigg|_{\text{coll.}} = \int d^3\vec{p}_2\, d^2\Omega\, \left|\frac{d\sigma}{d\Omega}\right| |\vec{v}_1 - \vec{v}_2| \left[f_2(\vec{p}_1{}', \vec{q}_1, \vec{p}_2{}', \vec{b}, -; t) \right.$$
$$\left. - f_2(\vec{p}_1, \vec{q}_1, \vec{p}_2, \vec{b}, -; t) \right].$$

(3.38)

The Jacobian of this transformation, $|d\sigma/d\Omega|$, has dimensions of area, and is known as the *differential cross-section*. It is equal to the area presented to an incoming beam that scatters into the solid angle Ω. The outgoing momenta $\vec{p}_1{}'$ and $\vec{p}_2{}'$ in Eq. (3.38) are now obtained from the two conditions $\vec{p}_1{}' + \vec{p}_2{}' = \vec{p}_1 + \vec{p}_2$ (conservation of momentum), and $\vec{p}_1{}' - \vec{p}_2{}' = |\vec{p}_1 - \vec{p}_2|\hat{\Omega}(\vec{b})$ (conservation of energy), as

$$\begin{cases} \vec{p}_1{}' = \left(\vec{p}_1 + \vec{p}_2 + |\vec{p}_1 - \vec{p}_2|\hat{\Omega}(\vec{b})\right)/2, \\ \vec{p}_2{}' = \left(\vec{p}_1 + \vec{p}_2 - |\vec{p}_1 - \vec{p}_2|\hat{\Omega}(\vec{b})\right)/2. \end{cases}$$

(3.39)

For the scattering of two *hard spheres* of diameter D, it is easy to show that the scattering angle is related to the impact parameter b by $\cos(\theta/2) = b/D$ for all ϕ. The differential cross-section is then obtained from

$$d^2\sigma = b\, db\, d\phi = D\cos\left(\frac{\theta}{2}\right) D\sin\left(\frac{\theta}{2}\right)\frac{d\theta}{2}\, d\phi = \frac{D^2}{4}\sin\theta\, d\theta\, d\phi = \frac{D^2}{4} d^2\Omega.$$

(Note that the solid angle in three dimensions is given by $d^2\Omega = \sin\theta\, d\theta\, d\phi$.) Integrating over all angles leads to the total cross-section of $\sigma = \pi D^2$, which is evidently correct. The differential cross-section for hard spheres is *independent* of both θ and $|\vec{P}|$. This is not the case for soft potentials. For example, the Coulomb potential $\mathcal{V} = e^2/|\vec{Q}|$ leads to

$$\left|\frac{d\sigma}{d\Omega}\right| = \left(\frac{me^2}{2|\vec{P}|^2 \sin^2(\theta/2)}\right)^2.$$

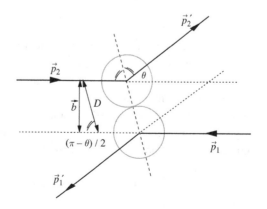

Fig. 3.7 The scattering of two hard spheres of diameter D, at an impact parameter $b = |\vec{b}|$.

(The dependence on $|\vec{P}|$ can be justified by obtaining a distance of closest approach from $|\vec{P}|^2/m + e^2/b \approx 0$.)

The Boltzmann equation is obtained from Eq. (3.38) after the substitution

$$f_2(\vec{p}_1, \vec{q}_1, \vec{p}_2, \vec{b}, -; t) = f_1(\vec{p}_1, \vec{q}_1, t) \cdot f_1(\vec{p}_2, \vec{q}_1, t), \tag{3.40}$$

known as the *assumption of molecular chaos*. Note that even if one starts with an uncorrelated initial probability distribution for particles, there is no guarantee that correlations are not generated as a result of collisions. The final result is the following closed form equation for f_1:

$$\left[\frac{\partial}{\partial t} - \frac{\partial U}{\partial \vec{q}_1} \cdot \frac{\partial}{\partial \vec{p}_1} + \frac{\vec{p}_1}{m} \cdot \frac{\partial}{\partial \vec{q}_1}\right] f_1 =$$

$$- \int d^3\vec{p}_2 \, d^2\Omega \, \left|\frac{d\sigma}{d\Omega}\right| |\vec{v}_1 - \vec{v}_2| \left[f_1(\vec{p}_1, \vec{q}_1, t) f_1(\vec{p}_2, \vec{q}_1, t) - f_1(\vec{p}_1', \vec{q}_1, t) f_1(\vec{p}_2', \vec{q}_1, t)\right].$$

$$\tag{3.41}$$

Given the complexity of the above "derivation" of the Boltzmann equation, it is appropriate to provide a heuristic explanation. The streaming terms on the left-hand side of the equation describe the motion of a single particle in the external potential U. The collision terms on the right-hand side have a simple physical interpretation: the probability of finding a particle of momentum $\vec{p}_1$ at $\vec{q}_1$ is suddenly altered if it undergoes a collision with another particle of momentum $\vec{p}_2$. The probability of such a collision is the product of kinematic factors described by the differential cross-section $|d\sigma/d\Omega|$, the "flux" of incident particles proportional to $|\vec{v}_2 - \vec{v}_1|$, and the joint probability of finding the two particles, approximated by $f_1(\vec{p}_1)f_1(\vec{p}_2)$. The first term on the right-hand side of Eq. (3.41) subtracts this probability and integrates over all possible momenta and solid angles describing the collision. The second term represents an *addition* to the probability that results from the inverse process: a particle can suddenly appear with coordinates $(\vec{p}_1, \vec{q}_1)$ as a result of a collision between two particles initially with momenta $\vec{p}_1'$ and $\vec{p}_2'$. The cross-section, and the momenta $(\vec{p}_1', \vec{p}_2')$, may have a complicated dependence on $(\vec{p}_1, \vec{p}_2)$ and Ω, determined by the specific form of the potential $\mathcal{V}$. Remarkably, various equilibrium properties of the gas are quite independent of this potential.

3.5 The H-theorem and irreversibility

The second question posed at the beginning of this chapter was whether a collection of particles naturally evolves toward an equilibrium state. While it is possible to obtain steady state solutions for the full phase space density ρ_N, because of time reversal symmetry these solutions are not attractors of generic non-equilibrium densities. Does the unconditional one-particle PDF ρ_1 suffer the same problem? While the exact density ρ_1 must necessarily reflect this property of ρ_N, the H-theorem proves that an approximate ρ_1, governed by the

Boltzmann equation, does in fact non-reversibly approach an equilibrium form. This theorem states that:

If $f_1(\vec{p}, \vec{q}, t)$ satisfies the Boltzmann equation, then $\mathrm{d}H/\mathrm{d}t \leq 0$, *where*

$$H(t) = \int \mathrm{d}^3 \vec{p}\, \mathrm{d}^3 \vec{q}\; f_1(\vec{p}, \vec{q}, t) \ln f_1(\vec{p}, \vec{q}, t). \tag{3.42}$$

The function $H(t)$ is related to the information content of the one-particle PDF. Up to an overall constant, the information content of $\rho_1 = f_1/N$ is given by $I[\rho_1] = \langle \ln \rho_1 \rangle$, which is clearly similar to $H(t)$.

Proof.   The time derivative of H is

$$\frac{\mathrm{d}H}{\mathrm{d}t} = \int \mathrm{d}^3 \vec{p}_1\, \mathrm{d}^3 \vec{q}_1\, \frac{\partial f_1}{\partial t}(\ln f_1 + 1) = \int \mathrm{d}^3 \vec{p}_1\, \mathrm{d}^3 \vec{q}_1\; \ln f_1 \frac{\partial f_1}{\partial t}, \tag{3.43}$$

since $\int \mathrm{d}V_1 f_1 = N \int \mathrm{d}\Gamma \rho = N$ is time-independent. Using Eq. (3.41), we obtain

$$\begin{aligned}
\frac{\mathrm{d}H}{\mathrm{d}t} = &\int \mathrm{d}^3 \vec{p}_1\, \mathrm{d}^3 \vec{q}_1\; \ln f_1 \left(\frac{\partial U}{\partial \vec{q}_1} \cdot \frac{\partial f_1}{\partial \vec{p}_1} - \frac{\vec{p}_1}{m} \cdot \frac{\partial f_1}{\partial \vec{q}_1} \right) \\
&- \int \mathrm{d}^3 \vec{p}_1\, \mathrm{d}^3 \vec{q}_1\, \mathrm{d}^3 \vec{p}_2\, \mathrm{d}^2 \sigma |\vec{v}_1 - \vec{v}_2| \left[f_1(\vec{p}_1, \vec{q}_1) f_1(\vec{p}_2, \vec{q}_1) \right. \\
&\left. - f_1(\vec{p}_1{'}, \vec{q}_1) f_1(\vec{p}_2{'}, \vec{q}_1) \right] \ln f_1(\vec{p}_1, \vec{q}_1),
\end{aligned} \tag{3.44}$$

where we shall interchangeably use $\mathrm{d}^2\sigma$, $\mathrm{d}^2\vec{b}$, or $\mathrm{d}^2\Omega |\mathrm{d}\sigma/\mathrm{d}\Omega|$ for the differential cross-section. The streaming terms in the above expression are zero, as shown through successive integrations by parts,

$$\begin{aligned}
\int \mathrm{d}^3 \vec{p}_1\, \mathrm{d}^3 \vec{q}_1\; \ln f_1 \frac{\partial U}{\partial \vec{q}_1} \cdot \frac{\partial f_1}{\partial \vec{p}_1} &= -\int \mathrm{d}^3 \vec{p}_1\, \mathrm{d}^3 \vec{q}_1 f_1 \frac{\partial U}{\partial \vec{q}_1} \cdot \frac{1}{f_1} \frac{\partial f_1}{\partial \vec{p}_1} \\
&= \int \mathrm{d}^3 \vec{p}_1\, \mathrm{d}^3 \vec{q}_1 f_1 \frac{\partial}{\partial \vec{p}_1} \cdot \frac{\partial U}{\partial \vec{q}_1} = 0,
\end{aligned}$$

and

$$\int \mathrm{d}^3 \vec{p}_1\, \mathrm{d}^3 \vec{q}_1\; \ln f_1 \frac{\vec{p}_1}{m} \cdot \frac{\partial f_1}{\partial \vec{q}_1} = -\int \mathrm{d}^3 \vec{p}_1\, \mathrm{d}^3 \vec{q}_1\, f_1 \frac{\vec{p}_1}{m} \cdot \frac{1}{f_1} \frac{\partial f_1}{\partial \vec{q}_1} = \int \mathrm{d}^3 \vec{p}_1\, \mathrm{d}^3 \vec{q}_1 f_1 \frac{\partial}{\partial \vec{q}_1} \cdot \frac{\vec{p}_1}{m} = 0.$$

The collision term in Eq. (3.44) involves integrations over dummy variables $\vec{p}_1$ and $\vec{p}_2$. The labels (1) and (2) can thus be exchanged without any change in the value of the integral. Averaging the resulting two expressions gives

$$\begin{aligned}
\frac{\mathrm{d}H}{\mathrm{d}t} = &-\frac{1}{2} \int \mathrm{d}^3 \vec{q}\, \mathrm{d}^3 \vec{p}_1\, \mathrm{d}^3 \vec{p}_2\, \mathrm{d}^2 \vec{b} |\vec{v}_1 - \vec{v}_2| \left[f_1(\vec{p}_1) f_1(\vec{p}_2) \right. \\
&\left. - f_1(\vec{p}_1{'}) f_1(\vec{p}_2{'}) \right] \ln \left(f_1(\vec{p}_1) f_1(\vec{p}_2) \right).
\end{aligned} \tag{3.45}$$

(The arguments, $\vec{q}$ and t, of f_1 are suppressed for ease of notation.) We would now like to change the variables of integration from the coordinates describing the initiators of the collision, $(\vec{p}_1, \vec{p}_2, \vec{b})$, to those of their products,

$(\vec{p}_1', \vec{p}_2', \vec{b}')$. The explicit functional forms describing this transformation are complicated because of the dependence of the solid angle $\hat{\Omega}$ in Eq. (3.39) on $\vec{b}$ and $|\vec{p}_2 - \vec{p}_1|$. However, we are assured that the Jacobian of the transformation is unity because of time reversal symmetry; since for every collision there is an inverse one obtained by reversing the momenta of the products. In terms of the new coordinates

$$\frac{\mathrm{d}H}{\mathrm{d}t} = -\frac{1}{2} \int \mathrm{d}^3\vec{q} \, \mathrm{d}^3\vec{p}_1' \, \mathrm{d}^3\vec{p}_2' \, \mathrm{d}^2\vec{b}' |\vec{v}_1 - \vec{v}_2| \left[f_1(\vec{p}_1) f_1(\vec{p}_2) \right. \tag{3.46}$$
$$\left. - f_1(\vec{p}_1') f_1(\vec{p}_2') \right] \ln \left(f_1(\vec{p}_1) f_1(\vec{p}_2) \right),$$

where we should now regard $(\vec{p}_1, \vec{p}_2)$ in the above equation as functions of the integration variables $(\vec{p}_1', \vec{p}_2', \vec{b}')$ as in Eq. (3.39). As noted earlier, $|\vec{v}_1 - \vec{v}_2| = |\vec{v}_1' - \vec{v}_2'|$ for any elastic collision, and we can use these quantities interchangeably. Finally, we relabel the dummy integration variables such that the primes are removed. Noting that the functional dependence of $(\vec{p}_1, \vec{p}_2, \vec{b})$ on $(\vec{p}_1', \vec{p}_2', \vec{b}')$ is exactly the same as its inverse, we obtain

$$\frac{\mathrm{d}H}{\mathrm{d}t} = -\frac{1}{2} \int \mathrm{d}^3\vec{q} \, \mathrm{d}^3\vec{p}_1 \, \mathrm{d}^3\vec{p}_2 \, \mathrm{d}^2\vec{b} \, |\vec{v}_1 - \vec{v}_2| \left[f_1(\vec{p}_1') f_1(\vec{p}_2') \right. \tag{3.47}$$
$$\left. - f_1(\vec{p}_1) f_1(\vec{p}_2) \right] \ln \left(f_1(\vec{p}_1') f_1(\vec{p}_2') \right).$$

Averaging Eqs. (3.45) and (3.47) results in

$$\frac{\mathrm{d}H}{\mathrm{d}t} = -\frac{1}{4} \int \mathrm{d}^3\vec{q} \, \mathrm{d}^3\vec{p}_1 \, \mathrm{d}^3\vec{p}_2 \, \mathrm{d}^2\vec{b} \, |\vec{v}_1 - \vec{v}_2| \left[f_1(\vec{p}_1) f_1(\vec{p}_2) - f_1(\vec{p}_1') f_1(\vec{p}_2') \right] \tag{3.48}$$
$$\left[\ln \left(f_1(\vec{p}_1) f_1(\vec{p}_2) \right) - \ln \left(f_1(\vec{p}_1') f_1(\vec{p}_2') \right) \right].$$

The integrand of the above expression is always positive. If $f_1(\vec{p}_1) f_1(\vec{p}_2) > f_1(\vec{p}_1') f_1(\vec{p}_2')$, both terms in square brackets are positive, while both are negative if $f_1(\vec{p}_1) f_1(\vec{p}_2) < f_1(\vec{p}_1') f_1(\vec{p}_2')$. In either case, their product is positive. The positivity of the integrand establishes the validity of the H-theorem,

$$\frac{\mathrm{d}H}{\mathrm{d}t} \leq 0. \tag{3.49}$$

Irreversibility: the second law is an *empirical* formulation of the vast number of everyday observations that support the existence of an *arrow of time*. Reconciling the reversibility of laws of physics governing the microscopic domain with the observed irreversibility of macroscopic phenomena is a fundamental problem. Of course, not all microscopic laws of physics are reversible: weak nuclear interactions violate time reversal symmetry, and the collapse of the quantum wave function in the act of observation is irreversible. The former interactions in fact do not play any significant role in everyday observations that lead to the second law. The irreversible collapse of the wave function may itself be an artifact of treating macroscopic observers and microscopic observables distinctly.[1] There are proponents of the view that the reversibility of the currently accepted microscopic equations of motion (classical or quantum) is indicative of their inadequacy. However, the advent of powerful computers has made it possible to simulate the evolution of collections of large numbers of particles, governed by *classical, reversible* equations of motion. Although simulations are currently limited to relatively small numbers of particles (10^6), they do exhibit the irreversible macroscopic behaviors similar to those observed in nature (typically involving 10^{23} particles). For example, particles initially occupying one half of a box proceed to irreversibly, and uniformly, occupy the whole box. (This has nothing to do with limitations of computational accuracy; the same macroscopic irreversibility is observed in exactly reversible integer-based simulations, such as with cellular automata.) Thus the origin of the observed irreversibilities should be sought in the classical evolution of large collections of particles.

The Boltzmann equation is the first formula we have encountered that is clearly not time reversible, as indicated by Eq. (3.49). We can thus ask the question of how we obtained this result from the Hamiltonian equations of motion. The key to this, of course, resides in the physically motivated approximations used to obtain Eq. (3.41). The first steps of the approximation were dropping the three-body collision term on the right-hand side of Eq. (3.30), and the implicit coarse graining of the resolution in the spatial and temporal scales. Neither of these steps explicitly violates time reversal symmetry, and the collision term in Eq. (3.37) retains this property. The next step in getting to Eq. (3.41) is to replace the two-body density $f_2(-)$, *evaluated before the collision*, with the product of two one-body densities according to Eq. (3.32). This treats the two-body densities before and after the collision differently. We could alternatively have expressed Eq. (3.37) in terms of the two-body densities $f_2(+)$ *evaluated after the collision*. Replacing $f_2(+)$ with the product of two one-particle densities would then lead to the opposite conclusion, with $dH/dt \geq 0$! For a system in equilibrium, it is hard to justify one choice over the other. However, once the system is out of equilibrium, for example, as in the situation of Fig. 3.4, the coordinates after the collision are more likely to be correlated, and hence the substitution of Eq. (3.32) for $f_2(+)$ does not make sense. Time reversal symmetry implies that there should also be subtle correlations in $f_2(-)$ that are ignored in the so-called *assumption of molecular chaos*.

While the assumption of molecular chaos before (but not after) collisions is the key to the irreversibility of the Boltzmann equation, the resulting loss of

---

[1] The time-dependent Schrödinger equation is fully time reversible. If it is possible to write a complicated wave function that includes the observing apparatus (possibly the whole Universe), it is hard to see how any irreversibility may occur.

information is best justified in terms of the coarse graining of space and time: the Liouville equation and its descendants contain precise information about the evolution of a pure state. This information, however, is inevitably transported to shorter scales. A useful image is that of mixing two immiscible fluids. While the two fluids remain distinct at each point, the transitions in space from one to the next occur at finer resolution on subsequent mixing. At some point, a finite resolution in any measuring apparatus will prevent keeping track of the two components. In the Boltzmann equation, the precise information of the pure state is lost at the scale of collisions. The resulting one-body density only describes space and time resolutions longer than those of a two-body collision, becoming more and more probabilistic as further information is lost.

3.6 Equilibrium properties

What is the nature of the equilibrium state described by f_1, for a homogeneous gas?

(1) *The equilibrium distribution*: after the gas has reached equilibrium, the function H should no longer decrease with time. Since the integrand in Eq. (3.48) is always positive, a *necessary* condition for $dH/dt = 0$ is that

$$f_1(\vec{p}_1, \vec{q}_1) f_1(\vec{p}_2, \vec{q}_1) - f_1(\vec{p}_1{}', \vec{q}_1) f_1(\vec{p}_2{}', \vec{q}_1) = 0, \qquad (3.50)$$

that is, *at each point* $\vec{q}$ we must have

$$\ln f_1(\vec{p}_1, \vec{q}) + \ln f_1(\vec{p}_2, \vec{q}) = \ln f_1(\vec{p}_1{}', \vec{q}) + \ln f_1(\vec{p}_2{}', \vec{q}). \qquad (3.51)$$

The left-hand side of the above equation refers to the momenta before a two-body collision, and the right-hand side to those after the collision. The equality is thus satisfied by any *additive* quantity that is conserved during the collision. There are five such conserved quantities for an elastic collision: the particle number, the three components of the net momentum, and the kinetic energy. Hence, a general solution for f_1 is

$$\ln f_1 = a(\vec{q}) - \vec{\alpha}(\vec{q}) \cdot \vec{p} - \beta(\vec{q}) \left(\frac{\vec{p}^{\,2}}{2m} \right). \qquad (3.52)$$

We can easily accommodate the potential energy $U(\vec{q})$ in the above form, and set

$$f_1(\vec{p}, \vec{q}) = \mathcal{N}(\vec{q}) \exp \left[-\vec{\alpha}(\vec{q}) \cdot \vec{p} - \beta(\vec{q}) \left(\frac{p^2}{2m} + U(\vec{q}) \right) \right]. \qquad (3.53)$$

We shall refer to the above distribution as describing *local equilibrium*. While this form is preserved *during collisions*, it will evolve in time *away from collisions*, due to the streaming terms, unless $\{\mathcal{H}_1, f_1\} = 0$. The latter condition is satisfied for any function f_1 that depends only on $\mathcal{H}_1$, or any other quantity that is conserved by it. Clearly, the above density satisfies this requirement as long as $\mathcal{N}$ and β are independent of $\vec{q}$, and $\vec{\alpha} = 0$.

According to Eq. (3.16), the appropriate normalization for f_1 is

$$\int d^3\vec{p}\, d^3\vec{q}\, f_1(\vec{p}, \vec{q}) = N. \tag{3.54}$$

For particles in a box of volume V, the potential $U(\vec{q})$ is zero inside the box, and infinite on the outside. The normalization factor in Eq. (3.53) can be obtained from Eq. (3.54) as

$$N = \mathcal{N}V\left[\int_{-\infty}^{\infty} dp_i \exp\left(-\alpha_i p_i - \frac{\beta p_i^2}{2m}\right)\right]^3 = \mathcal{N}V\left(\frac{2\pi m}{\beta}\right)^{3/2} \exp\left(\frac{m\alpha^2}{2\beta}\right). \tag{3.55}$$

Hence, the properly normalized Gaussian distribution for momenta is

$$f_1(\vec{p}, \vec{q}) = n\left(\frac{\beta}{2\pi m}\right)^{3/2} \exp\left[-\frac{\beta(\vec{p} - \vec{p}_0)^2}{2m}\right], \tag{3.56}$$

where $\vec{p}_0 = \langle \vec{p} \rangle = m\vec{\alpha}/\beta$ is the mean value for the momentum of the gas, which is zero for a stationary box, and $n = N/V$ is the particle density. From the Gaussian form of the distribution it can be easily concluded that the variance of each component of the momentum is $\langle p_i^2 \rangle = m/\beta$, and

$$\langle p^2 \rangle = \langle p_x^2 + p_y^2 + p_z^2 \rangle = \frac{3m}{\beta}. \tag{3.57}$$

(2) *Equilibrium between two gases*: consider two different gases (a) and (b), moving in the same potential U, and subject to a two-body interaction $\mathcal{V}_{ab}(\vec{q}^{(a)} - \vec{q}^{(b)})$. We can define one-particle densities, $f_1^{(a)}$ and $f_1^{(b)}$, for the two gases, respectively. In terms of a generalized collision integral

$$C_{\alpha,\beta} = -\int d^3\vec{p}_2\, d^2\Omega \left|\frac{d\sigma_{\alpha,\beta}}{d\Omega}\right| |\vec{v}_1 - \vec{v}_2| \left[f_1^{(a)}(\vec{p}_1, \vec{q}_1) f_1^{(\beta)}(\vec{p}_2, \vec{q}_1)\right.$$
$$\left. - f_1^{(a)}(\vec{p}_1\,', \vec{q}_1) f_1^{(\beta)}(\vec{p}_2\,', \vec{q}_1)\right], \tag{3.58}$$

the evolution of these densities is governed by a simple generalization of the Boltzmann equation to

$$\begin{cases} \dfrac{\partial f_1^{(a)}}{\partial t} = -\left\{f_1^{(a)}, \mathcal{H}_1^{(a)}\right\} + C_{a,a} + C_{a,b} \\[2mm] \dfrac{\partial f_1^{(b)}}{\partial t} = -\left\{f_1^{(b)}, \mathcal{H}_1^{(b)}\right\} + C_{b,a} + C_{b,b} \end{cases}. \tag{3.59}$$

Stationary distributions can be obtained if all six terms on the right-hand side of Eqs. (3.59) are zero. In the absence of interspecies collisions, that is, for $C_{a,b} = C_{b,a} = 0$, we can obtain independent stationary distributions $f_1^{(a)} \propto \exp\left(-\beta_a \mathcal{H}_1^{(a)}\right)$ and $f_1^{(b)} \propto \exp\left(-\beta_b \mathcal{H}_1^{(b)}\right)$. Requiring the vanishing of $C_{a,b}$ leads to the additional constraint,

$$f_1^{(a)}(\vec{p}_1) f_1^{(b)}(\vec{p}_2) - f_1^{(a)}(\vec{p}_1\,') f_1^{(b)}(\vec{p}_2\,') = 0, \implies$$
$$\beta_a \mathcal{H}_1^{(a)}(\vec{p}_1) + \beta_b \mathcal{H}_1^{(b)}(\vec{p}_2) = \beta_a \mathcal{H}_1^{(a)}(\vec{p}_1\,') + \beta_b \mathcal{H}_1^{(b)}(\vec{p}_2\,'). \tag{3.60}$$

Since the total energy $\mathcal{H}_1^{(a)} + \mathcal{H}_1^{(b)}$ is conserved in a collision, the above equation can be satisfied for $\beta_a = \beta_b = \beta$. From Eq. (3.57) this condition implies the equality of the kinetic energies of the two species,

$$\left\langle \frac{p_a^2}{2m_a} \right\rangle = \left\langle \frac{p_b^2}{2m_b} \right\rangle = \frac{3}{2\beta}. \qquad (3.61)$$

The parameter β thus plays the role of an *empirical temperature* describing the equilibrium of gases.

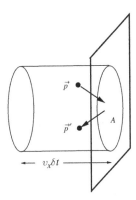

Fig. 3.8 The pressure exerted on the wall due to impact of particles.

(3) *The equation of state*: to complete the identification of β with temperature T, consider a gas of N particles confined to a box of volume V. The gas pressure results from the force exerted by the particles colliding with the walls of the container. Consider a wall element of area A perpendicular to the x direction. The number of particles impacting this area, with momenta in the interval $[\vec{p}, \vec{p} + d\vec{p}]$, over a time period δt, is

$$d\mathcal{N}(\vec{p}) = \left(f_1(\vec{p}) d^3\vec{p} \right) \left(A\, v_x\, \delta t \right). \qquad (3.62)$$

The final factor in the above expression is the volume of a cylinder of height $v_x \delta t$ perpendicular to the area element A. Only particles within this cylinder are close enough to impact the wall during δt. As each collision imparts a momentum $2p_x$ to the wall, the net force exerted is

$$F = \frac{1}{\delta t} \int_{-\infty}^{0} dp_x \int_{-\infty}^{\infty} dp_y \int_{-\infty}^{\infty} dp_z f_1(\vec{p}) \left(A\, \frac{p_x}{m}\, \delta t \right) (2p_x). \qquad (3.63)$$

As only particles with velocities directed toward the wall will hit it, the first integral is over half of the range of p_x. Since the integrand is even in p_x, this restriction can be removed by dividing the full integral by 2. The pressure P is then obtained from the force per unit area as

$$P = \frac{F}{A} = \int d^3\vec{p}\, f_1(\vec{p}) \frac{p_x^2}{m} = \frac{1}{m} \int d^3\vec{p}\, p_x^2\, n \left(\frac{\beta}{2\pi m} \right)^{3/2} \exp\left(-\frac{\beta p^2}{2m} \right) = \frac{n}{\beta}, \qquad (3.64)$$

where Eq. (3.56) is used for the equilibrium form of f_1. Comparing with the standard equation of state, $PV = Nk_BT$, for an ideal gas, leads to the identification, $\beta = 1/k_BT$.

(4) *Entropy*: as discussed earlier, the Boltzmann H-function is closely related to the information content of the one-particle PDF ρ_1. We can also define a corresponding Boltzmann entropy,

$$S_B(t) = -k_B \text{H}(t), \tag{3.65}$$

where the constant k_B reflects the historical origins of entropy. The H-theorem implies that S_B can only increase with time in approaching equilibrium. It has the further advantage of being defined through Eq. (3.42) for situations that are clearly out of equilibrium. For a gas in equilibrium in a box of volume V, from Eq. (3.56), we compute

$$\text{H} = V \int \text{d}^3 \vec{p} \, f_1(\vec{p}) \ln f_1(\vec{p})$$

$$= V \int \text{d}^3 \vec{p} \, \frac{N}{V} (2\pi m k_B T)^{-3/2} \exp\left(-\frac{p^2}{2mk_BT}\right) \left[\ln\left(\frac{n}{(2\pi m k_B T)^{3/2}}\right) - \frac{p^2}{2mk_BT}\right]$$

$$= N\left[\ln\left(\frac{n}{(2\pi m k_B T)^{3/2}}\right) - \frac{3}{2}\right]. \tag{3.66}$$

The entropy is now identified as

$$S_B = -k_B \text{H} = N k_B \left[\frac{3}{2} + \frac{3}{2} \ln(2\pi m k_B T) - \ln\left(\frac{N}{V}\right)\right]. \tag{3.67}$$

The thermodynamic relation, $T\text{d}S_B = \text{d}E + P\text{d}V$, implies

$$\left.\frac{\partial E}{\partial T}\right|_V = T \left.\frac{\partial S_B}{\partial T}\right|_V = \frac{3}{2} N k_B,$$

$$P + \left.\frac{\partial E}{\partial V}\right|_T = T \left.\frac{\partial S_B}{\partial V}\right|_T = \frac{N k_B T}{V}. \tag{3.68}$$

The usual properties of a monatomic ideal gas, $PV = Nk_BT$, and $E = 3Nk_BT/2$, can now be obtained from the above equations. Also note that for this classical gas, the zero temperature limit of the entropy in Eq. (3.67) *is not* independent of the density n, in violation of the third law of thermodynamics.

3.7 Conservation laws

Approach to equilibrium: we now address the third question posed in the introduction, of *how* the gas reaches its final equilibrium. Consider a situation in which the gas is perturbed from the equilibrium form described by Eq. (3.56), and follow its relaxation to equilibrium. There is a hierarchy of mechanisms that operate at different time scales.

(i) The fastest processes are the two-body collisions of particles in immediate vicinity. Over a time scale of the order of τ_c, $f_2(\vec{q}_1, \vec{q}_2, t)$ relaxes to $f_1(\vec{q}_1, t)f_1(\vec{q}_2, t)$ for separations $|\vec{q}_1 - \vec{q}_2| \gg d$. Similar relaxations occur for the higher-order densities f_s.

(ii) At the next stage, f_1 relaxes to a *local equilibrium* form, as in Eq. (3.53), over the time scale of the mean free time $\tau_\times$. This is the intrinsic scale set by the collision term on the right-hand side of the Boltzmann equation. After this time interval, quantities conserved in collisions achieve a state of local equilibrium. We can then define at each point a (time-dependent) local density by integrating over all momenta as

$$n(\vec{q}, t) = \int d^3\vec{p}\, f_1(\vec{p}, \vec{q}, t),\qquad (3.69)$$

as well as a local expectation value for any operator $\mathcal{O}(\vec{p}, \vec{q}, t)$,

$$\langle \mathcal{O}(\vec{q}, t)\rangle = \frac{1}{n(\vec{q}, t)} \int d^3\vec{p}\, f_1(\vec{p}, \vec{q}, t)\mathcal{O}(\vec{p}, \vec{q}, t).\qquad (3.70)$$

(iii) After the densities and expectation values have relaxed to their local equilibrium forms in the intrinsic time scales τ_c and $\tau_\times$, there is a subsequent slower relaxation to global equilibrium over extrinsic time and length scales. This final stage is governed by the smaller streaming terms on the left-hand side of the Boltzmann equation. It is most conveniently expressed in terms of the time evolution of conserved quantities according to *hydrodynamic equations*.

Conserved quantities are left unchanged by the two-body collisions, that is, they satisfy

$$\chi(\vec{p}_1, \vec{q}, t) + \chi(\vec{p}_2, \vec{q}, t) = \chi(\vec{p}_1{}', \vec{q}, t) + \chi(\vec{p}_2{}', \vec{q}, t),\qquad (3.71)$$

where $(\vec{p}_1, \vec{p}_2)$ and $(\vec{p}_1{}', \vec{p}_2{}')$ refer to the momenta before and after a collision, respectively. For such quantities, we have

$$J_\chi(\vec{q}, t) = \int d^3\vec{p}\, \chi(\vec{p}, \vec{q}, t) \left.\frac{df_1}{dt}\right|_{\text{coll.}} (\vec{p}, \vec{q}, t) = 0.\qquad (3.72)$$

Proof. Using the form of the collision integral, we have

$$J_\chi = -\int d^3\vec{p}_1\, d^3\vec{p}_2\, d^2\vec{b}\, |\vec{v}_1 - \vec{v}_2| \left[f_1(\vec{p}_1)f_1(\vec{p}_2) - f_1(\vec{p}_1{}')f_1(\vec{p}_2{}') \right] \chi(\vec{p}_1).\qquad (3.73)$$

(The implicit arguments $(\vec{q}, t)$ are left out for ease of notation.) We now perform the same set of changes of variables that were used in the proof of the H-theorem. The first step is averaging after exchange of the dummy variables $\vec{p}_1$ and $\vec{p}_2$, leading to

$$J_\chi = -\frac{1}{2}\int d^3\vec{p}_1\, d^3\vec{p}_2\, d^2\vec{b}\, |\vec{v}_1 - \vec{v}_2| \left[f_1(\vec{p}_1)f_1(\vec{p}_2) - f_1(\vec{p}_1{}')f_1(\vec{p}_2{}') \right] \left[\chi(\vec{p}_1) + \chi(\vec{p}_2) \right].$$
$$(3.74)$$

Next, change variables from the originators $(\vec{p}_1, \vec{p}_2, \vec{b})$ to the products $(\vec{p}_1{}', \vec{p}_2{}', \vec{b}')$ of the collision. After relabeling the integration variables, the above equation is transformed to

$$J_\chi = -\frac{1}{2}\int d^3\vec{p}_1\, d^3\vec{p}_2\, d^2\vec{b}\, |\vec{v}_1 - \vec{v}_2| \left[f_1(\vec{p}_1{}')f_1(\vec{p}_2{}') \right.$$
$$\left. - f_1(\vec{p}_1)f_1(\vec{p}_2) \right] \left[\chi(\vec{p}_1{}') + \chi(\vec{p}_2{}') \right].\qquad (3.75)$$

Averaging the last two equations leads to

$$J_\chi = -\frac{1}{4} \int d^3\vec{p}_1 \, d^3\vec{p}_2 \, d^2\vec{b} \, |\vec{v}_1 - \vec{v}_2| \left[f_1(\vec{p}_1) f_1(\vec{p}_2) - f_1(\vec{p}_1{}') f_1(\vec{p}_2{}') \right]$$
$$\left[\chi(\vec{p}_1) + \chi(\vec{p}_2) - \chi(\vec{p}_1{}') - \chi(\vec{p}_2{}') \right], \tag{3.76}$$

which is zero from Eq. (3.71).

Let us explore the consequences of this result for the evolution of expectation values involving χ. Substituting for the collision term in Eq. (3.72) the streaming terms on the left-hand side of the Boltzmann equation lead to

$$J_\chi(\vec{q}, t) = \int d^3\vec{p} \, \chi(\vec{p}, \vec{q}, t) \left[\partial_t + \frac{p_\alpha}{m} \partial_\alpha + F_\alpha \frac{\partial}{\partial p_\alpha} \right] f_1(\vec{p}, \vec{q}, t) = 0, \tag{3.77}$$

where we have introduced the notations $\partial_t \equiv \partial/\partial t$, $\partial_\alpha \equiv \partial/\partial q_\alpha$, and $F_\alpha = -\partial U/\partial q_\alpha$. We can manipulate the above equation into the form

$$\int d^3\vec{p} \left\{ \left[\partial_t + \frac{p_\alpha}{m} \partial_\alpha + F_\alpha \frac{\partial}{\partial p_\alpha} \right] (\chi f_1) - f_1 \left[\partial_t + \frac{p_\alpha}{m} \partial_\alpha + F_\alpha \frac{\partial}{\partial p_\alpha} \right] \chi \right\} = 0. \tag{3.78}$$

The third term is zero, as it is a complete derivative. Using the definition of expectation values in Eq. (3.70), the remaining terms can be rearranged into

$$\partial_t \left(n \langle \chi \rangle \right) + \partial_\alpha \left(n \left\langle \frac{p_\alpha}{m} \chi \right\rangle \right) - n \langle \partial_t \chi \rangle - n \left\langle \frac{p_\alpha}{m} \partial_\alpha \chi \right\rangle - n F_\alpha \left\langle \frac{\partial \chi}{\partial p_\alpha} \right\rangle = 0. \tag{3.79}$$

As discussed earlier, for elastic collisions, there are five conserved quantities: particle number, the three components of momentum, and kinetic energy. Each leads to a corresponding hydrodynamic equation, as constructed below.

(a) *Particle number*: setting $\chi = 1$ in Eq. (3.79) leads to

$$\partial_t n + \partial_\alpha \left(n u_\alpha \right) = 0, \tag{3.80}$$

where we have introduced the local velocity

$$\vec{u} \equiv \left\langle \frac{\vec{p}}{m} \right\rangle. \tag{3.81}$$

This equation simply states that the time variation of the local particle density is due to a particle current $\vec{J}_n = n\vec{u}$.

(b) *Momentum*: any linear function of the momentum $\vec{p}$ is conserved in the collision, and we shall explore the consequences of the conservation of

$$\vec{c} \equiv \frac{\vec{p}}{m} - \vec{u}. \tag{3.82}$$

Substituting c_α into Eq. (3.79) leads to

$$\partial_\beta \left(n \langle (u_\beta + c_\beta) c_\alpha \rangle \right) + n \partial_t u_\alpha + n \partial_\beta u_\alpha \langle u_\beta + c_\beta \rangle - n \frac{F_\alpha}{m} = 0. \tag{3.83}$$

Taking advantage of $\langle c_\alpha \rangle = 0$, from Eqs. (3.81) and (3.82), leads to

$$\partial_t u_\alpha + u_\beta \partial_\beta u_\alpha = \frac{F_\alpha}{m} - \frac{1}{mn} \partial_\beta P_{\alpha\beta}, \tag{3.84}$$

where we have introduced the *pressure tensor*,

$$P_{\alpha\beta} \equiv mn\langle c_\alpha c_\beta\rangle. \tag{3.85}$$

The left-hand side of the equation is the acceleration of an element of the fluid $d\vec{u}/dt$, which should equal $\vec{F}_{\text{net}}/m$ according to Newton's equation. The net force includes an additional component due to the variations in the pressure tensor across the fluid.

(c) *Kinetic energy*: we first introduce an average local kinetic energy

$$\varepsilon \equiv \left\langle\frac{mc^2}{2}\right\rangle = \left\langle\frac{p^2}{2m} - \vec{p}\cdot\vec{u} + \frac{mu^2}{2}\right\rangle, \tag{3.86}$$

and then examine the conservation law obtained by setting χ equal to $mc^2/2$ in Eq. (3.79). Since for space and time derivatives $\partial\varepsilon = mc_\beta\partial c_\beta = -mc_\beta\partial u_\beta$, we obtain

$$\partial_t(n\varepsilon) + \partial_\alpha\left(n\left\langle(u_\alpha + c_\alpha)\frac{mc^2}{2}\right\rangle\right) + nm\partial_t u_\beta\langle c_\beta\rangle + nm\partial_\alpha u_\beta\langle(u_\alpha + c_\alpha)c_\beta\rangle \tag{3.87}$$
$$- nF_\alpha\langle c_\alpha\rangle = 0.$$

Taking advantage of $\langle c_\alpha\rangle = 0$, the above equation is simplified to

$$\partial_t(n\varepsilon) + \partial_\alpha(nu_\alpha\varepsilon) + \partial_\alpha\left(n\left\langle c_\alpha\frac{mc^2}{2}\right\rangle\right) + P_{\alpha\beta}\partial_\alpha u_\beta = 0. \tag{3.88}$$

We next take out the dependence on n in the first two terms of the above equation, finding

$$\varepsilon\partial_t n + n\partial_t\varepsilon + \varepsilon\partial_\alpha(nu_\alpha) + nu_\alpha\partial_\alpha\varepsilon + \partial_\alpha h_\alpha + P_{\alpha\beta}u_{\alpha\beta} = 0, \tag{3.89}$$

where we have also introduced the local *heat flux*

$$\vec{h} \equiv \frac{nm}{2}\langle c_\alpha c^2\rangle, \tag{3.90}$$

and the *rate of strain tensor*

$$u_{\alpha\beta} = \frac{1}{2}\left(\partial_\alpha u_\beta + \partial_\beta u_\alpha\right). \tag{3.91}$$

Eliminating the first and third terms in Eq. (3.89) with the aid of Eq. (3.80) leads to

$$\partial_t\varepsilon + u_\alpha\partial_\alpha\varepsilon = -\frac{1}{n}\partial_\alpha h_\alpha - \frac{1}{n}P_{\alpha\beta}u_{\alpha\beta}. \tag{3.92}$$

Clearly, to solve the hydrodynamic equations for n, $\vec{u}$, and ε, we need expressions for $P_{\alpha\beta}$ and $\vec{h}$, which are either given phenomenologically, or calculated from the density f_1, as in the next sections.

3.8 Zeroth-order hydrodynamics

As a first approximation, we shall assume that in *local equilibrium*, the density f_1 at each point in space can be represented as in Eq. (3.56), that is,

$$f_1^0(\vec{p}, \vec{q}, t) = \frac{n(\vec{q}, t)}{\left(2\pi m k_B T(\vec{q}, t)\right)^{3/2}} \exp\left[-\frac{\left(\vec{p} - m\vec{u}(\vec{q}, t)\right)^2}{2m k_B T(\vec{q}, t)}\right]. \tag{3.93}$$

The choice of parameters clearly enforces $\int d^3\vec{p}\, f_1^0 = n$, and $\langle \vec{p}/m \rangle^0 = \vec{u}$, as required. Average values are easily calculated from this Gaussian weight; in particular,

$$\langle c_\alpha c_\beta \rangle^0 = \frac{k_B T}{m}\delta_{\alpha\beta}, \tag{3.94}$$

leading to

$$P_{\alpha\beta}^0 = n k_B T \delta_{\alpha\beta}, \quad \text{and} \quad \varepsilon = \frac{3}{2} k_B T. \tag{3.95}$$

Since the density f_1^0 is even in $\vec{c}$, all odd expectation values vanish, and in particular

$$\vec{h}^0 = 0. \tag{3.96}$$

The conservation laws in this approximation take the simple forms

$$\begin{cases} D_t n = -n\partial_\alpha u_\alpha \\[2mm] m D_t u_\alpha = F_\alpha - \dfrac{1}{n}\partial_\alpha\left(n k_B T\right) \\[2mm] D_t T = -\dfrac{2}{3} T \partial_\alpha u_\alpha \end{cases}. \tag{3.97}$$

In the above expression, we have introduced the *material derivative*

$$D_t \equiv \left[\partial_t + u_\beta \partial_\beta\right], \tag{3.98}$$

which measures the time variations of any quantity as it moves along the stream-lines set up by the average velocity field $\vec{u}$. By combining the first and third equations, it is easy to get

$$D_t \ln\left(n T^{-3/2}\right) = 0. \tag{3.99}$$

The quantity $\ln\left(n T^{-3/2}\right)$ is like a local entropy for the gas (see Eq. (3.67)), which according to the above equation is not changed along stream-lines. The zeroth-order hydrodynamics thus predicts that the gas flow is adiabatic. This prevents the local equilibrium solution of Eq. (3.93) from reaching a true global equilibrium form, which necessitates an increase in entropy.

To demonstrate that Eqs. (3.97) do not describe a satisfactory approach to equilibrium, examine the evolution of small deformations about a stationary ($\vec{u}_0 = 0$) state, in a uniform box ($\vec{F} = 0$), by setting

$$\begin{cases} n(\vec{q}, t) = \overline{n} + \nu(\vec{q}, t) \\[2mm] T(\vec{q}, t) = \overline{T} + \theta(\vec{q}, t) \end{cases}. \tag{3.100}$$

We shall next expand Eqs. (3.97) to *first order* in the deviations $(\nu, \theta, \vec{u})$. Note that to lowest order, $D_t = \partial_t + O(u)$, leading to the linearized zeroth-order hydrodynamic equations

$$
\begin{cases}
\partial_t \nu = -\overline{n}\partial_\alpha u_\alpha \\[2mm]
m\partial_t u_\alpha = -\dfrac{k_B \overline{T}}{\overline{n}}\partial_\alpha \nu - k_B \partial_\alpha \theta \;. \\[2mm]
\partial_t \theta = -\dfrac{2}{3}\overline{T}\partial_\alpha u_\alpha
\end{cases}
\tag{3.101}
$$

Normal modes of the system are obtained by Fourier transformations,

$$
\tilde{A}\left(\vec{k}, \omega\right) = \int \mathrm{d}^3 q\, \mathrm{d}t \, \exp\left[\mathrm{i}\left(\vec{k}\cdot\vec{q} - \omega t\right)\right] A\left(\vec{q}, t\right),
\tag{3.102}
$$

where A stands for any of the three fields $(\nu, \theta, \vec{u})$. The natural vibration frequencies are solutions to the matrix equation

$$
\omega \begin{pmatrix} \tilde{\nu} \\ \tilde{u}_\alpha \\ \tilde{\theta} \end{pmatrix} = \begin{pmatrix} 0 & \overline{n}k_\beta & 0 \\ \frac{k_B \overline{T}}{m\overline{n}}\delta_{\alpha\beta}k_\beta & 0 & \frac{k_B}{m}\delta_{\alpha\beta}k_\beta \\ 0 & \frac{2}{3}\overline{T}k_\beta & 0 \end{pmatrix} \begin{pmatrix} \tilde{\nu} \\ \tilde{u}_\beta \\ \tilde{\theta} \end{pmatrix}.
\tag{3.103}
$$

It is easy to check that this equation has the following modes, the first *three* with zero frequency:

(a) Two modes describe shear flows in a uniform $(n = \overline{n})$ and isothermal $(T = \overline{T})$ fluid, in which the velocity varies along a direction normal to its orientation (e.g., $\vec{u} = f(x, t)\hat{y}$). In terms of Fourier modes $\vec{k}\cdot\vec{\tilde{u}}_T(\vec{k}) = 0$, indicating *transverse* flows that are not relaxed in this zeroth-order approximation.

(b) A third zero-frequency mode describes a stationary fluid with *uniform pressure* $P = nk_B T$. While n and T may vary across space, their product is constant, insuring that the fluid will not start moving due to pressure variations. The corresponding eigenvector of Eq. (3.103) is

$$
\mathbf{v_e} = \begin{pmatrix} \overline{n} \\ 0 \\ -\overline{T} \end{pmatrix}.
\tag{3.104}
$$

(c) Finally, the *longitudinal* velocity $(\vec{u}_\ell \parallel \vec{k})$ combines with density and temperature variations in eigenmodes of the form

$$
\mathbf{v_l} = \begin{pmatrix} \overline{n}|\vec{k}| \\ \omega(\vec{k}) \\ \frac{2}{3}\overline{T}|\vec{k}| \end{pmatrix}, \qquad \text{with} \qquad \omega(\vec{k}) = \pm v_\ell |\vec{k}|,
\tag{3.105}
$$

where

$$
v_\ell = \sqrt{\frac{5}{3}\frac{k_B \overline{T}}{m}}
\tag{3.106}
$$

is the longitudinal sound velocity. Note that the density and temperature variations in this mode are *adiabatic*, that is, the local entropy (proportional to $\ln\left(nT^{-3/2}\right)$) is left unchanged.

We thus find that none of the conserved quantities relaxes to equilibrium in the zeroth-order approximation. Shear flow and entropy modes persist forever, while the two sound modes have undamped oscillations. This is a deficiency of the zeroth-order approximation that is removed by finding a better solution to the Boltzmann equation.

3.9 First-order hydrodynamics

While $f_1^0(\vec{p}, \vec{q}, t)$ of Eq. (3.93) does set the right-hand side of the Boltzmann equation to zero, it is not a full solution, as the left-hand side causes its form to vary. The left-hand side is a *linear* differential operator, which using the various notations introduced in the previous sections, can be written as

$$\mathcal{L}[f] \equiv \left[\partial_t + \frac{p_\alpha}{m}\partial_\alpha + F_\alpha \frac{\partial}{\partial p_\alpha}\right]f = \left[D_t + c_\alpha\partial_\alpha + \frac{F_\alpha}{m}\frac{\partial}{\partial c_\alpha}\right]f. \tag{3.107}$$

It is simpler to examine the effect of $\mathcal{L}$ on $\ln f_1^0$, which can be written as

$$\ln f_1^0 = \ln\left(nT^{-3/2}\right) - \frac{mc^2}{2k_BT} - \frac{3}{2}\ln\left(2\pi mk_B\right). \tag{3.108}$$

Using the relation $\partial(c^2/2) = c_\beta\partial c_\beta = -c_\beta\partial u_\beta$, we get

$$\mathcal{L}\left[\ln f_1^0\right] = D_t \ln\left(nT^{-3/2}\right) + \frac{mc^2}{2k_BT^2}D_tT + \frac{m}{k_BT}c_\alpha D_t u_\alpha$$
$$+ c_\alpha\left(\frac{\partial_\alpha n}{n} - \frac{3}{2}\frac{\partial_\alpha T}{T}\right) + \frac{mc^2}{2k_BT^2}c_\alpha\partial_\alpha T + \frac{m}{k_BT}c_\alpha c_\beta\partial_\alpha u_\beta - \frac{F_\alpha c_\alpha}{k_BT}. \tag{3.109}$$

If the fields n, T, and u_α, satisfy the zeroth-order hydrodynamic Eqs. (3.97), we can simplify the above equation to

$$\mathcal{L}\left[\ln f_1^0\right] = 0 - \frac{mc^2}{3k_BT}\partial_\alpha u_\alpha + c_\alpha\left[\left(\frac{F_\alpha}{k_BT} - \frac{\partial_\alpha n}{n} - \frac{\partial_\alpha T}{T}\right) + \left(\frac{\partial_\alpha n}{n} - \frac{3}{2}\frac{\partial_\alpha T}{T}\right) - \frac{F_\alpha}{k_BT}\right]$$
$$+ \frac{mc^2}{2k_BT^2}c_\alpha\partial_\alpha T + \frac{m}{k_BT}c_\alpha c_\beta u_{\alpha\beta}$$
$$= \frac{m}{k_BT}\left(c_\alpha c_\beta - \frac{\delta_{\alpha\beta}}{3}c^2\right)u_{\alpha\beta} + \left(\frac{mc^2}{2k_BT} - \frac{5}{2}\right)\frac{c_\alpha}{T}\partial_\alpha T. \tag{3.110}$$

The characteristic time scale τ_U for $\mathcal{L}$ is *extrinsic*, and can be made much larger than $\tau_\times$. The zeroth-order result is thus exact in the limit $(\tau_\times/\tau_U) \to 0$; and corrections can be constructed in a perturbation series in $(\tau_\times/\tau_U)$. To this purpose, we set $f_1 = f_1^0(1 + g)$, and linearize the collision operator as

$$C[f_1, f_1] = -\int d^3\vec{p}_2 d^2\vec{b}\,|\vec{v}_1 - \vec{v}_2|f_1^0(\vec{p}_1)f_1^0(\vec{p}_2)\left[g(\vec{p}_1) + g(\vec{p}_2) - g(\vec{p}_1') - g(\vec{p}_2')\right]$$
$$\equiv -f_1^0(\vec{p}_1)C_L[g]. \tag{3.111}$$

While linear, the above integral operator is still difficult to manipulate in general. As a first approximation, and noting its characteristic magnitude, we set

$$C_L[g] \approx \frac{g}{\tau_\times}. \tag{3.112}$$

This is known as the *single collision time* approximation, and from the linearized Boltzmann equation $\mathcal{L}[f_1] = -f_1^0 C_L[g]$, we obtain

$$g = -\tau_\times \frac{1}{f_1^0} \mathcal{L}[f_1] \approx -\tau_\times \mathcal{L}\left[\ln f_1^0\right], \tag{3.113}$$

where we have kept only the leading term. Thus the first-order solution is given by (using Eq. (3.110))

$$f_1^1(\vec{p}, \vec{q}, t) = f_1^0(\vec{p}, \vec{q}, t)\left[1 - \frac{\tau_\mu m}{k_B T}\left(c_\alpha c_\beta - \frac{\delta_{\alpha\beta}}{3}c^2\right)u_{\alpha\beta}\right.$$
$$\left. - \tau_K\left(\frac{mc^2}{2k_B T} - \frac{5}{2}\right)\frac{c_\alpha}{T}\partial_\alpha T\right], \tag{3.114}$$

where $\tau_\mu = \tau_K = \tau_\times$ in the single collision time approximation. However, in writing the above equation, we have anticipated the possibility of $\tau_\mu \neq \tau_K$, which arises in more sophisticated treatments (although both times are still of order of $\tau_\times$).

It is easy to check that $\int d^3\vec{p} f_1^1 = \int d^3\vec{p} f_1^0 = n$, and thus various local expectation values are calculated to first order as

$$\langle \mathcal{O} \rangle^1 = \frac{1}{n}\int d^3\vec{p}\, \mathcal{O} f_1^0(1+g) = \langle \mathcal{O} \rangle^0 + \langle g\mathcal{O} \rangle^0. \tag{3.115}$$

The calculation of averages over products of c_α's, distributed according to the Gaussian weight of f_1^0, is greatly simplified by the use of *Wick's theorem*, which states that expectation value of the product is the sum over all possible products of paired expectation values, for example

$$\langle c_\alpha c_\beta c_\gamma c_\delta \rangle_0 = \left(\frac{k_B T}{m}\right)^2 \left(\delta_{\alpha\beta}\delta_{\gamma\delta} + \delta_{\alpha\gamma}\delta_{\beta\delta} + \delta_{\alpha\delta}\delta_{\beta\gamma}\right). \tag{3.116}$$

(Expectation values involving a product of an odd number of c_α's are zero by symmetry.) Using this result, it is easy to verify that

$$\left\langle \frac{p_\alpha}{m} \right\rangle^1 = u_\alpha - \tau_K \frac{\partial_\beta T}{T}\left\langle \left(\frac{mc^2}{2k_B T} - \frac{5}{2}\right)c_\alpha c_\beta \right\rangle^0 = u_\alpha. \tag{3.117}$$

The pressure tensor at first order is given by

$$P_{\alpha\beta}^1 = nm\langle c_\alpha c_\beta \rangle^1 = nm\left[\langle c_\alpha c_\beta \rangle^0 - \frac{\tau_\mu m}{k_B T}\left\langle c_\alpha c_\beta\left(c_\mu c_\nu - \frac{\delta_{\mu\nu}}{3}c^2\right)\right\rangle^0 u_{\mu\nu}\right]$$
$$= nk_B T\delta_{\alpha\beta} - 2nk_B T\tau_\mu\left(u_{\alpha\beta} - \frac{\delta_{\alpha\beta}}{3}u_{\gamma\gamma}\right). \tag{3.118}$$

(Using the above result, we can further verify that $\varepsilon^1 = \langle mc^2/2 \rangle^1 = 3k_B T/2$, as before.) Finally, the heat flux is given by

$$
\begin{aligned}
h_\alpha^1 &= n \left\langle c_\alpha \frac{mc^2}{2} \right\rangle^1 = -\frac{nm\tau_K}{2} \frac{\partial_\beta T}{T} \left\langle \left(\frac{mc^2}{2k_B T} - \frac{5}{2} \right) c_\alpha c_\beta c^2 \right\rangle^0 \\
&= -\frac{5}{2} \frac{nk_B^2 T \tau_K}{m} \partial_\alpha T.
\end{aligned}
\tag{3.119}
$$

At this order, we find that spatial variations in temperature generate a heat flow that tends to smooth them out, while shear flows are opposed by the off-diagonal terms in the pressure tensor. These effects are sufficient to cause relaxation to equilibrium, as can be seen by examining the modified behavior of the modes discussed previously.

(a) The pressure tensor now has an off-diagonal term

$$
P_{\alpha \neq \beta}^1 = -2nk_B T \tau_\mu u_{\alpha\beta} \equiv -\mu \left(\partial_\alpha u_\beta + \partial_\beta u_\alpha \right),
\tag{3.120}
$$

where $\mu \equiv nk_B T \tau_\mu$ is the *viscosity coefficient*. A shearing of the fluid (e.g., described by a velocity $u_y(x, t)$) now leads to a viscous force that opposes it (proportional to $\mu \partial_x^2 u_y$), causing its diffusive relaxation as discussed below.

(b) Similarly, a temperature gradient leads to a heat flux

$$
\vec{h} = -K\nabla T,
\tag{3.121}
$$

where $K = (5nk_B^2 T \tau_K)/(2m)$ is the coefficient of *thermal conductivity* of the gas. If the gas is at rest ($\vec{u} = 0$, and uniform $P = nk_B T$), variations in temperature now satisfy

$$
n\partial_t \varepsilon = \frac{3}{2} nk_B \partial_t T = -\partial_\alpha (-K\partial_\alpha T), \quad \Rightarrow \quad \partial_t T = \frac{2K}{3nk_B} \nabla^2 T.
\tag{3.122}
$$

This is the *Fourier equation* and shows that temperature variations relax by diffusion.

We can discuss the behavior of all the modes by linearizing the equations of motion. The first-order contribution to $D_t u_\alpha \approx \partial_t u_\alpha$ is

$$
\delta^1 (\partial_t u_\alpha) \equiv \frac{1}{mn} \partial_\beta \delta^1 P_{\alpha\beta} \approx -\frac{\mu}{mn} \left(\frac{1}{3} \partial_\alpha \partial_\beta + \delta_{\alpha\beta} \partial_\gamma \partial_\gamma \right) u_\beta,
\tag{3.123}
$$

where $\mu \equiv \bar{n} k_B \overline{T} \tau_\mu$. Similarly, the correction for $D_t T \approx \partial_t \theta$ is given by

$$
\delta^1 (\partial_t \theta) \equiv -\frac{2}{3k_B n} \partial_\alpha h_\alpha \approx -\frac{2K}{3k_B \bar{n}} \partial_\alpha \partial_\alpha \theta,
\tag{3.124}
$$

with $K = (5\bar{n} k_B^2 \overline{T} \tau_K)/(2m)$. After Fourier transformation, the matrix equation (3.103) is modified to

$$
\omega \begin{pmatrix} \tilde{\nu} \\ \tilde{u}_\alpha \\ \tilde{\theta} \end{pmatrix} = \begin{pmatrix} 0 & \bar{n}\delta_{\alpha\beta}k_\beta & 0 \\ \frac{k_B \overline{T}}{m\bar{n}}\delta_{\alpha\beta}k_\beta & -i\frac{\mu}{m\bar{n}}\left(k^2\delta_{\alpha\beta} + \frac{k_\alpha k_\beta}{3}\right) & \frac{k_B}{m}\delta_{\alpha\beta}k_\beta \\ 0 & \frac{2}{3}\overline{T}\delta_{\alpha\beta}k_\beta & -i\frac{2Kk^2}{3k_B\bar{n}} \end{pmatrix} \begin{pmatrix} \tilde{\nu} \\ \tilde{u}_\beta \\ \tilde{\theta} \end{pmatrix}.
\tag{3.125}
$$

We can ask how the normal mode frequencies calculated in the zeroth-order approximation are modified at this order. It is simple to verify that the transverse (shear) normal models $(\vec{k} \cdot \tilde{\vec{u}}_T = 0)$ now have a frequency

$$\omega_T = -\mathrm{i}\frac{\mu}{m\overline{n}}k^2. \tag{3.126}$$

The imaginary frequency implies that these modes are damped over a characteristic time $\tau_T(k) \sim 1/|\omega_T| \sim (\lambda)^2/(\tau_\mu \overline{v}^2)$, where λ is the corresponding wavelength, and $\overline{v} \sim \sqrt{k_B T/m}$ is a typical gas particle velocity. We see that the characteristic time scales grow as the square of the wavelength, which is the signature of *diffusive* processes.

In the remaining normal modes the velocity is parallel to $\vec{k}$, and Eq. (3.125) reduces to

$$\omega \begin{pmatrix} \tilde{\nu} \\ \tilde{u}_\ell \\ \tilde{\theta} \end{pmatrix} = \begin{pmatrix} 0 & \overline{n}k & 0 \\ \frac{k_B \overline{T}}{m\overline{n}}k & -\mathrm{i}\frac{4\mu k^2}{3m\overline{n}} & \frac{k_B}{m}k \\ 0 & \frac{2}{3}\overline{T}k & -\mathrm{i}\frac{2Kk^2}{3k_B\overline{n}} \end{pmatrix} \begin{pmatrix} \tilde{\nu} \\ \tilde{u}_\beta \\ \tilde{\theta} \end{pmatrix}. \tag{3.127}$$

The determinant of the dynamical matrix is the product of the three eigenfrequencies, and to lowest order is given by

$$\det(M) = \mathrm{i}\frac{2Kk^2}{3k_B\overline{n}} \cdot \overline{n}k \cdot \frac{k_B \overline{T}k}{m\overline{n}} + \mathcal{O}(\tau_\times^2). \tag{3.128}$$

At zeroth order the two sound modes have $\omega_\pm^0(k) = \pm v_\ell k$, and hence the frequency of the isobaric mode is

$$\omega_e^1(k) \approx \frac{\det(M)}{-v_\ell^2 k^2} = -\mathrm{i}\frac{2Kk^2}{5k_B\overline{n}} + \mathcal{O}(\tau_\times^2). \tag{3.129}$$

At first order, the longitudinal sound modes also turn into damped oscillations with frequencies $\omega_\pm^1(k) = \pm v_\ell k - \mathrm{i}\gamma$. The simplest way to obtain the decay rates is to note that the trace of the dynamical matrix is equal to the sum of the eigenvalues, and hence

$$\omega_\pm^1(k) = \pm v_\ell k - \mathrm{i}k^2 \left(\frac{2\mu}{3m\overline{n}} + \frac{2K}{15k_B\overline{n}} \right) + \mathcal{O}(\tau_\times^2). \tag{3.130}$$

The damping of all normal modes guarantees the, albeit slow, approach of the gas to its final uniform and stationary equilibrium state.

Problems for chapter 3

1. *One-dimensional gas*: a thermalized gas particle is suddenly confined to a one-dimensional trap. The corresponding mixed state is described by an initial density function $\rho(q, p, t = 0) = \delta(q)f(p)$, where $f(p) = \exp(-p^2/2mk_B T)/\sqrt{2\pi mk_B T}$.

   (a) Starting from Liouville's equation, derive $\rho(q, p, t)$ and sketch it in the (q, p) plane.
   (b) Derive the expressions for the averages $\langle q^2 \rangle$ and $\langle p^2 \rangle$ at $t > 0$.
   (c) Suppose that hard walls are placed at $q = \pm Q$. Describe $\rho(q, p, t \gg \tau)$, where τ is an appropriately large relaxation time.

(d) A "coarse-grained" density $\tilde{\rho}$ is obtained by ignoring variations of ρ below some small resolution in the (q, p) plane; for example, by averaging ρ over cells of the resolution area. Find $\tilde{\rho}(q, p)$ for the situation in part (c), and show that it is stationary.

****

2. *Evolution of entropy*: the normalized ensemble density is a probability in the phase space Γ. This probability has an associated entropy $S(t) = -\int d\Gamma \rho(\Gamma, t) \ln \rho(\Gamma, t)$.

   (a) Show that if $\rho(\Gamma, t)$ satisfies Liouville's equation for a Hamiltonian $\mathcal{H}$, $dS/dt = 0$.

   (b) Using the method of Lagrange multipliers, find the function $\rho_{max}(\Gamma)$ that maximizes the functional $S[\rho]$, subject to the constraint of fixed average energy, $\langle \mathcal{H} \rangle = \int d\Gamma \rho \mathcal{H} = E$.

   (c) Show that the solution to part (b) is stationary, that is, $\partial \rho_{max}/\partial t = 0$.

   (d) How can one reconcile the result in (a) with the observed increase in entropy as the system approaches the equilibrium density in (b)? (*Hint.* Think of the situation encountered in the previous problem.)

****

3. *The Vlasov equation* is obtained in the limit of high particle density $n = N/V$, or large interparticle interaction range λ, such that $n\lambda^3 \gg 1$. In this limit, the collision terms are dropped from the left-hand side of the equations in the BBGKY hierarchy.

   The BBGKY hierarchy

   $$\left[\frac{\partial}{\partial t} + \sum_{n=1}^{s} \frac{\vec{p}_n}{m} \cdot \frac{\partial U}{\partial q_n} - \sum_{n=1}^{s} \left(\frac{\partial U}{\vec{q}_n} + \sum_{l} \frac{\partial \mathcal{V}(\vec{q}_n - \vec{q}_l)}{\vec{q}_n} \right) \cdot \frac{\partial}{\vec{p}_n} \right] f_s$$
   $$= \sum_{n=1}^{s} \int dV_{s+1} \frac{\partial \mathcal{V}(\vec{q}_n - \vec{q}_{s+1})}{\vec{q}_n} \cdot \frac{\partial f_{s+1}}{\partial \vec{p}_n}$$

   has the characteristic time scales

   $$\begin{cases} \dfrac{1}{\tau_U} \sim \dfrac{\partial U}{\partial \vec{q}} \cdot \dfrac{\partial}{\partial \vec{p}} \sim \dfrac{v}{L}, \\[2mm] \dfrac{1}{\tau_c} \sim \dfrac{\partial \mathcal{V}}{\partial \vec{q}} \cdot \dfrac{\partial}{\partial \vec{p}} \sim \dfrac{v}{\lambda}, \\[2mm] \dfrac{1}{\tau_\times} \sim \int dx \dfrac{\partial \mathcal{V}}{\vec{q}} \cdot \dfrac{\partial}{\vec{p}} \dfrac{f_{s+1}}{f_s} \sim \dfrac{1}{\tau_c} \cdot n\lambda^3, \end{cases}$$

   where $n\lambda^3$ is the number of particles within the interaction range λ, and v is a typical velocity. The Boltzmann equation is obtained in the dilute limit, $n\lambda^3 \ll 1$, by disregarding terms of order $1/\tau_\times \ll 1/\tau_c$. The Vlasov equation is obtained in the dense limit of $n\lambda^3 \gg 1$ by ignoring terms of order $1/\tau_c \ll 1/\tau_\times$.

   (a) Assume that the N-body density is a product of one-particle densities, that is, $\rho = \prod_{i=1}^{N} \rho_1(\mathbf{x}_i, t)$, where $\mathbf{x}_i \equiv (\vec{p}_i, \vec{q}_i)$. Calculate the densities f_s, and their normalizations.

(b) Show that once the collision terms are eliminated, all the equations in the BBGKY hierarchy are equivalent to the single equation

$$\left[\frac{\partial}{\partial t} + \frac{\vec{p}}{m} \cdot \frac{\partial}{\partial \vec{q}} - \frac{\partial U_{\text{eff}}}{\partial \vec{q}} \cdot \frac{\partial}{\partial \vec{p}}\right] f_1(\vec{p}, \vec{q}, t) = 0,$$

where

$$U_{\text{eff}}(\vec{q}, t) = U(\vec{q}) + \int d\mathbf{x}' \mathcal{V}(\vec{q} - \vec{q}') f_1(\mathbf{x}', t).$$

(c) Now consider N particles confined to a box of volume V, with no additional potential. Show that $f_1(\vec{q}, \vec{p}) = g(\vec{p})/V$ is a stationary solution to the Vlasov equation *for any* $g(\vec{p})$. Why is there no relaxation toward equilibrium for $g(\vec{p})$?

****

4. *Two-component plasma*: consider a *neutral* mixture of N ions of charge $+e$ and mass m_+, and N electrons of charge $-e$ and mass m_-, in a volume $V = N/n_0$.

(a) Show that the Vlasov equations for this two-component system are

$$\begin{cases} \left[\dfrac{\partial}{\partial t} + \dfrac{\vec{p}}{m_+} \cdot \dfrac{\partial}{\partial \vec{q}} + e\dfrac{\partial \Phi_{\text{eff}}}{\partial \vec{q}} \cdot \dfrac{\partial}{\partial \vec{p}}\right] f_+(\vec{p}, \vec{q}, t) = 0 \\[2ex] \left[\dfrac{\partial}{\partial t} + \dfrac{\vec{p}}{m_-} \cdot \dfrac{\partial}{\partial \vec{q}} - e\dfrac{\partial \Phi_{\text{eff}}}{\partial \vec{q}} \cdot \dfrac{\partial}{\partial \vec{p}}\right] f_-(\vec{p}, \vec{q}, t) = 0 \end{cases},$$

where the effective Coulomb potential is given by

$$\Phi_{\text{eff}}(\vec{q}, t) = \Phi_{\text{ext}}(\vec{q}) + e \int d\mathbf{x}' C(\vec{q} - \vec{q}') \left[f_+(\mathbf{x}', t) - f_-(\mathbf{x}', t)\right].$$

Here, Φ_{ext} is the potential set up by the external charges, and the Coulomb potential $C(\vec{q})$ satisfies the differential equation $\nabla^2 C = 4\pi\delta^3(\vec{q})$.

(b) Assume that the one-particle densities have the stationary forms $f_\pm = g_\pm(\vec{p})n_\pm(\vec{q})$. Show that the effective potential satisfies the equation

$$\nabla^2 \Phi_{\text{eff}} = 4\pi\rho_{\text{ext}} + 4\pi e \left(n_+(\vec{q}) - n_-(\vec{q})\right),$$

where ρ_{ext} is the external charge density.

(c) Further assuming that the densities relax to the equilibrium Boltzmann weights $n_\pm(\vec{q}) = n_0 \exp\left[\pm\beta e\Phi_{\text{eff}}(\vec{q})\right]$ leads to the self-consistency condition

$$\nabla^2 \Phi_{\text{eff}} = 4\pi \left[\rho_{\text{ext}} + n_0 e \left(e^{\beta e\Phi_{\text{eff}}} - e^{-\beta e\Phi_{\text{eff}}}\right)\right],$$

known as the *Poisson–Boltzmann equation*. Due to its non-linear form, it is generally not possible to solve the Poisson–Boltzmann equation. By linearizing the exponentials, one obtains the simpler *Debye equation*

$$\nabla^2 \Phi_{\text{eff}} = 4\pi\rho_{\text{ext}} + \Phi_{\text{eff}}/\lambda^2.$$

Give the expression for the *Debye screening length* λ.

(d) Show that the Debye equation has the general solution

$$\Phi_{\text{eff}}(\vec{q}) = \int d^3\vec{q}' G(\vec{q} - \vec{q}')\rho_{\text{ext}}(\vec{q}'),$$

where $G(\vec{q}) = \exp(-|\vec{q}|/\lambda)/|\vec{q}|$ is the screened Coulomb potential.

(e) Give the condition for the self-consistency of the Vlasov approximation, and interpret it in terms of the interparticle spacing.

(f) Show that the characteristic relaxation time $(\tau \approx \lambda/v)$ is temperature-independent. What property of the plasma is it related to?

****

5. *Two-dimensional electron gas in a magnetic field*: when donor atoms (such as P or As) are added to a semiconductor (e.g., Si or Ge), their conduction electrons can be thermally excited to move freely in the host lattice. By growing layers of different materials, it is possible to generate a spatially varying potential (work-function) that traps electrons at the boundaries between layers. In the following, we shall treat the trapped electrons as a gas of classical particles *in two dimensions*.

   If the layer of electrons is sufficiently separated from the donors, the main source of scattering is from electron–electron collisions.

(a) The Hamiltonian for non-interacting free electrons in a magnetic field has the form

$$\mathcal{H} = \sum_i \left[\frac{\left(\vec{p}_i - e\vec{A}\right)^2}{2m} \pm \mu_B |\vec{B}| \right].$$

(The two signs correspond to electron spins parallel or anti-parallel to the field.) The vector potential $\vec{A} = \vec{B} \times \vec{q}/2$ describes a uniform magnetic field $\vec{B}$. Obtain the classical equations of motion, and show that they describe rotation of electrons in cyclotron orbits in a plane orthogonal to $\vec{B}$.

(b) Write down heuristically (i.e., not through a step-by-step derivation) the Boltzmann equations for the densities $f_\uparrow(\vec{p}, \vec{q}, t)$ and $f_\downarrow(\vec{p}, \vec{q}, t)$ of electrons with up and down spins, in terms of the two cross-sections $\sigma \equiv \sigma_{\uparrow\uparrow} = \sigma_{\downarrow\downarrow}$, and $\sigma_\times \equiv \sigma_{\uparrow\downarrow}$, of *spin conserving* collisions.

(c) Show that $dH/dt \leq 0$, where $H = H_\uparrow + H_\downarrow$ is the sum of the corresponding H-functions.

(d) Show that $dH/dt = 0$ for any $\ln f$ that is, *at each location*, a linear combination of quantities conserved in the collisions.

(e) Show that the streaming terms in the Boltzmann equation are zero for any function that depends only on the quantities conserved by the one-body Hamiltonians.

(f) Show that angular momentum $\vec{L} = \vec{q} \times \vec{p}$ is conserved during, and away from, collisions.

(g) Write down the most general form for the equilibrium distribution functions for particles confined to a circularly symmetric potential.

(h) How is the result in part (g) modified by including scattering from magnetic and non-magnetic impurities?

(i) Do conservation of spin and angular momentum lead to new hydrodynamic equations?

****

6. *The Lorentz gas* describes non-interacting particles colliding with a fixed set of scatterers. It is a good model for scattering of electrons from donor impurities. Consider a uniform two-dimensional density n_0 of fixed impurities, which are hard circles of radius a.

(a) Show that the differential cross-section of a hard circle scattering through an angle θ is

$$d\sigma = \frac{a}{2} \sin \frac{\theta}{2} d\theta,$$

and calculate the total cross-section.

(b) Write down the Boltzmann equation for the one-particle density $f(\vec{q}, \vec{p}, t)$ of the Lorentz gas (including only collisions with the fixed impurities). *(Ignore the electron spin.)*

(c) Find the eigenfunctions and eigenvalues of the collision operator.

(d) Using the definitions $\vec{F} \equiv -\partial U / \partial \vec{q}$, and

$$n(\vec{q}, t) = \int d^2 \vec{p} f(\vec{q}, \vec{p}, t), \quad \text{and} \quad \langle g(\vec{q}, t) \rangle = \frac{1}{n(\vec{q}, t)} \int d^2 \vec{p} f(\vec{q}, \vec{p}, t) g(\vec{q}, t),$$

show that for any function $\chi(|\vec{p}|)$, we have

$$\frac{\partial}{\partial t} (n \langle \chi \rangle) + \frac{\partial}{\partial \vec{q}} \cdot \left(n \left\langle \frac{\vec{p}}{m} \chi \right\rangle \right) = \vec{F} \cdot \left(n \left\langle \frac{\partial \chi}{\partial \vec{p}} \right\rangle \right).$$

(e) Derive the conservation equation for local density $\rho \equiv m n(\vec{q}, t)$, in terms of the local velocity $\vec{u} \equiv \langle \vec{p}/m \rangle$.

(f) Since the magnitude of particle momentum is unchanged by impurity scattering, the Lorentz gas has an infinity of conserved quantities $|\vec{p}|^m$. This unrealistic feature is removed upon inclusion of particle–particle collisions. For the rest of this problem focus only on $p^2/2m$ as a conserved quantity. Derive the conservation equation for the energy density

$$\epsilon(\vec{q}, t) \equiv \frac{\rho}{2} \langle c^2 \rangle, \quad \text{where} \quad \vec{c} \equiv \frac{\vec{p}}{m} - \vec{u},$$

in terms of the energy flux $\vec{h} \equiv \rho \langle \vec{c} \, c^2 \rangle / 2$, and the pressure tensor $P_{\alpha\beta} \equiv \rho \langle c_\alpha c_\beta \rangle$.

(g) Starting with a one-particle density

$$f^0(\vec{p}, \vec{q}, t) = n(\vec{q}, t) \exp \left[-\frac{p^2}{2 m k_B T(\vec{q}, t)} \right] \frac{1}{2\pi m k_B T(\vec{q}, t)},$$

reflecting local equilibrium conditions, calculate $\vec{u}$, $\vec{h}$, and $P_{\alpha\beta}$. Hence obtain the zeroth-order hydrodynamic equations.

(h) Show that in the single collision time approximation to the collision term in the Boltzmann equation, the first-order solution is

$$f^1(\vec{p}, \vec{q}, t) = f^0(\vec{p}, \vec{q}, t) \left[1 - \tau \frac{\vec{p}}{m} \cdot \left(\frac{\partial \ln \rho}{\partial \vec{q}} - \frac{\partial \ln T}{\partial \vec{q}} + \frac{p^2}{2mk_B T^2} \frac{\partial T}{\partial \vec{q}} - \frac{\vec{F}}{k_B T} \right) \right].$$

(i) Show that using the first-order expression for f, we obtain

$$\rho \vec{u} = n\tau \left[\vec{F} - k_B T \nabla \ln (\rho T) \right].$$

(j) From the above equation, calculate the velocity response function $\chi_{\alpha\beta} = \partial u_\alpha / \partial F_\beta$.

(k) Calculate $P_{\alpha\beta}$, and $\vec{h}$, and hence write down the first-order hydrodynamic equations.

****

7. *Thermal conductivity*: consider a classical gas between two plates separated by a distance w. One plate at $y = 0$ is maintained at a temperature T_1, while the other plate at $y = w$ is at a different temperature T_2. The gas velocity is zero, so that the initial zeroth-order approximation to the one-particle density is

$$f_1^0(\vec{p}, x, y, z) = \frac{n(y)}{[2\pi mk_B T(y)]^{3/2}} \exp \left[-\frac{\vec{p} \cdot \vec{p}}{2mk_B T(y)} \right].$$

(a) What is the necessary relation between $n(y)$ and $T(y)$ to ensure that the gas velocity $\vec{u}$ remains zero? (Use this relation between $n(y)$ and $T(y)$ in the remainder of this problem.)

(b) Using Wick's theorem, or otherwise, show that

$$\langle p^2 \rangle^0 \equiv \langle p_\alpha p_\alpha \rangle^0 = 3 (mk_B T), \quad \text{and} \quad \langle p^4 \rangle^0 \equiv \langle p_\alpha p_\alpha p_\beta p_\beta \rangle^0 = 15 (mk_B T)^2,$$

where $\langle \mathcal{O} \rangle^0$ indicates local averages with the Gaussian weight f_1^0. Use the result $\langle p^6 \rangle^0 = 105(mk_B T)^3$ in conjunction with symmetry arguments to conclude

$$\langle p_y^2 p^4 \rangle^0 = 35 (mk_B T)^3.$$

(c) The zeroth-order approximation does not lead to relaxation of temperature/density variations related as in part (a). Find a better (time-independent) approximation $f_1^1(\vec{p}, y)$, by linearizing the Boltzmann equation in the single collision time approximation to

$$\mathcal{L}[f_1^1] \approx \left[\frac{\partial}{\partial t} + \frac{p_y}{m} \frac{\partial}{\partial y} \right] f_1^0 \approx -\frac{f_1^1 - f_1^0}{\tau_K},$$

where τ_K is of the order of the mean time between collisions.

(d) Use f_1^1, along with the averages obtained in part (b), to calculate h_y, the y component of the heat transfer vector, and hence find K, the coefficient of thermal conductivity.

(e) What is the temperature profile, $T(y)$, of the gas in steady state?

$$********$$

---

8. *Zeroth-order hydrodynamics*: the hydrodynamic equations resulting from the conservation of particle number, momentum, and energy in collisions are (in a uniform box):

$$
\begin{cases}
\partial_t n + \partial_\alpha (n u_\alpha) = 0 \\[2mm]
\partial_t u_\alpha + u_\beta \partial_\beta u_\alpha = -\dfrac{1}{mn} \partial_\beta P_{\alpha\beta} \quad , \\[2mm]
\partial_t \varepsilon + u_\alpha \partial_\alpha \varepsilon = -\dfrac{1}{n} \partial_\alpha h_\alpha - \dfrac{1}{n} P_{\alpha\beta} u_{\alpha\beta}
\end{cases}
$$

where n is the local density, $\vec{u} = \langle \vec{p}/m \rangle$, $u_{\alpha\beta} = (\partial_\alpha u_\beta + \partial_\beta u_\alpha)/2$, and $\varepsilon = \langle mc^2/2 \rangle$, with $\vec{c} = \vec{p}/m - \vec{u}$.

(a) For the zeroth-order density

$$
f_1^0(\vec{p}, \vec{q}, t) = \frac{n(\vec{q}, t)}{\left(2\pi m k_B T(\vec{q}, t)\right)^{3/2}} \exp\left[-\frac{\left(\vec{p} - m\vec{u}(\vec{q}, t)\right)^2}{2m k_B T(\vec{q}, t)} \right],
$$

calculate the pressure tensor $P_{\alpha\beta}^0 = mn \langle c_\alpha c_\beta \rangle^0$, and the heat flux $h_\alpha^0 = nm \langle c_\alpha c^2/2 \rangle^0$.

(b) Obtain the zeroth-order hydrodynamic equations governing the evolution of $n(\vec{q}, t)$, $\vec{u}(\vec{q}, t)$, and $T(\vec{q}, t)$.

(c) Show that the above equations imply $D_t \ln \left(nT^{-3/2}\right) = 0$, where $D_t = \partial_t + u_\beta \partial_\beta$ is the material derivative along stream-lines.

(d) Write down the expression for the function $H^0(t) = \int d^3\vec{q} d^3\vec{p} f_1^0(\vec{p}, \vec{q}, t) \ln f_1^0(\vec{p}, \vec{q}, t)$, after performing the integrations over $\vec{p}$, in terms of $n(\vec{q}, t)$, $\vec{u}(\vec{q}, t)$, and $T(\vec{q}, t)$.

(e) Using the hydrodynamic equations in (b), calculate dH^0/dt.

(f) Discuss the implications of the result in (e) for approach to equilibrium.

$$********$$

9. *Viscosity*: consider a classical gas between two plates separated by a distance w. One plate at $y = 0$ is stationary, while the other at $y = w$ moves with a constant velocity $v_x = u$. A zeroth-order approximation to the one-particle density is

$$
f_1^0(\vec{p}, \vec{q}) = \frac{n}{(2\pi m k_B T)^{3/2}} \exp\left[-\frac{1}{2m k_B T} \left((p_x - m\alpha y)^2 + p_y^2 + p_z^2 \right) \right],
$$

obtained from the *uniform* Maxwell–Boltzmann distribution by substituting the average value of the velocity at each point. ($\alpha = u/w$ is the velocity gradient.)

(a) The above approximation does not satisfy the Boltzmann equation as the collision term vanishes, while $df_1^0/dt \neq 0$. Find a better approximation, $f_1^1(\vec{p})$, by linearizing the Boltzmann equation, in the single collision time approximation, to

$$
\mathcal{L}\left[f_1^1 \right] \approx \left[\frac{\partial}{\partial t} + \frac{\vec{p}}{m} \cdot \frac{\partial}{\partial \vec{q}} \right] f_1^0 \approx -\frac{f_1^1 - f_1^0}{\tau_\times},
$$

where $\tau_\times$ is a characteristic mean time between collisions.

(b) Calculate the net transfer Π_{xy} of the x component of the momentum, of particles passing through a plane at y, per unit area and in unit time.

(c) Note that the answer to (b) is independent of y, indicating a uniform transverse force $F_x = -\Pi_{xy}$, exerted by the gas on each plate. Find the coefficient of viscosity, defined by $\eta = F_x/\alpha$.

****

10. *Light and matter*: in this problem we use kinetic theory to explore the equilibrium between atoms and radiation.

(a) The atoms are assumed to be either in their ground state a_0, or in an excited state a_1, which has a higher energy ε. By considering the atoms as a collection of N fixed two-state systems of energy E (i.e., ignoring their coordinates and momenta), calculate the ratio n_1/n_0 of densities of atoms in the two states as a function of temperature T.

Consider photons γ of frequency $\omega = \varepsilon/\hbar$ and momentum $|\vec{p}| = \hbar\omega/c$, which can interact with the atoms through the following processes:

  (i) *Spontaneous emission: $a_1 \to a_0 + \gamma$.*
  (ii) *Adsorption: $a_0 + \gamma \to a_1$.*
  (iii) *Stimulated emission: $a_1 + \gamma \to a_0 + \gamma + \gamma$.*

Assume that spontaneous emission occurs with a probability σ_{sp}, and that adsorption and stimulated emission have constant (angle-independent) differential cross-sections of $\sigma_{ad}/4\pi$ and $\sigma_{st}/4\pi$, respectively.

(b) Write down the Boltzmann equation governing the density f of the photon gas, treating the atoms as fixed scatterers of densities n_0 and n_1.

(c) Find the equilibrium density $f_{eq.}$ for the photons of the above frequency.

(d) According to Planck's law, the density of photons at a temperature T depends on their frequency ω as $f_{eq.} = [\exp(\hbar\omega/k_B T) - 1]^{-1}/h^3$. What does this imply about the above cross-sections?

(e) Consider a situation in which light shines along the x axis on a collection of atoms whose boundary coincides with the $x = 0$ plane, as illustrated in the figure.

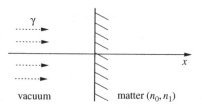

Clearly, f will depend on x (and p_x), but will be independent of y and z. Adapt the Boltzmann equation you propose in part (b) to the case of a uniform incoming flux of photons with momentum $\vec{p} = \hbar \omega \hat{x}/c$. What is the *penetration length* across which the incoming flux decays?

$$********$$

11. *Equilibrium density*: consider a gas of N particles of mass m, in an external potential $U(\vec{q})$. Assume that the one-body density $\rho_1(\vec{p}, \vec{q}, t)$ satisfies the Boltzmann equation. For a stationary solution, $\partial \rho_1/\partial t = 0$, it is *sufficient* from Liouville's theorem for ρ_1 to satisfy $\rho_1 \propto \exp\left[-\beta\left(p^2/2m + U(\vec{q})\right)\right]$. Prove that this condition is also *necessary* by using the H-theorem as follows.

(a) Find $\rho_1(\vec{p}, \vec{q})$ that minimizes $H = N \int d^3\vec{p} d^3\vec{q} \rho_1(\vec{p}, \vec{q}) \ln \rho_1(\vec{p}, \vec{q})$, subject to the constraint that the total energy $E = \langle \mathcal{H} \rangle$ is constant. (*Hint.* Use the method of Lagrange multipliers to impose the constraint.)

(b) For a mixture of two gases (particles of masses m_a and m_b) find the distributions $\rho_1^{(a)}$ and $\rho_1^{(b)}$ that minimize $H = H^{(a)} + H^{(b)}$ subject to the constraint of constant total energy. Hence show that the kinetic energy per particle can serve as an empirical temperature.

$$********$$

12. *Moments of momentum*: consider a gas of N classical particles of mass m in thermal equilibrium at a temperature T, in a box of volume V.

(a) Write down the equilibrium one-particle density $f_{eq.}(\vec{p}, \vec{q})$, for coordinate $\vec{q}$, and momentum $\vec{p}$.

(b) Calculate the joint characteristic function, $\left\langle \exp\left(-i\vec{k}\cdot\vec{p}\right)\right\rangle$, for momentum.

(c) Find all the joint cumulants $\left\langle p_x^\ell p_y^m p_z^n \right\rangle_c$.

(d) Calculate the joint moment $\left\langle p_\alpha p_\beta \left(\vec{p}\cdot\vec{p}\right)\right\rangle$.

$$********$$

13. *Generalized ideal gas*: consider a gas of N particles confined to a box of volume V in d dimensions. The energy, ϵ, and momentum, $\mathbf{p}$, of each particle are related by $\epsilon = p^s$, where $p = |\mathbf{p}|$. (For classical particles $s = 2$, while for highly relativistic ones $s = 1$.) Let $f(v)dv$ denote the probability of finding particles with speeds between v and $v+dv$, and $n = N/V$.

(a) Calculate the number of impacts of gas molecules per unit area of the wall of the box, and per unit time as follows:

(i) Show that the number of particles hitting area A in a time dt arriving from a specific direction $\vec{\Omega}$, with a speed v, is proportional to $A \cdot vdt \cos\theta \cdot nf(v)dv$, where θ is the angle between the direction $\vec{\Omega}$ and the normal to the wall.

(ii) Summing over all directions $\vec{\Omega}$ with the same polar angle θ, demonstrate that

$$dN(\theta, v) = A \cdot vdt \cos\theta \cdot nf(v)dv \cdot \frac{S_{d-1} \sin^{d-2}\theta \, d\theta}{S_d},$$

where $S_d = 2\pi^{d/2}/(d/2-1)!$ is the total solid angle in d dimensions.

(iii) By averaging over v and θ show that

$$\frac{N}{A\,dt} = \frac{S_{d-1}}{(d-1)S_d} \cdot n\bar{v}, \quad \text{where} \quad \bar{v} = \int vf(v)dv \quad \text{is the average speed.}$$

(b) Each (elastic) collision transfers a momentum $2p\cos\theta$ to the wall. By examining the net force on an element of area prove that the pressure P equals $\frac{s}{d} \cdot \frac{E}{V}$, where E is the average (kinetic) energy. (Note that the velocity $\mathbf{v}$ is not $\mathbf{p}/m$ but $\partial\epsilon/\partial\mathbf{p}$.) (*Hint.* Clearly, upon averaging over all directions $\langle\cos^2\theta\rangle = 1/d$.)

(c) Using thermodynamics and the result in (b), show that along an adiabatic curve PV^γ is constant, and calculate γ.

(d) According to the equipartition theorem, each degree of freedom that appears quadratically in the energy has an energy $k_B T/2$. Calculate the value of γ if each gas particle has ℓ such quadratic degrees of freedom in addition to its translational motion. What values of γ are expected for helium and hydrogen at room temperature?

(e) Consider the following experiment to test whether the motion of ants is random. 250 ants are placed inside a $10\,\text{cm} \times 10\,\text{cm}$ box. They cannot climb the wall, but can escape through an opening of size $5\,\text{mm}$ in the wall. If the motion of ants is indeed random, and they move with an average speed of $2\,\text{mm s}^{-1}$, how many are expected to escape the box in the first 30 seconds?

****

14. *Effusion*: a box contains a perfect gas at temperature T and density n.

(a) What is the one-particle density, $\rho_1(\vec{v})$, for particles with velocity $\vec{v}$?

A small hole is opened in the wall of the box for a short time to allow some particles to escape into a previously empty container.

(b) During the time that the hole is open what is the flux (number of particles per unit time and per unit area) of particles into the container? (Ignore the possibility of any particles returning to the box.)

(c) Show that the average kinetic energy of escaping particles is $2k_B T$. (*Hint.* Calculate contributions to kinetic energy of velocity components parallel and perpendicular to the wall separately.)

(d) The hole is closed and the container (now thermally insulated) is allowed to reach equilibrium. What is the final temperature of the gas in the container?

(e) A vessel partially filled with mercury (atomic weight 201), and closed except for a hole of area $0.1\,\text{mm}^2$ above the liquid level, is kept at $0°\,\text{C}$ in a continuously evacuated enclosure. After 30 days it is found that $24\,\text{mg}$ of mercury has been lost. What is the vapor pressure of mercury at $0°\,\text{C}$?

****

15. *Adsorbed particles*: consider a gas of classical particles of mass m in thermal equilibrium at a temperature T, and with a density n. A clean metal surface is introduced into the gas. Particles hitting this surface with normal velocity less than v_t are reflected back into the gas, while particles with normal velocity greater than v_t are absorbed by it.

(a) Find the average number of particles hitting one side of the surface per unit area and per unit time.

(b) Find the average number of particles absorbed by one side of the surface per unit area and per unit time.

****

16. *Electron emission*: when a metal is heated in vacuum, electrons are emitted from its surface. The metal is modeled as a classical gas of non-interacting electrons held in the solid by an abrupt potential well of depth ϕ (the work-function) relative to the vacuum.

(a) What is the relationship between the initial and final velocities of an escaping electron?

(b) In thermal equilibrium at temperature T, what is the probability density function for the velocity of electrons?

(c) If the number density of electrons is n, calculate the current density of thermally emitted electrons.

****

4

Classical statistical mechanics

4.1 General definitions

Statistical mechanics *is a probabilistic approach to equilibrium macroscopic properties of large numbers of degrees of freedom.*

As discussed in chapter 1, equilibrium properties of macroscopic bodies are phenomenologically described by the laws of thermodynamics. The macrostate M depends on a relatively small number of thermodynamic coordinates. To provide a more fundamental derivation of these properties, we can examine the dynamics of the many degrees of freedom comprising a macroscopic body. Description of each microstate μ requires an enormous amount of information, and the corresponding time evolution, governed by the Hamiltonian equations discussed in chapter 3, is usually quite complicated. Rather than following the evolution of an individual (pure) microstate, statistical mechanics examines an ensemble of microstates corresponding to a given (mixed) macrostate. It aims to provide the probabilities $p_M(\mu)$ for the equilibrium ensemble. Liouville's theorem justifies the assumption that all accessible microstates are equally likely in an equilibrium ensemble. As explained in chapter 2, such assignment of probabilities is subjective. In this chapter we shall provide unbiased estimates of $p_M(\mu)$ for a number of different equilibrium ensembles. A central conclusion is that in the thermodynamic limit, with large numbers of degrees of freedom, all these ensembles are in fact equivalent. In contrast to kinetic theory, equilibrium statistical mechanics leaves out the question of *how* various systems evolve to a state of equilibrium.

4.2 The microcanonical ensemble

Our starting point in thermodynamics is a mechanically and adiabatically isolated system. In the absence of heat or work input to the system, the internal energy E, and the generalized coordinates $\mathbf{x}$, are fixed, specifying a macrostate $M \equiv (E, \mathbf{x})$. The corresponding set of mixed microstates form the *microcanonical ensemble*. In classical statistical mechanics, these microstates are defined by points in phase space, their time evolution governed by a Hamiltonian $\mathcal{H}(\mu)$, as discussed in chapter 3. Since the Hamiltonian equations (3.1) conserve the total energy of a given system, all microstates are confined to the surface $\mathcal{H}(\mu) = E$ in phase space. Let us assume that there are no other conserved quantities, so that all points on this surface are mutually accessible. The central postulate of statistical mechanics is that the equilibrium probability distribution is given by

$$p_{(E,\mathbf{x})}(\mu) = \frac{1}{\Omega(E,\mathbf{x})} \cdot \begin{cases} 1 & \text{for } \mathcal{H}(\mu) = E \\ 0 & \text{otherwise} \end{cases}. \tag{4.1}$$

Some remarks and clarification are in order:

(1) *Boltzmann's assumption of equal a priori equilibrium probabilities* refers to the above postulate, which is in fact the unbiased probability estimate in phase space subject to the constraint of constant energy. This assignment is consistent with, but not required by, Liouville's theorem. Note that the phase space specifying the microstates μ must be composed of *canonically conjugate pairs*. Under a *canonical* change of variables, $\mu \to \mu'$, volumes in phase space are left invariant. The Jacobian of such transformations is unity, and the transformed probability, $p(\mu') = p(\mu) |\partial\mu/\partial\mu'|$, is again uniform on the surface of constant energy.

(2) *The normalization factor* $\Omega(E,\mathbf{x})$ is the area of the surface of constant energy E in phase space. To avoid subtleties associated with densities that are non-zero only on a surface, it is sometimes more convenient to define the microcanonical ensemble by requiring $E - \Delta \leq \mathcal{H}(\mu) \leq E + \Delta$, that is, constraining the energy of the ensemble up to an uncertainty of Δ. In this case, the accessible phase space forms a shell of thickness 2Δ around the surface of energy E. The normalization is now the volume of the shell, $\Omega' \approx 2\Delta\Omega$. Since Ω typically depends exponentially on E, as long as $\Delta \sim \mathcal{O}(E^0)$ (or even $\mathcal{O}(E^1)$), the difference between the surface and volume of the shell is negligible in the $E \propto N \to \infty$ limit, and we can use Ω and Ω' interchangeably.

(3) *The entropy* of this uniform probability distribution is given by

$$S(E,\mathbf{x}) = k_B \ln \Omega(E,\mathbf{x}). \tag{4.2}$$

An additional factor of k_B is introduced compared with the definition of Eq. (2.70), so that the entropy has the correct dimensions of energy per degree Kelvin, used in thermodynamics. Ω and S are not changed by a canonical change of coordinates in phase space. For a collection of independent systems, the overall allowed phase space is the product of individual ones, that is, $\Omega_{\text{Total}} = \prod_i \Omega_i$. The resulting entropy is thus additive, as expected for an extensive quantity.

Various results in thermodynamics now follow from Eq. (4.1), provided that we consider macroscopic systems with many degrees of freedom.

Fig. 4.1 The exchange of energy between two isolated systems.

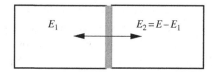

The zeroth law: equilibrium properties are discussed in thermodynamics by placing two previously isolated systems in contact, and allowing them to exchange heat. We can similarly bring together two microcanonical systems, and allow them to exchange energy, but not work. If the original systems have energies E_1 and E_2, respectively, the combined system has energy $E = E_1 + E_2$. Assuming that interactions between the two parts are small, each microstate of the joint system corresponds to a pair of microstates of the two components, that is, $\mu = \mu_1 \otimes \mu_2$, and $\mathcal{H}(\mu_1 \otimes \mu_2) = \mathcal{H}_1(\mu_1) + \mathcal{H}_2(\mu_2)$. As the joint system is in a microcanonical ensemble of energy $E = E_1 + E_2$, in equilibrium

$$p_E(\mu_1 \otimes \mu_2) = \frac{1}{\Omega(E)} \cdot \begin{cases} 1 & \text{for } \mathcal{H}_1(\mu_1) + \mathcal{H}_2(\mu_2) = E \\ 0 & \text{otherwise} \end{cases}. \tag{4.3}$$

Since only the overall energy is fixed, the total allowed phase space is computed from

$$\Omega(E) = \int dE_1 \, \Omega_1(E_1) \Omega_2(E - E_1) = \int dE_1 \exp\left[\frac{S_1(E_1) + S_2(E - E_1)}{k_B}\right]. \tag{4.4}$$

Fig. 4.2 The joint number of states of two systems in contact is overwhelmingly larger at the "equilibrium" energies E_1^* and E_2^*.

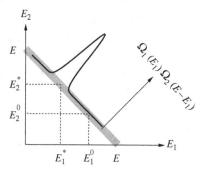

The properties of the two systems in the new joint equilibrium state are implicit in Eq. (4.3). We can make them explicit by examining the entropy that follows from Eq. (4.4). Extensivity of entropy suggests that S_1 and S_2 are proportional to the numbers of particles in each system, making the integrand in Eq. (4.4) an exponentially large quantity. Hence the integral can be equated by the saddle point method to the maximum value of the integrand, obtained for energies E_1^* and $E_2^* = E - E_1^*$, that is,

$$S(E) = k_B \ln \Omega(E) \approx S_1(E_1^*) + S_2(E_2^*). \tag{4.5}$$

The position of the maximum is obtained by extremizing the exponent in Eq. (4.4) with respect to E_1, resulting in the condition

$$\left.\frac{\partial S_1}{\partial E_1}\right|_{\mathbf{x}_1} = \left.\frac{\partial S_2}{\partial E_2}\right|_{\mathbf{x}_2}. \tag{4.6}$$

Although all joint microstates are equally likely, the above results indicate that there is an exponentially larger number of states in the vicinity of (E_1^*, E_2^*). Originally, the joint system starts in the vicinity of some point (E_1^0, E_2^0). After the exchange of energy takes place, the combined system explores a whole set of new microstates. The probabilistic arguments provide no information on the dynamics of evolution amongst these microstates, or on the amount of time needed to establish equilibrium. However, once sufficient time has elapsed so that the assumption of equal a priori probabilities is again valid, the system is overwhelmingly likely to be at a state with internal energies (E_1^*, E_2^*). At this *equilibrium* point, condition (4.6) is satisfied, specifying a relation between two functions of state. These state functions are thus equivalent to empirical temperatures, and, indeed, consistent with the fundamental result of thermodynamics, we have

$$\left.\frac{\partial S}{\partial E}\right|_{\mathbf{x}} = \frac{1}{T}. \tag{4.7}$$

The first law: we next inquire about the variations of $S(E, \mathbf{x})$ with $\mathbf{x}$, by changing the coordinates *reversibly* by $\delta\mathbf{x}$. This results in doing work on the system by an amount $đW = \mathbf{J} \cdot \delta\mathbf{x}$, which changes the internal energy to $E + \mathbf{J} \cdot \delta\mathbf{x}$. The first-order change in entropy is given by

$$\delta S = S(E + \mathbf{J} \cdot \delta\mathbf{x}, \mathbf{x} + \delta\mathbf{x}) - S(E, \mathbf{x}) = \left(\left.\frac{\partial S}{\partial E}\right|_{\mathbf{x}} \mathbf{J} + \left.\frac{\partial S}{\partial \mathbf{x}}\right|_{E}\right) \cdot \delta\mathbf{x}. \tag{4.8}$$

This change will occur spontaneously, taking the system into a more probable state, unless the quantity in brackets is zero. Using Eq. (4.7), this allows us to identify the derivatives

$$\left.\frac{\partial S}{\partial x_i}\right|_{E, x_{j \neq i}} = -\frac{J_i}{T}. \tag{4.9}$$

Having thus identified all variations of S, we have

$$dS(E, \mathbf{x}) = \frac{dE}{T} - \frac{\mathbf{J} \cdot d\mathbf{x}}{T}, \implies dE = T dS + \mathbf{J} \cdot d\mathbf{x}, \tag{4.10}$$

allowing us to identify the heat input $đQ = T dS$.

The second law: clearly, the above statistical definition of equilibrium rests on the presence of many degrees of freedom $N \gg 1$, which make it exponentially unlikely in N that the combined systems are found with component energies different from (E_1^*, E_2^*). By this construction, the equilibrium point has a larger number of accessible states than any starting point, that is,

$$\Omega_1(E_1^*, \mathbf{x}_1)\Omega_2(E_2^*, \mathbf{x}_2) \geq \Omega_1(E_1, \mathbf{x}_1)\Omega_2(E_2, \mathbf{x}_2). \tag{4.11}$$

In the process of evolving to the more likely (and more densely populated) regions, there is an irreversible loss of information, accompanied by an increase in entropy,

$$\delta S = S_1(E_1^*) + S_2(E_2^*) - S_1(E_1) - S_2(E_2) \geq 0, \qquad (4.12)$$

as required by the second law of thermodynamics. When the two bodies are first brought into contact, the equality in Eq. (4.6) does not hold. The change in entropy is such that

$$\delta S = \left(\frac{\partial S_1}{\partial E_1}\bigg|_{\mathbf{x_1}} - \frac{\partial S_2}{\partial E_2}\bigg|_{\mathbf{x_2}} \right) \delta E_1 = \left(\frac{1}{T_1} - \frac{1}{T_2} \right) \delta E_1 \geq 0, \qquad (4.13)$$

that is, heat (energy) flows from the hotter to the colder body, as in Clausius's statement of the second law.

Stability conditions: since the point (E_1^*, E_2^*) is a maximum, the second derivative of $S_1(E_1) + S_2(E_2)$ must be negative at this point, that is,

$$\frac{\partial^2 S_1}{\partial E_1^2}\bigg|_{\mathbf{x_1}} + \frac{\partial^2 S_2}{\partial E_2^2}\bigg|_{\mathbf{x_2}} \leq 0. \qquad (4.14)$$

Applying the above condition to two parts of the same system, the condition of thermal stability, $C_{\mathbf{x}} \geq 0$, as discussed in Section 1.9, is regained. Similarly, the second-order changes in Eq. (4.8) must be negative, requiring that the matrtix $\partial^2 S / \partial x_i \partial x_j \big|_E$ be positive definite.

4.3 Two-level systems

Consider N impurity atoms trapped in a solid matrix. Each impurity can be in one of two states, with energies 0 and ϵ, respectively. This example is somewhat different from the situations considered so far, in that the allowed microstates are discrete. Liouville's theorem applies to Hamiltonian evolution in a continuous phase space. Although there is less ambiguity in enumeration of discrete states, the dynamics that ensures that all allowed microstates are equally accessed will remain unspecified for the moment. (An example from quantum mechanical evolution will be presented later on.)

The microstates of the two-level system are specified by the set of *occupation numbers* $\{n_i\}$, where $n_i = 0$ or 1 depending on whether the ith impurity is in its ground state or excited. The overall energy is

$$\mathcal{H}(\{n_i\}) = \epsilon \sum_{i=1}^{N} n_i \equiv \epsilon N_1, \qquad (4.15)$$

where N_1 is the total number of excited impurities. The macrostate is specified by the total energy E, and the number of impurities N. The microcanonical probability is thus

$$p(\{n_i\}) = \frac{1}{\Omega(E, N)} \delta_{\epsilon \sum_i n_i, E.} \qquad (4.16)$$

As there are $N_1 = E/\epsilon$ excited impurities, the normalization Ω is the number of ways of choosing N_1 excited levels among the available N, and given by the binomial coefficient

$$\Omega(E, N) = \frac{N!}{N_1!(N-N_1)!}. \tag{4.17}$$

The entropy

$$S(E, N) = k_B \ln \frac{N!}{N_1!(N-N_1)!} \tag{4.18}$$

can be simplified by Stirling's formula in the limit of $N_1, N \gg 1$ to

$$\begin{aligned} S(E, N) &\approx -Nk_B \left[\frac{N_1}{N} \ln \frac{N_1}{N} + \frac{N-N_1}{N} \ln \frac{N-N_1}{N} \right] \\ &= -Nk_B \left[\left(\frac{E}{N\epsilon}\right) \ln \left(\frac{E}{N\epsilon}\right) + \left(1 - \frac{E}{N\epsilon}\right) \ln \left(1 - \frac{E}{N\epsilon}\right) \right]. \end{aligned} \tag{4.19}$$

The equilibrium temperature can now be calculated from Eq. (4.7) as

$$\frac{1}{T} = \left. \frac{\partial S}{\partial E} \right|_N = -\frac{k_B}{\epsilon} \ln \left(\frac{E}{N\epsilon - E} \right). \tag{4.20}$$

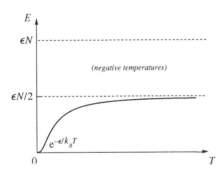

Fig. 4.3 The internal energy of a two-level system, as a function of its temperature T.

Alternatively, the internal energy at a temperature T is given by

$$E(T) = \frac{N\epsilon}{\exp\left(\frac{\epsilon}{k_B T}\right) + 1}. \tag{4.21}$$

The internal energy is a monotonic function of temperature, increasing from a minimum value of 0 at $T = 0$ to a maximum value of $N\epsilon/2$ at infinite temperature. It is, however, possible to start with energies larger than $N\epsilon/2$, which correspond to *negative temperatures* from Eq. (4.20). The origin of the negative temperature is the *decrease* in the number of microstates with increasing energy, the opposite of what happens in most systems. Two-level systems have an upper bound on their energy, and very few microstates close to this maximal energy. Hence, increased energy leads to more order in the system. However, once a negative temperature system is brought into contact with the rest of the Universe (or any portion of it without an upper bound in energy), it loses its excess energy and comes to equilibrium at a positive temperature. The world of negative temperatures is quite unusual in that systems can be cooled by adding heat, and heated by removing it. There are physical examples of systems temporarily prepared at a metastable equilibrium of negative temperature in lasers, and for magnetic spins.

The heat capacity of the system, given by

$$C = \frac{dE}{dT} = Nk_B \left(\frac{\epsilon}{k_B T} \right)^2 \exp\left(\frac{\epsilon}{k_B T} \right) \left[\exp\left(\frac{\epsilon}{k_B T} \right) + 1 \right]^{-2}, \qquad (4.22)$$

Fig. 4.4 The heat capacity of a two-level system goes to zero at both high and low temperatures T.

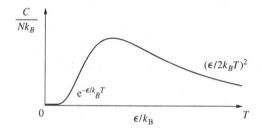

vanishes at both low and high temperatures. The vanishing of C as $\exp(-\epsilon/k_B T)$ at low temperatures is characteristic of all systems with an *energy gap* separating the ground state and lowest excited states. The vanishing of C at high temperatures is a *saturation* effect, common to systems with a maximum in the number of states as a function of energy. In between, the heat capacity exhibits a peak at a characteristic temperature of $T_\epsilon \propto \epsilon/k_B$.

Fig. 4.5 The probabilities for finding a single impurity in its ground state (top), or excited state (bottom), as a function of temperature T.

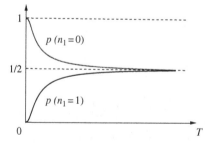

Statistical mechanics provides much more than just macroscopic information for quantities such as energy and heat capacity. Equation (4.16) is a complete joint probability distribution with considerable information on the microstates. For example, the unconditional probability for exciting a particular impurity is obtained from

$$p(n_1) = \sum_{\{n_2, \cdots, n_N\}} p(\{n_i\}) = \frac{\Omega(E - n_1\epsilon, N - 1)}{\Omega(E, N)}. \qquad (4.23)$$

The second equality is obtained by noting that once the energy taken by the first impurity is specified, the remaining energy must be distributed among the other $N-1$ impurities. Using Eq. (4.17),

$$p(n_1 = 0) = \frac{\Omega(E, N-1)}{\Omega(E, N)} = \frac{(N-1)!}{N_1!(N-N_1-1)!} \cdot \frac{N_1!(N-N_1)!}{N!} = 1 - \frac{N_1}{N}, \quad (4.24)$$

and $p(n_1 = 1) = 1 - p(n_1 = 0) = N_1/N$. Using $N_1 = E/\epsilon$, and Eq. (4.21), the occupation probabilities at a temperature T are

$$p(0) = \frac{1}{1 + \exp\left(-\frac{\epsilon}{k_B T}\right)}, \quad \text{and} \quad p(1) = \frac{\exp\left(-\frac{\epsilon}{k_B T}\right)}{1 + \exp\left(-\frac{\epsilon}{k_B T}\right)}. \quad (4.25)$$

4.4 The ideal gas

As discussed in chapter 3, microstates of a gas of N particles correspond to points $\mu \equiv \{\vec{p}_i, \vec{q}_i\}$ in the $6N$-dimensional phase space. Ignoring the potential energy of interactions, the particles are subject to a Hamiltonian

$$\mathcal{H} = \sum_{i=1}^{N} \left[\frac{\vec{p}_i^{\,2}}{2m} + U(\vec{q}_i) \right], \quad (4.26)$$

where $U(\vec{q})$ describes the potential imposed by a box of volume V. A microcanonical ensemble is specified by its energy, volume, and number of particles, $M \equiv (E, V, N)$. The joint PDF for a microstate is

$$p(\mu) = \frac{1}{\Omega(E, V, N)} \cdot \begin{cases} 1 & \text{for } \vec{q}_i \in \text{box, and } \sum_i \vec{p}_i^{\,2}/2m = E \quad (\pm\Delta_E) \\ 0 & \text{otherwise} \end{cases}. \quad (4.27)$$

In the allowed microstates, coordinates of the particles must be within the box, while the momenta are constrained to the surface of the (hyper-)sphere $\sum_{i=1}^{N} \vec{p}_i^{\,2} - 2mE$. The allowed phase space is thus the product of a contribution V^N from the coordinates, with the surface area of a $3N$-dimensional sphere of radius $\sqrt{2mE}$ from the momenta. (If the microstate energies are accepted in the energy interval $E \pm \Delta_E$, the corresponding volume in momentum space is that of a (hyper-)spherical shell of thickness $\Delta_R = \sqrt{2m/E}\Delta E$.) The area of a d-dimensional sphere is $\mathcal{A}_d = S_d R^{d-1}$, where S_d is the generalized solid angle.

A simple way to calculate the d-dimensional solid angle is to consider the product of d Gaussian integrals,

$$I_d \equiv \left(\int_{-\infty}^{\infty} \mathrm{d}x e^{-x^2} \right)^d = \pi^{d/2}. \quad (4.28)$$

Alternatively, we may consider I_d as an integral over the entire d-dimensional space, that is,

$$I_d = \int \prod_{i=1}^{d} \mathrm{d}x_i \exp\left(-x_i^2\right). \quad (4.29)$$

The integrand is spherically symmetric, and we can change coordinates to $R^2 = \sum_i x_i^2$. Noting that the corresponding volume element in these coordinates is $\mathrm{d}V_d = S_d R^{d-1} \mathrm{d}R$,

$$I_d = \int_0^\infty \mathrm{d}R S_d R^{d-1} \mathrm{e}^{-R^2} = \frac{S_d}{2} \int_0^\infty \mathrm{d}y\, y^{d/2-1} \mathrm{e}^{-y} = \frac{S_d}{2}(d/2-1)!, \qquad (4.30)$$

where we have first made a change of variables to $y = R^2$, and then used the integral representation of $n!$ in Eq. (2.63). Equating expressions (4.28) and (4.30) for I_d gives the final result for the solid angle,

$$S_d = \frac{2\pi^{d/2}}{(d/2-1)!}. \qquad (4.31)$$

The volume of the available phase space is thus given by

$$\Omega(E, V, N) = V^N \frac{2\pi^{3N/2}}{(3N/2-1)!}(2mE)^{(3N-1)/2}\Delta_R. \qquad (4.32)$$

The entropy is obtained from the logarithm of the above expression. Using Stirling's formula, and neglecting terms of order of 1 or $\ln E \sim \ln N$ in the large N limit, results in

$$\begin{aligned} S(E, V, N) &= k_B \left[N \ln V + \frac{3N}{2}\ln(2\pi mE) - \frac{3N}{2}\ln\frac{3N}{2} + \frac{3N}{2} \right] \\ &= Nk_B \ln\left[V\left(\frac{4\pi emE}{3N}\right)^{3/2} \right]. \end{aligned} \qquad (4.33)$$

Properties of the ideal gas can now be recovered from $T\mathrm{d}S = \mathrm{d}E + P\mathrm{d}V - \mu\mathrm{d}N$,

$$\frac{1}{T} = \left.\frac{\partial S}{\partial E}\right|_{N,V} = \frac{3}{2}\frac{Nk_B}{E}. \qquad (4.34)$$

The internal energy $E = 3Nk_BT/2$ is only a function of T, and the heat capacity $C_V = 3Nk_B/2$ is a constant. The equation of state is obtained from

$$\frac{P}{T} = \left.\frac{\partial S}{\partial V}\right|_{N,E} = \frac{Nk_B}{V}, \quad \Longrightarrow \quad PV = Nk_BT. \qquad (4.35)$$

The unconditional probability of finding a particle of momentum $\vec{p}_1$ in the gas can be calculated from the joint PDF in Eq. (4.27), by integrating over all other variables,

$$\begin{aligned} p(\vec{p}_1) &= \int \mathrm{d}^3\vec{q}_1 \prod_{i=2}^N \mathrm{d}^3\vec{q}_i \mathrm{d}^3\vec{p}_i\, p(\{\vec{q}_i, \vec{p}_i\}) \\ &= \frac{V\,\Omega(E - \vec{p}_1^{\,2}/2m, V, N-1)}{\Omega(E, V, N)}. \end{aligned} \qquad (4.36)$$

The final expression indicates that once the kinetic energy of one particle is specified, the remaining energy must be shared amongst the other $N - 1$. Using Eq. (4.32),

$$
\begin{aligned}
p(\vec{p}_1) &= \frac{V^N \pi^{3(N-1)/2}(2mE - \vec{p}_1{}^2)^{(3N-4)/2}}{\big(3(N-1)/2 - 1\big)!} \cdot \frac{(3N/2 - 1)!}{V^N \pi^{3N/2}(2mE)^{(3N-1)/2}} \\
&= \left(1 - \frac{\vec{p}_1{}^2}{2mE}\right)^{3N/2 - 2} \frac{1}{(2\pi mE)^{3/2}} \frac{(3N/2 - 1)!}{\big(3(N-1)/2 - 1\big)!}.
\end{aligned}
\tag{4.37}
$$

From Stirling's formula, the ratio of $(3N/2 - 1)!$ to $\big(3(N-1)/2 - 1\big)!$ is approximately $(3N/2)^{3/2}$, and in the large E limit,

$$
p(\vec{p}_1) = \left(\frac{3N}{4\pi mE}\right)^{3/2} \exp\left(-\frac{3N}{2}\frac{\vec{p}_1{}^2}{2mE}\right).
\tag{4.38}
$$

This is a properly normalized Maxwell–Boltzmann distribution, which can be displayed in its more familiar form after the substitution $E = 3Nk_BT/2$,

$$
p(\vec{p}_1) = \frac{1}{(2\pi mk_BT)^{3/2}} \exp\left(-\frac{\vec{p}_1{}^2}{2mk_BT}\right).
\tag{4.39}
$$

4.5 Mixing entropy and the Gibbs paradox

The expression in Eq. (4.33) for the entropy of the ideal gas has a major shortcoming in that *it is not extensive*. Under the transformation $(E, V, N) \to (\lambda E, \lambda V, \lambda N)$, the entropy changes to $\lambda(S + Nk_B \ln \lambda)$. The additional term comes from the contribution V^N of the coordinates to the available phase space. This difficulty is intimately related to the *mixing entropy* of two gases. Consider two distinct gases, initially occupying volumes V_1 and V_2 at the same temperature T. The partition between them is removed, and they are allowed to expand and occupy the combined volume $V = V_1 + V_2$. The mixing process is clearly irreversible, and must be accompanied by an increase in entropy, calculated as follows. According to Eq. (4.33), the initial entropy is

$$
S_i = S_1 + S_2 = N_1 k_B(\ln V_1 + \sigma_1) + N_2 k_B(\ln V_2 + \sigma_2),
\tag{4.40}
$$

where

$$
\sigma_\alpha = \ln\left(\frac{4\pi em_\alpha}{3} \cdot \frac{E_\alpha}{N_\alpha}\right)^{3/2}
\tag{4.41}
$$

is the momentum contribution to the entropy of the αth gas. Since $E_\alpha/N_\alpha = 3k_BT/2$ for a monotonic gas,

$$
\sigma_\alpha(T) = \frac{3}{2}\ln\left(2\pi em_\alpha k_BT\right).
\tag{4.42}
$$

The temperature of the gas is unchanged by mixing, since

$$
\frac{3}{2}k_BT_f = \frac{E_1 + E_2}{N_1 + N_2} = \frac{E_1}{N_1} = \frac{E_2}{N_2} = \frac{3}{2}k_BT.
\tag{4.43}
$$

The final entropy of the mixed gas is

$$S_f = N_1 k_B \ln(V_1 + V_2) + N_2 k_B \ln(V_1 + V_2) + k_B(N_1\sigma_1 + N_2\sigma_2). \tag{4.44}$$

There is no change in the contribution from the momenta, which depends only on temperature. The mixing entropy,

$$\Delta S_{\text{Mix}} = S_f - S_i = N_1 k_B \ln \frac{V}{V_1} + N_2 k_B \ln \frac{V}{V_2} = -N k_B \left[\frac{N_1}{N} \ln \frac{V_1}{V} + \frac{N_2}{N} \ln \frac{V_2}{V} \right], \tag{4.45}$$

is solely from the contribution of the coordinates. The above expression is easily generalized to the mixing of many components, with $\Delta S_{\text{Mix}} = -N k_B \sum_\alpha (N_\alpha/N) \ln(V_\alpha/V)$.

Fig. 4.6 A mixing entropy results from removing the partition separating two gases.

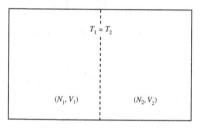

The Gibbs paradox is related to what happens when the two gases, initially on the two sides of the partition, are identical with the same density, $n = N_1/V_1 = N_2/V_2$. Since removing or inserting the partition does not change the state of the system, there should be no entropy of mixing, while Eq. (4.45) does predict such a change. For the resolution of this paradox, note that while after removing and reinserting the partition, the macroscopic system does return to its initial configuration, the actual particles that occupy the two chambers are not the same. But as the particles are by assumption *identical*, these configurations cannot be distinguished. In other words, while the exchange of *distinct* particles leads to two configurations

$$\frac{\bullet \mid \circ}{A \mid B} \quad \text{and} \quad \frac{\circ \mid \bullet}{A \mid B},$$

a similar exchange has no effect on *identical* particles, as in

$$\frac{\bullet \mid \bullet}{A \mid B} \quad \text{and} \quad \frac{\bullet \mid \bullet}{A \mid B}.$$

Therefore, we have over-counted the phase space associated with N identical particles by the number of possible permutations. As there are $N!$ permutations leading to indistinguishable microstates, Eq. (4.32) should be corrected to

$$\Omega(N, E, V) = \frac{V^N}{N!} \frac{2\pi^{3N/2}}{(3N/2 - 1)!} (2mE)^{(3N-1)/2} \Delta_R, \tag{4.46}$$

resulting in a modified entropy,

$$S = k_B \ln \Omega = k_B[N \ln V - N \ln N + N \ln e] + N k_B \sigma = N k_B \left[\ln \frac{eV}{N} + \sigma \right]. \quad (4.47)$$

As the argument of the logarithm has changed from V to V/N, the final expression is now properly extensive. The mixing entropies can be recalculated using Eq. (4.47). For the mixing of distinct gases,

$$\begin{aligned}
\Delta S_{\text{Mix}} = S_f - S_i &= N_1 k_B \ln \frac{V}{N_1} + N_2 k_B \ln \frac{V}{N_2} - N_1 k_B \ln \frac{V_1}{N_1} - N_2 k_B \ln \frac{V_2}{N_2} \\
&= N_1 k_B \ln \left(\frac{V}{N_1} \cdot \frac{N_1}{V_1} \right) + N_2 k_B \ln \left(\frac{V}{N_2} \cdot \frac{N_2}{V_2} \right) \qquad (4.48) \\
&= -N k_B \left[\frac{N_1}{N} \ln \frac{V_1}{V} + \frac{N_2}{N} \ln \frac{V_2}{V} \right],
\end{aligned}$$

exactly as obtained before in Eq. (4.45). For the "mixing" of two identical gases, with $N_1/V_1 = N_2/V_2 = (N_1 + N_2)/(V_1 + V_2)$,

$$\Delta S_{\text{Mix}} = S_f - S_i = (N_1 + N_2) k_B \ln \frac{V_1 + V_2}{N_1 + N_2} - N_1 k_B \ln \frac{V_1}{N_1} - N_2 k_B \ln \frac{V_2}{N_2} = 0. \quad (4.49)$$

Note that after taking the permutations of identical particles into account, the available coordinate volume in the final state is $V^{N_1+N_2}/N_1!N_2!$ for distinct particles, and $V^{N_1+N_2}/(N_1 + N_2)!$ for identical particles.

Additional comments on the microcanonical entropy:

1. In the example of two-level impurities in a solid matrix (Section 4.3), there is no need for the additional factor of $N!$, as the defects can be distinguished by their locations.
2. The corrected formula for the ideal gas entropy in Eq. (4.47) does not affect the computations of energy and pressure in Eqs. (4.34) and (4.35). It is essential to obtaining an intensive chemical potential,

$$\frac{\mu}{T} = -\frac{\partial S}{\partial N}\bigg|_{E,V} = -\frac{S}{N} + \frac{5}{2} k_B = k_B \ln \left[\frac{V}{N} \left(\frac{4\pi m E}{3N} \right)^{3/2} \right]. \quad (4.50)$$

3. The above treatment of identical particles is somewhat artificial. This is because the concept of identical particles does not easily fit within the framework of classical mechanics. To implement the Hamiltonian equations of motion on a computer, one has to keep track of the coordinates of the N particles. The computer will have no difficulty in distinguishing exchanged particles. The indistinguishability of their phase spaces is in a sense an additional postulate of classical statistical mechanics. This problem is elegantly resolved within the framework of quantum statistical mechanics. Description of identical particles in quantum mechanics requires proper symmetrization of the wave function. The corresponding quantum microstates naturally yield the $N!$ factor, as will be shown later on.

4. Yet another difficulty with the expression (4.47), resolved in quantum statistical mechanics, is the arbitrary constant that appears in changing the units of measurement for q and p. The volume of phase space involves products pq, of coordinates and conjugate momenta, and hence has dimensions of (action)^N. Quantum mechanics provides the appropriate measure of action in Planck's constant h. Anticipating these quantum results, we shall henceforth set the measure of phase space for identical particles to

$$\mathrm{d}\Gamma_N = \frac{1}{h^{3N}N!} \prod_{i=1}^{N} \mathrm{d}^3\vec{q}_i \mathrm{d}^3\vec{p}_i. \tag{4.51}$$

4.6 The canonical ensemble

In the microcanonical ensemble, the energy E of a large macroscopic system is precisely specified, and its equilibrium temperature T emerges as a consequence (Eq. (4.7)). However, from a thermodynamic perspective, E and T are both functions of state and on the same footing. It is possible to construct a statistical mechanical formulation in which the temperature of the system is specified and its internal energy is then deduced. This is achieved in the *canonical ensemble* where the macrostates, specified by $M \equiv (T, \mathbf{x})$, allow the input of heat into the system, but no external work. The system S is maintained at a constant temperature through contact with a reservoir R. The *reservoir* is another macroscopic system that is sufficiently large so that its temperature is not changed due to interactions with S. To find the probabilities $p_{(T,\mathbf{x})}(\mu)$ of the various microstates of S, note that the combined system R $\oplus$ S belongs to a microcanonical ensemble of energy $E_{\mathrm{Tot}} \gg E_S$. As in Eq. (4.3), the joint probability of microstates $(\mu_S \otimes \mu_R)$ is

$$p(\mu_S \otimes \mu_R) = \frac{1}{\Omega_{S\oplus R}(E_{\mathrm{Tot}})} \cdot \begin{cases} 1 & \text{for } \mathcal{H}_S(\mu_S) + \mathcal{H}_R(\mu_R) = E_{\mathrm{Tot}} \\ 0 & \text{otherwise} \end{cases}. \tag{4.52}$$

Fig. 4.7 The system S can be maintained at a temperature T through heat exchanges with the reservoir R.

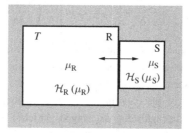

The unconditional probability for microstates of S is now obtained from

$$p(\mu_S) = \sum_{\{\mu_R\}} p(\mu_S \otimes \mu_R). \tag{4.53}$$

Once μ_S is specified, the above sum is restricted to microstates of the reservoir with energy $E_{Tot} - \mathcal{H}_S(\mu_S)$. The number of such states is related to the entropy of the reservoir, and leads to

$$p(\mu_S) = \frac{\Omega_R\left(E_{Tot} - \mathcal{H}_S(\mu_S)\right)}{\Omega_{S \oplus R}(E_{Tot})} \propto \exp\left[\frac{1}{k_B} S_R\left(E_{Tot} - \mathcal{H}_S(\mu_S)\right)\right]. \tag{4.54}$$

Since, by assumption, the energy of the system is insignificant compared with that of the reservoir,

$$S_R\left(E_{Tot} - \mathcal{H}_S(\mu_S)\right) \approx S_R(E_{Tot}) - \mathcal{H}_S(\mu_S)\frac{\partial S_R}{\partial E_R} = S_R(E_{Tot}) - \frac{\mathcal{H}_S(\mu_S)}{T}. \tag{4.55}$$

Dropping the subscript S, the normalized probabilities are given by

$$p_{(T,\mathbf{x})}(\mu) = \frac{e^{-\beta\mathcal{H}(\mu)}}{Z(T,\mathbf{x})}. \tag{4.56}$$

The normalization

$$Z(T, \mathbf{x}) = \sum_{\{\mu\}} e^{-\beta\mathcal{H}(\mu)} \tag{4.57}$$

is known as the *partition function*, and $\beta \equiv 1/k_B T$. (Note that probabilities similar to Eq. (4.56) were already obtained in Eqs. (4.25) and (4.39), when considering a portion of the system in equilibrium with the rest of it.)

Is the internal energy E of the system S well defined? Unlike in a microcanonical ensemble, the energy of a system exchanging heat with a reservoir is a random variable. Its probability distribution $p(\mathcal{E})$ is obtained by changing variables from μ to $\mathcal{H}(\mu)$ in $p(\mu)$, resulting in

$$p(\mathcal{E}) = \sum_{\{\mu\}} p(\mu)\,\delta\left(\mathcal{H}(\mu) - \mathcal{E}\right) = \frac{e^{-\beta\mathcal{E}}}{Z}\sum_{\{\mu\}}\delta\left(\mathcal{H}(\mu) - \mathcal{E}\right). \tag{4.58}$$

Since the restricted sum is just the number $\Omega(\mathcal{E})$ of microstates of appropriate energy,

$$p(\mathcal{E}) = \frac{\Omega(\mathcal{E})e^{-\beta\mathcal{E}}}{Z} = \frac{1}{Z}\exp\left[\frac{S(\mathcal{E})}{k_B} - \frac{\mathcal{E}}{k_B T}\right] = \frac{1}{Z}\exp\left[-\frac{F(\mathcal{E})}{k_B T}\right], \tag{4.59}$$

where we have set $F = \mathcal{E} - TS(\mathcal{E})$, in anticipation of its relation to the *Helmholtz free energy*. The probability $p(\mathcal{E})$ is sharply peaked at a *most probable energy* E^*, which minimizes $F(\mathcal{E})$. Using the result in Section 3.6 for sums over exponentials,

$$Z = \sum_{\{\mu\}} e^{-\beta\mathcal{H}(\mu)} = \sum_{\mathcal{E}} e^{-\beta F(\mathcal{E})} \approx e^{-\beta F(E^*)}. \tag{4.60}$$

The *average* energy computed from the distribution in Eq. (4.59) is

$$\langle\mathcal{H}\rangle = \sum_{\mu}\mathcal{H}(\mu)\frac{e^{-\beta\mathcal{H}(\mu)}}{Z} = -\frac{1}{Z}\frac{\partial}{\partial\beta}\sum_{\mu}e^{-\beta\mathcal{H}} = -\frac{\partial\ln Z}{\partial\beta}. \tag{4.61}$$

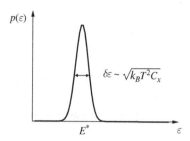

In thermodynamics, a similar expression was encountered for the energy
(Eq. (1.37)),

$$E = F + TS = F - T \left.\frac{\partial F}{\partial T}\right|_{\mathbf{x}} = -T^2 \frac{\partial}{\partial T}\left(\frac{F}{T}\right) = \frac{\partial(\beta F)}{\partial \beta}. \tag{4.62}$$

Equations (4.60) and (4.61) both suggest identifying

$$F(T, \mathbf{x}) = -k_B T \ln Z(T, \mathbf{x}). \tag{4.63}$$

However, note that Eq. (4.60) refers to the most likely energy, while the average
energy appears in Eq. (4.61). How close are these two values of the energy? We
can get an idea of the width of the probability distribution $p(\mathcal{E})$, by computing
the variance $\langle \mathcal{H}^2 \rangle_c$. This is most easily accomplished by noting that $Z(\beta)$ is
related to the characteristic function for $\mathcal{H}$ (with β replacing ik) and

$$-\frac{\partial Z}{\partial \beta} = \sum_\mu \mathcal{H} e^{-\beta \mathcal{H}}, \quad \text{and} \quad \frac{\partial^2 Z}{\partial \beta^2} = \sum_\mu \mathcal{H}^2 e^{-\beta \mathcal{H}}. \tag{4.64}$$

Cumulants of $\mathcal{H}$ are generated by $\ln Z(\beta)$,

$$\langle \mathcal{H} \rangle_c = \frac{1}{Z} \sum_\mu \mathcal{H} e^{-\beta \mathcal{H}} = -\frac{1}{Z}\frac{\partial Z}{\partial \beta} = -\frac{\partial \ln Z}{\partial \beta}, \tag{4.65}$$

and

$$\langle \mathcal{H}^2 \rangle_c = \langle \mathcal{H}^2 \rangle - \langle \mathcal{H} \rangle^2 = \frac{1}{Z} \sum_\mu \mathcal{H}^2 e^{-\beta \mathcal{H}} - \frac{1}{Z^2}\left(\sum_\mu \mathcal{H} e^{-\beta \mathcal{H}}\right)^2 = \frac{\partial^2 \ln Z}{\partial \beta^2} = -\frac{\partial \langle \mathcal{H} \rangle}{\partial \beta}. \tag{4.66}$$

More generally, the nth cumulant of $\mathcal{H}$ is given by

$$\langle \mathcal{H}^n \rangle_c = (-1)^n \frac{\partial^n \ln Z}{\partial \beta^n}. \tag{4.67}$$

From Eq. (4.66),

$$\langle \mathcal{H}^2 \rangle_c = -\frac{\partial \langle \mathcal{H} \rangle}{\partial (1/k_B T)} = k_B T^2 \left.\frac{\partial \langle \mathcal{H} \rangle}{\partial T}\right|_{\mathbf{x}}, \quad \Rightarrow \quad \langle \mathcal{H}^2 \rangle_c = k_B T^2 C_{\mathbf{x}}, \tag{4.68}$$

where we have identified the heat capacity with the thermal derivative of the
average energy $\langle \mathcal{H} \rangle$. Equation (4.68) shows that it is justified to treat the mean
and most likely energies interchangeably, since the width of the distribution

Table 4.1 *Comparison of canonical and microcanonical ensembles*

Ensemble	Macrostate	$p(\mu)$	Normalization
Microcanonical	$(E, \mathbf{x})$	$\delta_\Delta\big(\mathcal{H}(\mu) - E\big)/\Omega$	$S(E, \mathbf{x}) = k_B \ln \Omega$
Canonical	$(T, \mathbf{x})$	$\exp\big(-\beta\mathcal{H}(\mu)\big)/Z$	$F(T, \mathbf{x}) = -k_B T \ln Z$

$p(\mathcal{E})$ only grows as $\sqrt{\langle\mathcal{H}^2\rangle_c} \propto N^{1/2}$. The relative error, $\sqrt{\langle\mathcal{H}^2\rangle_c}/\langle\mathcal{H}\rangle_c$, vanishes in the thermodynamic limit as $1/\sqrt{N}$. (In fact, Eq. (4.67) shows that all cumulants of $\mathcal{H}$ are proportional to N.) The PDF for energy in a canonical ensemble can thus be approximated by

$$p(\mathcal{E}) = \frac{1}{Z} e^{-\beta F(\mathcal{E})} \approx \exp\left(-\frac{(\mathcal{E} - \langle\mathcal{H}\rangle)^2}{2k_B T^2 C_{\mathbf{x}}}\right) \frac{1}{\sqrt{2\pi k_B T^2 C_{\mathbf{x}}}}. \tag{4.69}$$

The above distribution is sufficiently sharp to make the internal energy in a canonical ensemble unambiguous in the $N \to \infty$ limit. Some care is necessary if the heat capacity $C_{\mathbf{x}}$ is divergent, as is the case in some continuous phase transitions.

The canonical probabilities in Eq. (4.56) are unbiased estimates obtained (as in Section 2.7) by constraining the *average energy*. The entropy of the canonical ensemble can also be calculated directly from Eq. (4.56) (using Eq. (2.68)) as

$$S = -k_B \langle \ln p(\mu)\rangle = -k_B \langle(-\beta\mathcal{H} - \ln Z)\rangle = \frac{E - F}{T}, \tag{4.70}$$

again using the identification of $\ln Z$ with the free energy from Eq. (4.63). For any finite system, the canonical and microcanonical properties are distinct. However, in the so-called *thermodynamic limit* of $N \to \infty$, the canonical energy probability is so sharply peaked around the average energy that the ensemble becomes essentially indistinguishable from the microcanonical ensemble at that energy. Table 4.1 compares the prescriptions used in the two ensembles.

4.7 Canonical examples

The two examples of Sections 4.3 and 4.4 are now re-examined in the canonical ensemble.

(1) *Two-level systems*: the N impurities are described by a macrostate $M \equiv (T, N)$. Subject to the Hamiltonian $\mathcal{H} = \epsilon \sum_{i=1}^{N} n_i$, the canonical probabilities of the microstates $\mu \equiv \{n_i\}$ are given by

$$p(\{n_i\}) = \frac{1}{Z} \exp\left[-\beta\epsilon \sum_{i=1}^{N} n_i\right]. \tag{4.71}$$

From the partition function,

$$Z(T,N) = \sum_{\{n_i\}} \exp\left[-\beta\epsilon \sum_{i=1}^{N} n_i\right] = \left(\sum_{n_1=0}^{1} e^{-\beta\epsilon n_1}\right) \cdots \left(\sum_{n_N=0}^{1} e^{-\beta\epsilon n_N}\right)$$

$$= \left(1 + e^{-\beta\epsilon}\right)^N,$$

(4.72)

we obtain the free energy

$$F(T,N) = -k_B T \ln Z = -Nk_B T \ln\left[1 + e^{-\epsilon/(k_B T)}\right].$$

(4.73)

The entropy is now given by

$$S = -\left.\frac{\partial F}{\partial T}\right|_N = \underbrace{Nk_B \ln\left[1 + e^{-\epsilon/(k_B T)}\right]}_{-F/T} + Nk_B T \left(\frac{\epsilon}{k_B T^2}\right) \frac{e^{-\epsilon/(k_B T)}}{1 + e^{-\epsilon/(k_B T)}}.$$

(4.74)

The internal energy,

$$E = F + TS = \frac{N\epsilon}{1 + e^{\epsilon/(k_B T)}},$$

(4.75)

can also be obtained from

$$E = -\frac{\partial \ln Z}{\partial \beta} = \frac{N\epsilon e^{-\beta\epsilon}}{1 + e^{-\beta\epsilon}}.$$

(4.76)

Since the joint probability in Eq. (4.71) is in the form of a product, $p = \prod_i p_i$, the excitations of different impurities are *independent* of each other, with the unconditional probabilities

$$p_i(n_i) = \frac{e^{-\beta\epsilon n_i}}{1 + e^{-\beta\epsilon}}.$$

(4.77)

This result coincides with Eqs. (4.25), obtained through a more elaborate analysis in the microcanonical ensemble. As expected, in the large N limit, the canonical and microcanonical ensembles describe exactly the same physics, both at the macroscopic and microscopic levels.

(2) *The ideal gas*: for the canonical macrostate $M \equiv (T, V, N)$, the joint PDF for the microstates $\mu \equiv \{\vec{p}_i, \vec{q}_i\}$ is

$$p(\{\vec{p}_i, \vec{q}_i\}) = \frac{1}{Z} \exp\left[-\beta \sum_{i=1}^{N} \frac{p_i^2}{2m}\right] \cdot \begin{cases} 1 & \text{for } \{\vec{q}_i\} \in \text{box} \\ 0 & \text{otherwise} \end{cases}.$$

(4.78)

Including the modifications to the phase space of identical particles in Eq. (4.51), the dimensionless partition function is computed as

$$Z(T,V,N) = \int \frac{1}{N!} \prod_{i=1}^{N} \frac{d^3\vec{q}_i d^3\vec{p}_i}{h^3} \exp\left[-\beta \sum_{i=1}^{N} \frac{p_i^2}{2m}\right]$$

$$= \frac{V^N}{N!} \left(\frac{2\pi m k_B T}{h^2}\right)^{3N/2} = \frac{1}{N!} \left(\frac{V}{\lambda(T)^3}\right)^N,$$

(4.79)

where

$$\lambda(T) = \frac{h}{\sqrt{2\pi m k_B T}}$$

(4.80)

is a characteristic length associated with the action h. It shall be demonstrated later on that this length scale controls the onset of quantum mechanical effects in an ideal gas.

The free energy is given by

$$F = -k_B T \ln Z = -N k_B T \ln V + N k_B T \ln N - N k_B T - \frac{3N}{2} k_B T \ln \left(\frac{2\pi m k_B T}{h^2} \right)$$

$$= -N k_B T \left[\ln \left(\frac{Ve}{N} \right) + \frac{3}{2} \ln \left(\frac{2\pi m k_B T}{h^2} \right) \right]. \qquad (4.81)$$

Various thermodynamic properties of the ideal gas can now be obtained from $dF = -SdT - PdV + \mu dN$. For example, from the entropy

$$-S = \left. \frac{\partial F}{\partial T} \right|_{V,N} = -N k_B \left[\ln \frac{Ve}{N} + \frac{3}{2} \ln \left(\frac{2\pi m k_B T}{h^2} \right) \right] - N k_B T \frac{3}{2T} = \frac{F - E}{T}, \qquad (4.82)$$

we obtain the internal energy $E = 3N k_B T/2$. The equation of state is obtained from

$$P = - \left. \frac{\partial F}{\partial V} \right|_{T,N} = \frac{N k_B T}{V}, \qquad \Longrightarrow \qquad PV = N k_B T, \qquad (4.83)$$

and the chemical potential is given by

$$\mu = \left. \frac{\partial F}{\partial N} \right|_{T,V} = \frac{F}{N} + k_B T = \frac{E - TS + PV}{N} = k_B T \ln \left(n \lambda^3 \right). \qquad (4.84)$$

Also, according to Eq. (4.78), the momenta of the N particles are taken from independent Maxwell–Boltzmann distributions, consistent with Eq. (4.39).

4.8 The Gibbs canonical ensemble

We can also define a generalized canonical ensemble in which the internal energy changes by the addition of both heat and work. The macrostates $M \equiv (T, \mathbf{J})$ are specified in terms of the external temperature and forces acting on the system; the thermodynamic coordinates $\mathbf{x}$ appear as additional random variables. The system is maintained at constant force through external elements (e.g., pistons or magnets). Including the work done against the forces, the energy of the combined system that includes these elements is $\mathcal{H} - \mathbf{J} \cdot \mathbf{x}$. Note that while the work done *on the system* is $+\mathbf{J} \cdot \mathbf{x}$, the energy change associated with the external elements with coordinates $\mathbf{x}$ has the opposite sign. The microstates of this combined system occur with the (canonical) probabilities

$$p(\mu_S, \mathbf{x}) = \exp \left[-\beta \mathcal{H}(\mu_S) + \beta \mathbf{J} \cdot \mathbf{x} \right] / \mathcal{Z}(T, N, \mathbf{J}), \qquad (4.85)$$

with the *Gibbs partition function*

$$\mathcal{Z}(N, T, \mathbf{J}) = \sum_{\mu_S, \mathbf{x}} e^{\beta \mathbf{J} \cdot \mathbf{x} - \beta \mathcal{H}(\mu_S)}. \qquad (4.86)$$

(Note that we have explicitly included the particle number N to emphasize that there is no chemical work. Chemical work is considered in the grand canonical ensemble, which is discussed next.)

Fig. 4.9 A system in
contact with a reservoir
at temperature T, and
maintained at a fixed
force J.

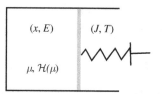

In this ensemble, the expectation value of the coordinates is obtained from

$$\langle \mathbf{x} \rangle = k_B T \frac{\partial \ln \mathcal{Z}}{\partial \mathbf{J}}, \tag{4.87}$$

which together with the thermodynamic identity $\mathbf{x} = -\partial G / \partial \mathbf{J}$, suggests the identification

$$G(N, T, \mathbf{J}) = -k_B T \ln \mathcal{Z}, \tag{4.88}$$

where $G = E - TS - \mathbf{x} \cdot \mathbf{J}$ is the *Gibbs free energy*. (The same conclusion can be reached by equating $\mathcal{Z}$ in Eq. (4.86) to the term that maximizes the probability with respect to $\mathbf{x}$.) The *enthalpy* $H \equiv E - \mathbf{x} \cdot \mathbf{J}$ is easily obtained in this ensemble from

$$-\frac{\partial \ln \mathcal{Z}}{\partial \beta} = \langle \mathcal{H} - \mathbf{x} \cdot \mathbf{J} \rangle = H. \tag{4.89}$$

Note that heat capacities at constant force (which include work done against the external forces) are obtained from the enthalpy as $C_\mathbf{J} = \partial H / \partial T$.

The following examples illustrate the use of the Gibbs canonical ensemble.

(1) *The ideal gas* in the *isobaric* ensemble is described by the macrostate $M \equiv (N, T, P)$. A microstate $\mu \equiv \{\vec{p}_i, \vec{q}_i\}$, with a volume V, occurs with the probability

$$p(\{\vec{p}_i, \vec{q}_i\}, V) = \frac{1}{\mathcal{Z}} \exp\left[-\beta \sum_{i=1}^{N} \frac{p_i^2}{2m} - \beta P V \right] \cdot \begin{cases} 1 & \text{for } \{\vec{q}_i\} \in \text{box of volume } V \\ 0 & \text{otherwise} \end{cases}. \tag{4.90}$$

The normalization factor is now

$$\mathcal{Z}(N, T, P) = \int_0^\infty \mathrm{d}V e^{-\beta PV} \int \frac{1}{N!} \prod_{i=1}^{N} \frac{\mathrm{d}^3 \vec{q}_i \mathrm{d}^3 \vec{p}_i}{h^3} \exp\left[-\beta \sum_{i=1}^{N} \frac{p_i^2}{2m} \right]$$

$$= \int_0^\infty \mathrm{d}V V^N e^{-\beta PV} \frac{1}{N! \lambda(T)^{3N}} = \frac{1}{(\beta P)^{N+1} \lambda(T)^{3N}}. \tag{4.91}$$

Ignoring non-extensive contributions, the Gibbs free energy is given by

$$G = -k_B T \ln \mathcal{Z} \approx N k_B T \left[\ln P - \frac{5}{2} \ln (k_B T) + \frac{3}{2} \ln \left(\frac{h^2}{2\pi m} \right) \right]. \tag{4.92}$$

Starting from $\mathrm{d}G = -S \mathrm{d}T + V \mathrm{d}P + \mu \mathrm{d}N$, the volume of the gas is obtained as

$$V = \left. \frac{\partial G}{\partial P} \right|_{T,N} = \frac{N k_B T}{P}, \qquad \Longrightarrow \quad PV = N k_B T. \tag{4.93}$$

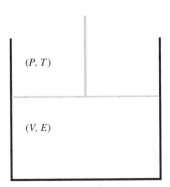

Fig. 4.10 In the isobaric
ensemble, the gas in the
piston is maintained at a
pressure P.

The *enthalpy* $H = \langle E + PV \rangle$ is easily calculated from

$$H = -\frac{\partial \ln \mathcal{Z}}{\partial \beta} = \frac{5}{2} N k_B T,$$

from which we get $C_P = dH/dT = 5/2 N k_B$.

(2) *Spins* in an external magnetic field $\vec{B}$ provide a common example for usage of the Gibbs canonical ensemble. Adding the work done against the magnetic field to the internal Hamiltonian $\mathcal{H}$ results in the Gibbs partition function

$$\mathcal{Z}(N, T, B) = \text{tr} \left[\exp \left(-\beta \mathcal{H} + \beta \vec{B} \cdot \vec{M} \right) \right],$$

where $\vec{M}$ is the net *magnetization*. The symbol "tr" is used to indicate the sum over all spin degrees of freedom, which in a quantum mechanical formulation are restricted to discrete values. The simplest case is spin of 1/2, with two possible projections of the spin along the magnetic field. A microstate of N spins is now described by the set of *Ising* variables $\{\sigma_i = \pm 1\}$. The corresponding magnetization along the field direction is given by $M = \mu_0 \sum_{i=1}^{N} \sigma_i$, where μ_0 is a microscopic magnetic moment. Assuming that there are no interactions between spins ($\mathcal{H} = 0$), the probability of a microstate is

$$p(\{\sigma_i\}) = \frac{1}{\mathcal{Z}} \exp \left[\beta B \mu_0 \sum_{i=1}^{N} \sigma_i \right]. \tag{4.94}$$

Clearly, this is closely related to the example of two-level systems discussed in the canonical ensemble, and we can easily obtain the Gibbs partition function

$$\mathcal{Z}(N, T, B) = [2 \cosh(\beta \mu_0 B)]^N, \tag{4.95}$$

and the Gibbs free energy

$$G = -k_B T \ln \mathcal{Z} = -N k_B T \ln [2 \cosh(\beta \mu_0 B)]. \tag{4.96}$$

The average magnetization is given by

$$M = -\frac{\partial G}{\partial B} = N \mu_0 \tanh(\beta \mu_0 B). \tag{4.97}$$

Expanding Eq. (4.97) for small B results in the well-known Curie law for magnetic *susceptibility* of non-interacting spins,

$$\chi(T) = \left.\frac{\partial M}{\partial B}\right|_{B=0} = \frac{N\mu_0^2}{k_B T}.$$ (4.98)

The enthalpy is simply $H = \langle \mathcal{H} - BM \rangle = -BM$, and $C_B = -B\partial M/\partial T$.

4.9 The grand canonical ensemble

The previous sections demonstrate that while the canonical and microcanonical ensembles are completely equivalent in the thermodynamic limit, it is frequently easier to perform statistical mechanical computations in the canonical framework. Sometimes it is more convenient to allow chemical work (by fixing the chemical potential μ, rather than at a fixed number of particles), but no mechanical work. The resulting macrostates $M \equiv (T, \mu, \mathbf{x})$ are described by the *grand canonical* ensemble. The corresponding microstates μ_S contain an indefinite number of particles $N(\mu_S)$. As in the case of the canonical ensemble, the system S can be maintained at a constant chemical potential through contact with a reservoir R, at temperature T and chemical potential μ. The probability distribution for the microstates of S is obtained by summing over all states of the reservoir, as in Eq. (4.53), and is given by

$$p(\mu_S) = \exp\left[\beta\mu N(\mu_S) - \beta\mathcal{H}(\mu_S)\right]/\mathcal{Q}.$$ (4.99)

The normalization factor is the *grand partition function*,

$$\mathcal{Q}(T, \mu, \mathbf{x}) = \sum_{\mu_S} e^{\beta\mu N(\mu_S) - \beta\mathcal{H}(\mu_S)}.$$ (4.100)

Fig. 4.11 A system S, in contact with a reservoir R, of temperature T and chemical potential μ.

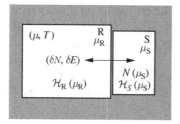

We can reorganize the above summation by grouping together all microstates with a given number of particles, that is,

$$\mathcal{Q}(T, \mu, \mathbf{x}) = \sum_{N=0}^{\infty} e^{\beta\mu N} \sum_{(\mu_S|N)} e^{-\beta\mathcal{H}_N(\mu_S)}.$$ (4.101)

The restricted sums in Eq. (4.101) are just the N-particle partition functions. As each term in $\mathcal{Q}$ is the total weight of all microstates of N particles, the unconditional probability of finding N particles in the system is

$$p(N) = \frac{e^{\beta\mu N} Z(T, N, \mathbf{x})}{\mathcal{Q}(T, \mu, \mathbf{x})}. \tag{4.102}$$

The average number of particles in the system is

$$\langle N \rangle = \frac{1}{\mathcal{Q}} \frac{\partial}{\partial(\beta\mu)} \mathcal{Q} = \frac{\partial}{\partial(\beta\mu)} \ln \mathcal{Q}, \tag{4.103}$$

while the number fluctuations are related to the variance

$$\langle N^2 \rangle_C = \langle N^2 \rangle - \langle N \rangle^2 = \frac{1}{\mathcal{Q}} \frac{\partial^2}{\partial(\beta\mu)^2} \ln \mathcal{Q} - \left(\frac{\partial}{\partial(\beta\mu)} \ln \mathcal{Q} \right)^2 = \frac{\partial^2}{\partial(\beta\mu)^2} \ln \mathcal{Q} = \frac{\partial \langle N \rangle}{\partial(\beta\mu)}. \tag{4.104}$$

The variance is thus proportional to N, and the *relative number fluctuations* vanish in the thermodynamic limit, establishing the equivalence of this ensemble to the previous ones.

Because of the sharpness of the distribution for N, the sum in Eq. (4.101) can be approximated by its largest term at $N = N^* \approx \langle N \rangle$, that is,

$$\mathcal{Q}(T, \mu, \mathbf{x}) = \lim_{N \to \infty} \sum_{N=0}^{\infty} e^{\beta\mu N} Z(T, N, \mathbf{x}) = e^{\beta\mu N^*} Z(T, N^*, \mathbf{x}) = e^{\beta\mu N^* - \beta F}$$
$$= e^{-\beta(-\mu N^* + E - TS)} = e^{-\beta\mathcal{G}}, \tag{4.105}$$

where

$$\mathcal{G}(T, \mu, \mathbf{x}) = E - TS - \mu N = -k_B T \ln \mathcal{Q} \tag{4.106}$$

is the *grand potential*. Thermodynamic information is obtained by using $d\mathcal{G} = -SdT - Nd\mu + \mathbf{J} \cdot d\mathbf{x}$, as

$$-S = \left. \frac{\partial \mathcal{G}}{\partial T} \right|_{\mu, \mathbf{x}}, \qquad N = -\left. \frac{\partial \mathcal{G}}{\partial \mu} \right|_{T, \mathbf{x}}, \qquad J_i = \left. \frac{\partial \mathcal{G}}{\partial x_i} \right|_{T, \mu}. \tag{4.107}$$

As a final example, we compute the properties of the ideal gas of non-interacting particles in the grand canonical ensemble. The macrostate is $M \equiv (T, \mu, V)$, and the corresponding microstates $\{\vec{p}_1, \vec{q}_1, \vec{p}_2, \vec{q}_2, \cdots\}$ have indefinite particle number. The grand partition function is given by

$$\mathcal{Q}(T, \mu, V) = \sum_{N=0}^{\infty} e^{\beta\mu N} \frac{1}{N!} \int \left(\prod_{i=1}^{N} \frac{d^3\vec{q}_i d^3\vec{p}_i}{h^3} \right) \exp\left[-\beta \sum_i \frac{p_i^2}{2m} \right]$$
$$= \sum_{N=0}^{\infty} \frac{e^{\beta\mu N}}{N!} \left(\frac{V}{\lambda^3} \right)^N \qquad \left(\text{with } \lambda = \frac{h}{\sqrt{2\pi m k_B T}} \right) \tag{4.108}$$
$$= \exp\left[e^{\beta\mu} \frac{V}{\lambda^3} \right],$$

and the grand potential is

$$\mathcal{G}(T, \mu, V) = -k_B T \ln \mathcal{Q} = -k_B T e^{\beta\mu} \frac{V}{\lambda^3}. \tag{4.109}$$

But, since $\mathcal{G} = E - TS - \mu N = -PV$, the gas pressure can be obtained directly as

$$P = -\frac{\mathcal{G}}{V} = -\frac{\partial \mathcal{G}}{\partial V}\bigg|_{\mu, T} = k_B T \frac{e^{\beta\mu}}{\lambda^3}. \tag{4.110}$$

The particle number and the chemical potential are related by

$$N = -\frac{\partial \mathcal{G}}{\partial \mu}\bigg|_{T, V} = \frac{e^{\beta\mu} V}{\lambda^3}. \tag{4.111}$$

The equation of state is obtained by comparing Eqs. (4.110) and (4.111), as $P = k_B T N / V$. Finally, the chemical potential is given by

$$\mu = k_B T \ln\left(\lambda^3 \frac{N}{V}\right) = k_B T \ln\left(\frac{P\lambda^3}{k_B T}\right). \tag{4.112}$$

Problems for chapter 4

1. *Classical harmonic oscillators*: consider N harmonic oscillators with coordinates and momenta $\{q_i, p_i\}$, and subject to a Hamiltonian

$$\mathcal{H}(\{q_i, p_i\}) = \sum_{i=1}^{N} \left[\frac{p_i^2}{2m} + \frac{m\omega^2 q_i^2}{2}\right].$$

   (a) Calculate the entropy S, as a function of the total energy E.
       (*Hint.* By appropriate change of scale, the surface of constant energy can be deformed into a sphere. You may then ignore the difference between the surface area and volume for $N \gg 1$. A more elegant method is to implement this deformation through a canonical transformation.)
   (b) Calculate the energy E, and heat capacity C, as functions of temperature T, and N.
   (c) Find the joint probability density $P(p, q)$ for a single oscillator. Hence calculate the mean kinetic energy, and mean potential energy for each oscillator.

$$*\,*\,*\,*\,*\,*\,*\,*$$

2. *Quantum harmonic oscillators*: consider N independent quantum oscillators subject to a Hamiltonian

$$\mathcal{H}(\{n_i\}) = \sum_{i=1}^{N} \hbar\omega\left(n_i + \frac{1}{2}\right),$$

   where $n_i = 0, 1, 2, \cdots$ is the quantum occupation number for the ith oscillator.

   (a) Calculate the entropy S, as a function of the total energy E.
       (*Hint.* $\Omega(E)$ can be regarded as the number of ways of rearranging $M = \sum_i n_i$ balls, and $N - 1$ partitions along a line.)

(b) Calculate the energy E, and heat capacity C, as functions of temperature T, and N.

(c) Find the probability $p(n)$ that a particular oscillator is in its nth quantum level.

(d) Comment on the difference between heat capacities for classical and quantum oscillators.

****

3. *Relativistic particles*: N *indistinguishable* relativistic particles move in *one dimension* subject to a Hamiltonian

$$\mathcal{H}\left(\{p_i, q_i\}\right) = \sum_{i=1}^{N} [c|p_i| + U(q_i)],$$

with $U(q_i) = 0$ for $0 \leq q_i \leq L$, and $U(q_i) = \infty$ otherwise. Consider a *microcanonical* ensemble of total energy E.

(a) Compute the contribution of the coordinates q_i to the available volume in phase space $\Omega(E, L, N)$.

(b) Compute the contribution of the momenta p_i to $\Omega(E, L, N)$.
 (*Hint.* The volume of the hyperpyramid defined by $\sum_{i=1}^{d} x_i \leq R$, and $x_i \geq 0$, in d dimensions is $R^d/d!$.)

(c) Compute the entropy $S(E, L, N)$.

(d) Calculate the one-dimensional pressure P.

(e) Obtain the heat capacities C_L and C_P.

(f) What is the probability $p(p_1)$ of finding a particle with momentum p_1?

****

4. *Hard sphere gas*: consider a gas of N hard spheres in a box. A single sphere excludes a volume ω around it, while its center of mass can explore a volume V (if the box is otherwise empty). There are no other interactions between the spheres, except for the constraints of hard-core exclusion.

(a) Calculate the entropy S, as a function of the total energy E.
 (*Hint.* $(V - a\omega)(V - (N-a)\omega) \approx (V - N\omega/2)^2$.)

(b) Calculate the equation of state of this gas.

(c) Show that the isothermal compressibility, $\kappa_T = -V^{-1} \, \partial V/\partial P|_T$, is always positive.

****

5. *Non-harmonic gas*: let us re-examine the generalized ideal gas introduced in the previous section, using statistical mechanics rather than kinetic theory. Consider a gas of N non-interacting atoms in a d-dimensional box of "volume" V, with a kinetic energy

$$\mathcal{H} = \sum_{i=1}^{N} A \left|\vec{p}_i\right|^s,$$

where $\vec{p}_i$ is the momentum of the ith particle.

(a) Calculate the classical partition function $Z(N, T)$ at a temperature T. (*You don't have to keep track of numerical constants in the integration.*)

(b) Calculate the pressure and the internal energy of this gas. (Note how the usual equipartition theorem is modified for non-quadratic degrees of freedom.)

(c) Now consider a diatomic gas of N molecules, each with energy

$$\mathcal{H}_i = A \left(\left| \vec{p}_i^{\,(1)} \right|^s + \left| \vec{p}_i^{\,(2)} \right|^s \right) + K \left| \vec{q}_i^{\,(1)} - \vec{q}_i^{\,(2)} \right|^t,$$

where the superscripts refer to the two particles in the molecule. (Note that this unrealistic potential allows the two atoms to occupy the same point.) Calculate the expectation value $\left\langle \left| \vec{q}_i^{\,(1)} - \vec{q}_i^{\,(2)} \right|^t \right\rangle$, at temperature T.

(d) Calculate the heat capacity ratio $\gamma = C_P/C_V$, for the above diatomic gas.

****

6. *Surfactant adsorption*: a dilute solution of surfactants can be regarded as an ideal three-dimensional gas. As surfactant molecules can reduce their energy by contact with air, a fraction of them migrate to the surface where they can be treated as a two-dimensional ideal gas. Surfactants are similarly adsorbed by other porous media such as polymers and gels with an affinity for them.

(a) Consider an ideal gas of classical particles of mass m in d dimensions, moving in a *uniform* attractive potential of strength ε_d. By calculating the partition function, or otherwise, show that the chemical potential at a temperature T and particle density n_d is given by

$$\mu_d = -\varepsilon_d + k_B T \ln \left[n_d \lambda(T)^d \right], \quad \text{where} \quad \lambda(T) = \frac{h}{\sqrt{2\pi m k_B T}}.$$

(b) If a surfactant lowers its energy by ε_0 in moving from the solution to the surface, calculate the concentration of floating surfactants as a function of the solution concentration $n \, (= n_3)$, at a temperature T.

(c) Gels are formed by cross-linking linear polymers. It has been suggested that the porous gel should be regarded as *fractal*, and the surfactants adsorbed on its surface treated as a gas in d_f-dimensional space, with a non-integer d_f. Can this assertion be tested by comparing the relative adsorption of surfactants to a gel, and to the individual polymers (presumably one-dimensional) before cross-linking, as a function of temperature?

****

7. *Molecular adsorption*: N diatomic molecules are stuck on a metal surface of square symmetry. Each molecule can either lie flat on the surface, in which case it must be aligned to one of two directions, x and y, or it can stand up along the z direction. There is an energy cost of $\varepsilon > 0$ associated with a molecule standing up, and zero energy for molecules lying flat along x or y directions.

(a) How many microstates have the smallest value of energy? What is the largest microstate energy?

(b) For *microcanonical* macrostates of energy E, calculate the number of states $\Omega(E, N)$, and the entropy $S(E, N)$.

(c) Calculate the heat capacity $C(T)$ and sketch it.

(d) What is the probability that a specific molecule is standing up?

(e) What is the largest possible value of the internal energy at any positive temperature?

****

8. *Curie susceptibility*: consider N non-interacting quantized spins in a magnetic field $\vec{B} = B\hat{z}$, and at a temperature T. The work done by the field is given by BM_z, with a magnetization $M_z = \mu \sum_{i=1}^{N} m_i$. For each spin, m_i takes only the $2s+1$ values $-s, -s+1, \cdots, s-1, s$.

(a) Calculate the Gibbs partition function $\mathcal{Z}(T, B)$. (Note that the ensemble corresponding to the macrostate (T, B) includes magnetic work.)

(b) Calculate the Gibbs free energy $G(T, B)$, and show that for small B,

$$G(B) = G(0) - \frac{N\mu^2 s(s+1)B^2}{6k_B T} + \mathcal{O}(B^4).$$

(c) Calculate the zero field susceptibility $\chi = \partial M_z/\partial B|_{B=0}$, and show that it satisfies *Curie's law*

$$\chi = c/T.$$

(d) Show that $C_B - C_M = cB^2/T^2$, where C_B and C_M are heat capacities at constant B and M, respectively.

****

9. *Langmuir isotherms*: an ideal gas of particles is in contact with the surface of a catalyst.

(a) Show that the chemical potential of the gas particles is related to their temperature and pressure via $\mu = k_B T \left[\ln\left(P/T^{5/2}\right) + A_0\right]$, where A_0 is a constant.

(b) If there are $\mathcal{N}$ distinct adsorption sites on the surface, and each adsorbed particle gains an energy ϵ upon adsorption, calculate the grand partition function for the two-dimensional gas with a chemical potential μ.

(c) In equilibrium, the gas and surface particles are at the same temperature and chemical potential. Show that the fraction of occupied surface sites is then given by $f(T, P) = P/(P + P_0(T))$. Find $P_0(T)$.

(d) In the grand canonical ensemble, the particle number N is a random variable. Calculate its characteristic function $\langle \exp(-ikN) \rangle$ in terms of $\mathcal{Q}(\beta\mu)$, and hence show that

$$\langle N^m \rangle_c = -(k_B T)^{m-1} \left.\frac{\partial^m \mathcal{G}}{\partial \mu^m}\right|_T,$$

where $\mathcal{G}$ is the grand potential.

(e) Using the characteristic function, show that

$$\left\langle N^2 \right\rangle_c = k_B T \left. \frac{\partial \langle N \rangle}{\partial \mu} \right|_T.$$

(f) Show that fluctuations in the number of adsorbed particles satisfy

$$\frac{\left\langle N^2 \right\rangle_c}{\langle N \rangle_c^2} = \frac{1-f}{\mathcal{N}f}.$$

****

10. *Molecular oxygen* has a net magnetic spin $\vec{S}$ of unity, that is, S^z is quantized to -1, 0, or $+1$. The Hamiltonian for an ideal gas of N such molecules in a magnetic field $\vec{B} \parallel \hat{z}$ is

$$\mathcal{H} = \sum_{i=1}^{N} \left[\frac{\vec{p}_i^{\,2}}{2m} - \mu B S_i^z \right],$$

where $\{\vec{p}_i\}$ are the center of mass momenta of the molecules. The corresponding coordinates $\{\vec{q}_i\}$ are confined to a volume V. (Ignore all other degrees of freedom.)

(a) Treating $\{\vec{p}_i, \vec{q}_i\}$ classically, but the spin degrees of freedom as quantized, calculate the partition function, $\tilde{Z}(T, N, V, B)$.
(b) What are the probabilities for S_i^z of a specific molecule to take on values of -1, 0, $+1$ at a temperature T?
(c) Find the average magnetic dipole moment, $\langle M \rangle / V$, where $M = \mu \sum_{i=1}^{N} S_i^z$.
(d) Calculate the zero field susceptibility $\chi = \partial \langle M \rangle / \partial B |_{B=0}$.

****

11. *One-dimensional polymer*: consider a polymer formed by connecting N disc-shaped molecules into a one-dimensional chain. Each molecule can align along either its long axis (of length $2a$) or short axis (length a). The energy of the monomer aligned along its shorter axis is higher by ε, that is, the total energy is $\mathcal{H} = \varepsilon U$, where U is the number of monomers standing up.

(a) Calculate the partition function, $Z(T, N)$, of the polymer.
(b) Find the relative probabilities for a monomer to be aligned along its short or long axis.
(c) Calculate the average length, $\langle L(T, N) \rangle$, of the polymer.
(d) Obtain the variance, $\left\langle L(T, N)^2 \right\rangle_c$.
(e) What does the central limit theorem say about the probability distribution for the length $L(T, N)$?

****

12. *Polar rods*: consider rod-shaped molecules with moment of inertia I, and a dipole moment μ. The contribution of the rotational degrees of freedom to the Hamiltonian is given by

$$\mathcal{H}_{\text{rot.}} = \frac{1}{2I}\left(p_\theta^2 + \frac{p_\phi^2}{\sin^2\theta}\right) - \mu E \cos\theta,$$

where E is an external electric field. ($\phi \in [0, 2\pi]$, $\theta \in [0, \pi]$ are the azimuthal and polar angles, and p_ϕ, p_θ are their conjugate momenta.)

(a) Calculate the contribution of the rotational degrees of freedom of each dipole to the *classical* partition function.

(b) Obtain the mean polarization $P = \langle \mu \cos\theta \rangle$ of each dipole.

(c) Find the *zero-field* polarizability

$$\chi_T = \left.\frac{\partial P}{\partial E}\right|_{E=0}.$$

(d) Calculate the rotational energy per particle (at finite E), and comment on its high- and low-temperature limits.

(e) Sketch the rotational heat capacity per dipole.

****

5

Interacting particles

5.1 The cumulant expansion

The examples studied in the previous chapter involve non-interacting particles. It is precisely the lack of interactions that renders these problems exactly solvable. Interactions, however, are responsible for the wealth of interesting materials and phases observed in nature. We would thus like to understand the role of interactions amongst particles, and learn how to treat them in statistical mechanics. For a general Hamiltonian,

$$\mathcal{H}_N = \sum_{i=1}^{N} \frac{\vec{p}_i^{\,2}}{2m} + \mathcal{U}(\vec{q}_1, \cdots, \vec{q}_N), \tag{5.1}$$

the partition function can be written as

$$Z(T, V, N) = \frac{1}{N!} \int \prod_{i=1}^{N} \left(\frac{\mathrm{d}^3 \vec{p}_i \mathrm{d}^3 \vec{q}_i}{h^3} \right) \exp\left[-\beta \sum_i \frac{\vec{p}_i^{\,2}}{2m} \right] \exp\left[-\beta \mathcal{U}(\vec{q}_1, \cdots, \vec{q}_N) \right]$$

$$= Z_0(T, V, N) \left\langle \exp\left[-\beta \mathcal{U}(\vec{q}_1, \cdots, \vec{q}_N) \right] \right\rangle^0, \tag{5.2}$$

where $Z_0(T, V, N) = \left(V/\lambda^3 \right)^N / N!$ is the partition function of the *ideal gas* (Eq. (4.73)), and $\langle \mathcal{O} \rangle^0$ denotes the expectation value of $\mathcal{O}$ computed with the probability distribution of the non-interacting system. In terms of the cumulants of the random variable $\mathcal{U}$, Eq. (5.2) can be recast as

$$\ln Z = \ln Z_0 + \sum_{\ell=1}^{\infty} \frac{(-\beta)^\ell}{\ell!} \left\langle \mathcal{U}^\ell \right\rangle_c^0. \tag{5.3}$$

The cumulants are related to the moments by the relations in Section 2.2. Since $\mathcal{U}$ depends only on $\{\vec{q}_i\}$, which are uniformly *and independently* distributed within the box of volume V, the moments are given by

$$\left\langle \mathcal{U}^\ell \right\rangle^0 = \int \left(\prod_{i=1}^{N} \frac{\mathrm{d}^3 \vec{q}_i}{V} \right) \mathcal{U}(\vec{q}_1, \cdots, \vec{q}_N)^\ell. \tag{5.4}$$

Various expectation values can also be calculated perturbatively, from

$$\langle \mathcal{O} \rangle = \frac{1}{Z} \frac{1}{N!} \int \prod_{i=1}^{N} \left(\frac{\mathrm{d}^3 \vec{p}_i \mathrm{d}^3 \vec{q}_i}{h^3} \right) \exp\left[-\beta \sum_i \frac{\vec{p}_i^{\,2}}{2m} \right] \exp\left[-\beta \mathcal{U}(\vec{q}_1, \cdots, \vec{q}_N) \right] \times \mathcal{O}$$

$$= \frac{\langle \mathcal{O} \exp[-\beta \mathcal{U}] \rangle^0}{\langle \exp[-\beta \mathcal{U}] \rangle^0} = i \frac{\partial}{\partial k} \ln \langle \exp[-ik\mathcal{O} - \beta \mathcal{U}] \rangle^0 \Big|_{k=0} . \tag{5.5}$$

The final expectation value generates the joint cumulants of the random variables $\mathcal{O}$ and $\mathcal{U}$, as

$$\ln \langle \exp[-ik\mathcal{O} - \beta \mathcal{U}] \rangle^0 \equiv \sum_{\ell, \ell' = 1}^{\infty} \frac{(-ik)^{\ell'}}{\ell'!} \frac{(-\beta)^{\ell}}{\ell!} \left\langle \mathcal{O}^{\ell'} * \mathcal{U}^{\ell} \right\rangle_c^0 , \tag{5.6}$$

in terms of which[1]

$$\langle \mathcal{O} \rangle = \sum_{\ell=0}^{\infty} \frac{(-\beta)^{\ell}}{\ell!} \left\langle \mathcal{O} * \mathcal{U}^{\ell} \right\rangle_c^0 . \tag{5.7}$$

The simplest system for treating interactions is again the dilute gas. As discussed in chapter 3, for a weakly interacting gas we can specialize to

$$\mathcal{U}(\vec{q}_1, \cdots, \vec{q}_N) = \sum_{i<j} \mathcal{V}(\vec{q}_i - \vec{q}_j) , \tag{5.8}$$

where $\mathcal{V}(\vec{q}_i - \vec{q}_j)$ is a pairwise interaction between particles. The first correction in Eq. (5.3) is

$$\langle \mathcal{U} \rangle_c^0 = \sum_{i<j} \int \frac{\mathrm{d}^3 \vec{q}_i}{V} \frac{\mathrm{d}^3 \vec{q}_j}{V} \mathcal{V}(\vec{q}_i - \vec{q}_j)$$

$$= \frac{N(N-1)}{2V} \int \mathrm{d}^3 \vec{q} \, \mathcal{V}(\vec{q}) . \tag{5.9}$$

The final result is obtained by performing the integrals over the relative and center of mass coordinates of $\vec{q}_i$ and $\vec{q}_j$ separately. (Each of the $N(N-1)/2$ pairs makes an identical contribution.)

The second-order correction,

$$\langle \mathcal{U}^2 \rangle_c^0 = \sum_{i<j, \, k<l} \left[\langle \mathcal{V}(\vec{q}_i - \vec{q}_j) \mathcal{V}(\vec{q}_k - \vec{q}_l) \rangle^0 - \langle \mathcal{V}(\vec{q}_i - \vec{q}_j) \rangle^0 \langle \mathcal{V}(\vec{q}_k - \vec{q}_l) \rangle^0 \right] , \tag{5.10}$$

is the sum of $[N(N-1)/2]^2$ terms that can be grouped as follows:

(i) There is no contribution from terms in which the four indices $\{i, j, k, l\}$ are different. This is because the different $\{\vec{q}_i\}$ are independently distributed, and $\langle \mathcal{V}(\vec{q}_i - \vec{q}_j) \mathcal{V}(\vec{q}_k - \vec{q}_l) \rangle^0$ equals $\langle \mathcal{V}(\vec{q}_i - \vec{q}_j) \rangle^0 \langle \mathcal{V}(\vec{q}_k - \vec{q}_l) \rangle^0$.

(ii) There is one common index between the two pairs, for example, $\{(i, j), (i, l)\}$. By changing coordinates to $\vec{q}_{ij} = \vec{q}_i - \vec{q}_j$ and $\vec{q}_{il} = \vec{q}_i - \vec{q}_l$, it again follows that $\langle \mathcal{V}(\vec{q}_i - \vec{q}_j) \mathcal{V}(\vec{q}_i - \vec{q}_l) \rangle^0$ equals $\langle \mathcal{V}(\vec{q}_i - \vec{q}_j) \rangle^0 \langle \mathcal{V}(\vec{q}_i - \vec{q}_l) \rangle^0$. The vanishing of these

---

[1] While this compact notation is convenient, it may be confusing. The joint cumulants are defined through Eq. (5.6), and for example $\langle 1 * \mathcal{U}^{\ell} \rangle_c = 0$ ($\neq \langle \mathcal{U}^{\ell} \rangle_c$), while $\langle 1 * \mathcal{U}^0 \rangle_c = 1$!

terms is a consequence of the *translational symmetry* of the problem in the absence of an external potential.

(iii) In the remaining $N(N-1)/2$ terms the pairs are identical, resulting in

$$\langle \mathcal{U}^2 \rangle_c^0 = \frac{N(N-1)}{2} \left[\int \frac{d^3 \vec{q}}{V} \mathcal{V}(\vec{q})^2 - \left(\int \frac{d^3 \vec{q}}{V} \mathcal{V}(\vec{q}) \right)^2 \right]. \qquad (5.11)$$

The second term in the above equation is smaller by a factor of d^3/V, where d is a characteristic range for the potential $\mathcal{V}$. For any reasonable potential that decays with distance, this term vanishes in the thermodynamic limit.

Similar groupings occur for higher-order terms in this *cumulant expansion*. It is helpful to visualize the terms in the expansion diagrammatically as follows:

(a) For a term of order p, draw p pairs of points (representing $\vec{q}_i$ and $\vec{q}_j$) connected by bonds, representing the interaction $\mathcal{V}_{ij} \equiv \mathcal{V}(\vec{q}_i - \vec{q}_j)$. An overall factor of $1/p!$ accompanies such graphs. At the third order, we have

$$\frac{1}{3!} \quad \sum_{i<j,\, k<l,\, m<n} \quad \underset{i \quad j}{\bullet \cdots \bullet} \quad \underset{k \quad l}{\bullet \cdots \bullet} \quad \underset{m \quad n}{\bullet \cdots \bullet} \quad .$$

(b) By multiple selections of the same index i, two or more bonds can be joined together forming a diagram of interconnected points. There is a factor S_p associated with the number of ways of assigning labels 1 through N to the different points of the graph. Ignoring the differences between N, $N-1$, etc., a diagram with s points makes a contribution proportional to N^s. There is typically also division by a *symmetry factor* that takes into account the number of equivalent assignments. For example, the diagrams involving a pair of points, calculated in Eqs. (5.9) and (5.11), have a symmetry factor of 1/2. A potential pairing at the third order is

$$\frac{1}{3!} \quad \frac{N^5}{2 \times 2} \quad \underset{1 \quad 2}{\bullet \cdots \bullet} \quad \underset{3 \quad 4 \quad 5}{\bullet \cdots \bullet \cdots \bullet} \quad .$$

(c) Apart from these numerical prefactors, the contribution of a diagram is an integral over all the s coordinates $\vec{q}_i$, of products of corresponding $-\beta \mathcal{V}_{ij}$. If the graph has n_c *disconnected clusters*, integration over the center of mass coordinates of the clusters gives a factor of V^{n_c}. Thus, the contribution of the above diagram is

$$\frac{(-\beta)^3}{3!} \quad \frac{N^5}{4V^5} \quad V^2 \left(\int d^3 \vec{q}_{12} \mathcal{V}(\vec{q}_{12}) \right) \left(\int d^3 \vec{q}_{34} d^3 \vec{q}_{45} \mathcal{V}(\vec{q}_{34}) \mathcal{V}(\vec{q}_{45}) \right).$$

Fortunately, many cancellations occur in calculating cumulants. In particular:

• When calculating the moment $\langle U^p \rangle^0$, the contribution of a *disconnected diagram* is simply the product of its *disjoint clusters*. The coordinates of these clusters are independent random variables, and make no contribution to the joint cumulant $\langle U^p \rangle_c^0$. This result also ensures the extensivity of $\ln Z$, as the surviving connected diagrams give a factor of V from their center of mass integration. (Disconnected clusters have more factors of V, and are non-extensive.)

- There are also *one-particle reducible* clusters, which are fully connected, yet fall to disjoint fragments if a *single* coordinate point is removed. By measuring all other coordinates relative to this special point, we observe that (in a translationally invariant system) the value of such a diagram is the product of its disjoint fragments. It can be shown that such diagrams are also canceled out in calculating the cumulant. Thus only *one-particle irreducible* clusters survive in this cumulant expansion. A cluster with s sites and p bonds makes a contribution of order of $V(N/V)^s(\beta\mathcal{V})^p$ to $\ln Z$. These results are summarized in the expansion

$$\ln Z = \ln Z_0 + V \sum_{p,s} \frac{(-\beta)^p}{p!} \frac{(N/V)^s}{s!} D(p, s), \qquad (5.12)$$

where $D(p, s)$ indicates the sum over the contributions of *all* one-particle irreducible clusters of s sites and p bonds. (The factor of $1/s!$, reflecting the symmetry of permuting the vertices of clusters, is included for later ease of comparison with the cluster expansion.)

At the second order in $(\beta\mathcal{V})$, the perturbative series for the free energy is

$$F(T, V, N) = F_0(T, V, N) + \frac{N^2}{2V} \left(\int d^3\vec{q}\, \mathcal{V}(\vec{q}) - \frac{\beta}{2} \int d^3\vec{q}\, \mathcal{V}(\vec{q})^2 + \mathcal{O}\left(\beta^2\mathcal{V}^3\right) \right)$$
$$+ \mathcal{O}\left(\frac{N^3\beta^2\mathcal{V}^3}{V^2}\right). \qquad (5.13)$$

From this expression we can proceed to calculate other modified state functions, for example, $P = -\partial F/\partial V|_{T,N}$. Unfortunately, the expansion in powers of $\beta\mathcal{V}$ is not particularly useful. The interatomic potential $\mathcal{V}(\vec{r})$ for most particles has an attractive tail due to van der Waals interaction that decays as $-1/r^6$ at large separations $r = |\vec{r}|$. At short distances the overlap of the electron clouds makes the potential strongly repulsive. Typically there is a minimum of depth a few hundred Kelvin, at a distance of a few angstroms. The infinity in $\mathcal{V}(\vec{r})$ due to the *hard core* at short distances makes it an unsuitable expansion parameter. This problem can be alleviated by a partial resummation of diagrams. For example, to get the correction at order of N^2/V, we need to sum over all two-point clusters, independent of the number of bonds. The resulting sum is actually quite trivial, and leads to

$$\ln Z = \ln Z_0 + \sum_{p=1}^{\infty} \frac{(-\beta)^p}{p!} \frac{N(N-1)}{2} \int \frac{d^3\vec{q}}{V} \mathcal{V}(\vec{q})^p + \mathcal{O}\left(\frac{N^3}{V^2}\right)$$
$$= \ln Z_0 + \frac{N(N-1)}{2V} \int d^3\vec{q} \left[\exp\left(-\beta\mathcal{V}(\vec{q})\right) - 1 \right] + \mathcal{O}\left(\frac{N^3}{V^2}\right). \qquad (5.14)$$

The resummation can be expressed diagrammatically by introducing a new bond

$$\bullet\!-\!\bullet \;\equiv\; \bullet\cdots\bullet \;+\; \frac{1}{2!}\; \bullet::::\bullet \;+\; \frac{1}{3!}\; \bullet::::\bullet \;+\cdots$$

The quantity $f(\vec{q}) = \exp\left(-\beta \mathcal{V}(\vec{q})\right) - 1$ is a much more convenient expansion parameter, which goes to -1 at short distances and rapidly vanishes for large separations. Clearly, such partial resummations can be performed for bonds connecting any pair of vertices, enabling us to remove the summation over p in Eq. (5.12), recasting it as a power series in density N/V. The precise form of the resulting series is discussed in the next section.

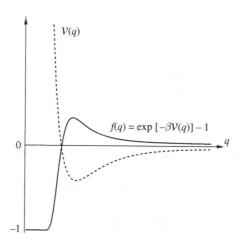

Fig. 5.1 A typical two-body potential $\mathcal{V}(q)$ (dashed curve), and the corresponding resummed function $f(q)$ (solid curve).

5.2 The cluster expansion

For short-range interactions, especially with a hard core, it is much better to replace the expansion parameter $\mathcal{V}(\vec{q})$ by $f(\vec{q}) = \exp\left(-\beta \mathcal{V}(\vec{q})\right) - 1$, which is obtained by summing over all possible numbers of bonds between two points on a cumulant graph. The resulting series is organized in powers of the density N/V, and is most suitable for obtaining a *virial expansion*, which expresses the deviations from the ideal gas equation of state in a power series

$$\frac{P}{k_B T} = \frac{N}{V}\left[1 + B_2(T)\frac{N}{V} + B_3(T)\left(\frac{N}{V}\right)^2 + \cdots\right]. \tag{5.15}$$

The temperature-dependent parameters, $B_i(T)$, are known as the *virial coefficients* and originate from the interparticle interactions. Our initial goal is to compute these coefficients from first principles.

To illustrate a different method of expansion, we shall perform computations in the grand canonical ensemble. With a macrostate $M \equiv (T, \mu, V)$, the grand partition function is given by

$$\mathcal{Q}(\mu, T, V) = \sum_{N=0}^{\infty} e^{\beta \mu N} Z(N, T, V) = \sum_{N=0}^{\infty} \frac{1}{N!}\left(\frac{e^{\beta \mu}}{\lambda^3}\right)^N \mathcal{S}_N, \tag{5.16}$$

where

$$\mathcal{S}_N = \int \prod_{i=1}^{N} \mathrm{d}^3 \vec{q}_i \prod_{i<j} (1 + f_{ij}), \tag{5.17}$$

and $f_{ij} = f(\vec{q}_i - \vec{q}_j)$.

The $2^{N(N-1)/2}$ terms in $\mathcal{S}_N$ can now be ordered in powers of f_{ij} as

$$\mathcal{S}_N = \int \prod_{i=1}^{N} \mathrm{d}^3 \vec{q}_i \left(1 + \sum_{i<j} f_{ij} + \sum_{i<j,k<l} f_{ij} f_{kl} + \cdots \right). \tag{5.18}$$

An efficient method for organizing the perturbation series is to represent the various contributions diagrammatically. In particular, we shall apply the following conventions:

(a) Draw N dots labeled by $i = 1, \cdots, N$ to represent the coordinates $\vec{q}_1$ through $\vec{q}_N$,

$$\begin{array}{cccc} \bullet & \bullet & \cdots & \bullet \\ 1 & 2 & & N \end{array}.$$

(b) Each term in Eq. (5.18) corresponds to a product of f_{ij}, represented by drawing lines connecting i and j for each f_{ij}. For example, the graph

$$\begin{array}{ccccccccc} \bullet & & \bullet\!-\!\bullet & & \bullet\!-\!\bullet\!-\!\bullet & \cdots & \bullet \\ 1 & & 2 \quad 3 & & 4 \quad 5 \quad 6 & & N \end{array}$$

represents the integral

$$\left(\int \mathrm{d}^3 \vec{q}_1 \right) \left(\int \mathrm{d}^3 \vec{q}_2 \mathrm{d}^3 \vec{q}_3 f_{23} \right) \left(\int \mathrm{d}^3 \vec{q}_4 \mathrm{d}^3 \vec{q}_5 \mathrm{d}^3 \vec{q}_6 f_{45} f_{56} \right) \cdots \left(\int \mathrm{d}^3 \vec{q}_N \right).$$

As the above example indicates, the value of each graph is the product of the contributions from its *linked clusters*. Since these clusters are more fundamental, we reformulate the sum in terms of them by defining a quantity b_ℓ, equal to the *sum over all ℓ-particle linked clusters* (one-particle irreducible or not). For example,

$$b_1 = \quad \bullet \quad = \int \mathrm{d}^3 \vec{q} = V, \tag{5.19}$$

and

$$b_2 = \bullet\!-\!\bullet = \int \mathrm{d}^3 \vec{q}_1 \mathrm{d}^3 \vec{q}_2 f(\vec{q}_1 - \vec{q}_2). \tag{5.20}$$

There are four diagrams contributing to b_3, leading to

$$b_3 = \overset{\bullet}{\diagup}\!\bullet\!-\!\bullet \; + \; \overset{\bullet}{\diagdown}\!\bullet\!-\!\bullet \; + \; \overset{\bullet}{\diagup\diagdown}\!\bullet\;\bullet \; + \; \overset{\bullet}{\diagup\diagdown}\!\bullet\!-\!\bullet$$

$$= \int \mathrm{d}^3 \vec{q}_1 \mathrm{d}^3 \vec{q}_2 \mathrm{d}^3 \vec{q}_3 \big[f(\vec{q}_1 - \vec{q}_2) f(\vec{q}_2 - \vec{q}_3) + f(\vec{q}_2 - \vec{q}_3) f(\vec{q}_3 - \vec{q}_1) + f(\vec{q}_3 - \vec{q}_1) f(\vec{q}_1 - \vec{q}_2)$$

$$+ f(\vec{q}_1 - \vec{q}_2) f(\vec{q}_2 - \vec{q}_3) f(\vec{q}_3 - \vec{q}_1) \big]. \tag{5.21}$$

A given N-particle graph can be decomposed into n_1 1-clusters, n_2 2-clusters, $\cdots$, n_ℓ ℓ-clusters, etc. Hence,

$$\mathcal{S}_N = \sum_{\{n_\ell\}'} \prod_\ell b_\ell^{n_\ell} W(\{n_\ell\}), \tag{5.22}$$

where the restricted sum is over all *distinct divisions of N points* into a set of clusters $\{n_\ell\}$, such that $\sum_\ell \ell n_\ell = N$. The coefficients $W(\{n_\ell\})$ are the number of ways of assigning N particle labels to groups of n_ℓ ℓ-clusters. For example, the divisions of three particles into a 1-cluster and a 2-cluster are

$$
\begin{array}{ccc}
\bullet & \bullet - \bullet & \\
1 & 2 \quad 3 \\
\end{array}
\quad , \quad
\begin{array}{cc}
\bullet & \bullet - \bullet \\
2 & 1 \quad 3 \\
\end{array}
\quad , \text{ and } \quad
\begin{array}{cc}
\bullet & \bullet - \bullet \\
3 & 2 \quad 1 \\
\end{array} \quad .
$$

All the above graphs have $n_1 = 1$ and $n_2 = 1$, and contribute a factor of $b_1 b_2$ to S_3; thus, $W(1, 1) = 3$. Other possible divisions of a three-particle system are to three 1-point clusters, or one 3-point cluster, leading to

$$
S_3 = \bullet \quad \bullet \quad \bullet + 3 \, \bullet\!\diagup \bullet\!-\!\bullet + \left(\bullet\!\diagup\!\bullet\!-\!\bullet \quad \bullet\!-\!\bullet\!\diagdown\!\bullet \quad \bullet\!\diagup\!\diagdown\!\bullet \quad \bullet\!\diagup\!\bullet\!-\!\bullet \right)
$$

$$
= b_1^3 + 3 b_1 b_2 + b_3.
$$

In general, $W(\{n_\ell\})$ is the number of distinct ways of grouping the labels $1, \cdots, N$ into bins of n_ℓ ℓ-clusters. It can be obtained from the total number of permutations, $N!$, after dividing by the number of equivalent assignments. Within each bin of ℓn_ℓ particles, equivalent assignments are obtained by: (i) permuting the ℓ labels in each subgroup in $\ell!$ ways, for a total of $(\ell!)^{n_\ell}$ permutations; and (ii) the $n_\ell!$ rearrangements of the n_ℓ subgroups. Hence,

$$
W(\{n_\ell\}) = \frac{N!}{\prod_\ell n_\ell! \, (\ell!)^{n_\ell}}. \tag{5.23}
$$

(We can indeed check that $W(1, 1) = 3!/(1!)(2!) = 3$ as obtained above.)

Using the above value of W, the expression for S_N in Eq. (5.22) can be evaluated. However, the restriction of the sum to configurations such that $\sum_\ell \ell n_\ell = N$ complicates the calculation. Fortunately, this restriction disappears in the expression for the grand partition function in Eq. (5.17),

$$
\mathcal{Q} = \sum_{N=0}^{\infty} \frac{1}{N!} \left(\frac{e^{\beta \mu}}{\lambda^3} \right)^N \sum_{\{n_\ell\}'} \frac{N!}{\prod_\ell n_\ell! \, (\ell!)^{n_\ell}} \prod_\ell b_\ell^{n_\ell}. \tag{5.24}
$$

The restriction in the second sum is now removed by noting that $\sum_{N=0}^{\infty} \sum_{\{n_\ell\}} \delta_{\sum_\ell \ell n_\ell, N} = \sum_{\{n_\ell\}}$. Therefore,

$$
\mathcal{Q} = \sum_{\{n_\ell\}} \left(\frac{e^{\beta \mu}}{\lambda^3} \right)^{\sum_\ell \ell n_\ell} \prod_\ell \frac{b_\ell^{n_\ell}}{n_\ell! \, (\ell!)^{n_\ell}} = \sum_{\{n_\ell\}} \prod_\ell \frac{1}{n_\ell!} \left(\frac{e^{\beta \mu \ell} b_\ell}{\lambda^{3\ell} \ell!} \right)^{n_\ell}
$$

$$
= \prod_\ell \sum_{n_\ell=0}^{\infty} \frac{1}{n_\ell!} \left[\left(\frac{e^{\beta \mu}}{\lambda^3} \right)^\ell \frac{b_\ell}{\ell!} \right]^{n_\ell} = \prod_\ell \exp\left[\left(\frac{e^{\beta \mu}}{\lambda^3} \right)^\ell \frac{b_\ell}{\ell!} \right] \tag{5.25}
$$

$$
= \exp\left[\sum_{\ell=1}^{\infty} \left(\frac{e^{\beta \mu}}{\lambda^3} \right)^\ell \frac{b_\ell}{\ell!} \right].
$$

The above result has the simple geometrical interpretation that the sum over all graphs, connected or not, equals the exponential of the sum over *connected clusters*. This is a quite general result that is also related to the graphical connection between moments and cumulants discussed in Section 2.1.

The grand potential is now obtained from

$$\ln \mathcal{Q} = -\beta \mathcal{G} = \frac{PV}{k_B T} = \sum_{\ell=1}^{\infty} \left(\frac{e^{\beta \mu}}{\lambda^3}\right)^{\ell} \frac{b_{\ell}}{l!}. \tag{5.26}$$

In Eq. (5.26), the extensivity condition is used to get $\mathcal{G} = E - TS - \mu N = -PV$. Thus, the terms on the right-hand side of the above equation must also be proportional to the volume V. This can be explicitly verified by noting that in evaluating each b_{ℓ} there is an integral over the center of mass coordinate that explores the whole volume. For example, $b_2 = \int \mathrm{d}^3 \vec{q}_1 \mathrm{d}^3 \vec{q}_2 f(\vec{q}_1 - \vec{q}_2) = V \int \mathrm{d}^3 \vec{q}_{12} f(\vec{q}_{12})$. Quite generally, we can set

$$\lim_{V \to \infty} b_{\ell} = V \bar{b}_{\ell}, \tag{5.27}$$

and the pressure is now obtained from

$$\frac{P}{k_B T} = \sum_{\ell=1}^{\infty} \left(\frac{e^{\beta \mu}}{\lambda^3}\right)^{\ell} \frac{\bar{b}_{\ell}}{\ell!}. \tag{5.28}$$

The linked cluster theorem ensures $\mathcal{G} \propto V$, since if any non-linked cluster had appeared in $\ln \mathcal{Q}$, it would have contributed a higher power of V.

Although an expansion for the gas pressure, Eq. (5.28) is quite different from Eq. (5.15) in that it involves powers of $e^{\beta \mu}$ rather than the density $n = N/V$. This difference can be removed by solving for the density in terms of the chemical potential, using

$$N = \frac{\partial \ln \mathcal{Q}}{\partial (\beta \mu)} = \sum_{\ell=1}^{\infty} \ell \left(\frac{e^{\beta \mu}}{\lambda^3}\right)^{\ell} \frac{V \bar{b}_{\ell}}{\ell!}. \tag{5.29}$$

The equation of state can be obtained by eliminating the *fugacity* $x = e^{\beta \mu}/\lambda^3$, between the equations

$$n = \sum_{\ell=1}^{\infty} \frac{x^{\ell}}{(\ell-1)!} \bar{b}_{\ell}, \quad \text{and} \quad \frac{P}{k_B T} = \sum_{\ell=1}^{\infty} \frac{x^{\ell}}{\ell!} \bar{b}_{\ell}, \tag{5.30}$$

using the following steps:

(a) Solve for $x(n)$ from ($\bar{b}_1 = \int \mathrm{d}^3 \vec{q}/V = 1$)

$$x = n - \bar{b}_2 x^2 - \frac{\bar{b}_3}{2} x^3 - \cdots. \tag{5.31}$$

The perturbative solution at each order is obtained by substituting the solution at the previous order in Eq. (5.31),

$$x_1 = n + \mathcal{O}(n^2),$$

$$x_2 = n - \bar{b}_2 n^2 + \mathcal{O}(n^3), \tag{5.32}$$

$$x_3 = n - \bar{b}_2 (n - \bar{b}_2 n)^2 - \frac{\bar{b}_3}{2} n^3 + \mathcal{O}(n^4) = n - \bar{b}_2 n^2 + \left(2 \bar{b}_2^2 - \frac{\bar{b}_3}{2}\right) n^3 + \mathcal{O}(n^4).$$

(b) Substitute the perturbative result for $x(n)$ into Eq. (5.30), yielding

$$
\begin{aligned}
\beta P &= x + \frac{b_2}{2} x^2 + \frac{b_3}{6} x^3 + \cdots \\
&= n - b_2 n^2 + \left(2b_2{}^2 - \frac{b_3}{2} \right) n^3 + \frac{b_2}{2} n^2 - b_2{}^2 n^3 + \frac{b_3}{6} n^3 + \cdots \\
&= n - \frac{b_2}{2} n^2 + \left(b_2{}^2 - \frac{b_3}{3} \right) n^3 + \mathcal{O}(n^4).
\end{aligned}
\tag{5.33}
$$

The final result is in the form of the virial expansion of Eq. (5.15),

$$
\beta P = n + \sum_{\ell=2}^{\infty} B_\ell(T) n^\ell.
$$

The first term in the series reproduces the ideal gas result. The next two corrections are

$$
B_2 = -\frac{b_2}{2} = -\frac{1}{2} \int \mathrm{d}^3 \vec{q} \left(\mathrm{e}^{-\beta \mathcal{V}(\vec{q})} - 1 \right),
\tag{5.34}
$$

and

$$
\begin{aligned}
B_3 &= b_2{}^2 - \frac{b_3}{3} \\
&= \left[\int \mathrm{d}^3 \vec{q} \left(\mathrm{e}^{-\beta \mathcal{V}(\vec{q})} - 1 \right) \right]^2 \\
&\quad - \frac{1}{3} \left[3 \int \mathrm{d}^3 \vec{q}_{12} \mathrm{d}^3 \vec{q}_{13} f(\vec{q}_{12}) f(\vec{q}_{13}) + \int \mathrm{d}^3 \vec{q}_{12} \mathrm{d}^3 \vec{q}_{13} f(\vec{q}_{12}) f(\vec{q}_{13}) f(\vec{q}_{12} - \vec{q}_{13}) \right] \\
&= -\frac{1}{3} \int \mathrm{d}^3 \vec{q}_{12} \mathrm{d}^3 \vec{q}_{13} f(\vec{q}_{12}) f(\vec{q}_{13}) f(\vec{q}_{12} - \vec{q}_{13}).
\end{aligned}
\tag{5.35}
$$

The above example demonstrates the cancellation of the one-particle reducible clusters that appear in b_3. While all clusters (reducible or not) appear in the sum for b_ℓ, as demonstrated in the previous section, only the *one-particle irreducible* ones can appear in an expansion in powers of density. The final expression for the ℓth virial coefficient is thus

$$
B_\ell(T) = -\frac{(\ell-1)}{\ell!} d_\ell,
\tag{5.36}
$$

where d_ℓ is defined as the sum over all one-particle irreducible clusters of ℓ points.

This result can also be obtained from the expansion of the partition function in Eq. (5.12). The key observation is that the summation over all powers of $\beta \mathcal{V}$ (leading to multiple connections between pairs of sites) in the diagrams considered in the previous section, results in precisely the clusters introduced in this section (with pairs of sites connected by f). Thus, $\sum_p (-\beta)^p/(p!) D(p, \ell) = d_\ell$, and

$$
\ln Z = \ln Z_0 + V \sum_{\ell=2}^{\infty} \frac{n^\ell}{\ell!} d_\ell,
\tag{5.37}
$$

reproducing the above virial expansion from $\beta P = \partial \ln Z / \partial V$.

5.3 The second virial coefficient and van der Waals equation

Let us study the second virial coefficient B_2 for a typical gas using Eq. (5.34). As discussed before, the two-body potential is characterized by a hard core repulsion at short distances and a van der Waals attraction at large distances. To make the computations easier, we shall use the following approximation for the potential:

$$\mathcal{V}(r) = \begin{cases} +\infty & \text{for } r < r_0, \\ -u_0 \left(r_0/r \right)^6 & \text{for } r > r_0, \end{cases} \tag{5.38}$$

which combines both features. The contributions of the two portions can then be calculated separately, as

$$\begin{aligned} b_2 &= \int_0^\infty \mathrm{d}^3\vec{r} \left(\mathrm{e}^{-\beta \mathcal{V}(r)} - 1 \right) \\ &= \int_0^{r_0} 4\pi r^2 \mathrm{d}r (-1) + \int_{r_0}^\infty 4\pi r^2 \mathrm{d}r \left[\mathrm{e}^{+\beta u_0 (r_0/r)^6} - 1 \right]. \end{aligned} \tag{5.39}$$

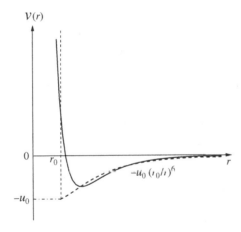

Fig. 5.2 The continuous two-body potential (solid curve) is replaced by a hard core and an attractive tail (dashed curve).

The second integrand can be approximated by $\beta u_0 (r_0/r)^6$ in the high-temperature limit, $\beta u_0 \gg 1$, and leads to

$$B_2 = -\frac{1}{2} \left[-\frac{4\pi r_0^3}{3} + 4\pi \beta u_0 r_0^6 \left(-\frac{r^{-3}}{3} \right) \Big|_{r_0}^\infty \right] = \frac{2\pi r_0^3}{3} (1 - \beta u_0). \tag{5.40}$$

We can define an *excluded volume* of $\Omega = 4\pi r_0^3/3$, which is 8 times the atomic volume (since the distance of minimum approach r_0 is twice an atomic radius), to get

$$B_2(T) = \frac{\Omega}{2} \left(1 - \frac{u_0}{k_B T} \right). \tag{5.41}$$

Remarks and observations:

(1) The tail of the attractive van der Waals interaction ($\propto r^{-6}$) extends to very long separations. Yet, its integral in Eq. (5.40) is dominated by contributions from the short scales r_0. In this limited context, the van der Waals potential is *short-ranged*, and results in corrections to the ideal gas behavior that are *analytical* in density n, leading to the virial series.

(2) By contrast, potentials that fall off with separation as $1/r^3$ or slower are *long-ranged*. The integral appearing in calculation of the second virial coefficient is then dominated by long distances, and is divergent. As a result, corrections to the ideal gas behavior cannot be written in the form of a virial series, and are in fact *non-analytic*. A good example is provided by the Coulomb plasma discussed in the problem section. The non-analytic corrections can be obtained by summing all the *ring diagrams* in the cumulant (or cluster) expansions.[2]

(3) The second virial coefficient has dimensions of volume and (for short-range potentials) is proportional to the atomic volume Ω. In the high-temperature limit, the importance of corrections to ideal gas behavior can be estimated by comparing the first two terms of Eq. (5.15),

$$\frac{B_2 n^2}{n} = \frac{B_2}{n^{-1}} \sim \frac{\text{atomic volume}}{\text{volume per particle in gas}} \sim \frac{\text{gas density}}{\text{liquid density}}. \tag{5.42}$$

This ratio is roughly 10^{-3} for air at room temperature and pressure. The corrections to ideal gas behavior are thus small at low densities. On dimensional grounds, a similar ratio is expected for the higher-order terms, $B_\ell n^\ell / B_{\ell-1} n^{\ell-1}$, in the virial series. We should thus be wary of the convergence of the series at high densities (when the gas may liquify).

(4) The virial expansion breaks down not only at high densities, but also at low temperatures. This is suggested by the divergences in Eqs. (5.41) and (5.39) as $T \to 0$, and reflects the fact that in the presence of attractive interactions the particles can lower their energy at low temperatures by condensing into a liquid state.

(5) The truncated virial expansion,

$$\frac{P}{k_B T} = n + \frac{\Omega}{2} \left(1 - \frac{u_0}{k_B T} \right) n^2 + \cdots, \tag{5.43}$$

can be rearranged as

$$\frac{1}{k_B T} \left(P + \frac{u_0 \Omega}{2} n^2 \right) = n \left(1 + n \frac{\Omega}{2} + \cdots \right) \approx \frac{n}{1 - n\Omega/2} = \frac{N}{V - N\Omega/2}. \tag{5.44}$$

---

[2] Consider a potential with typical strength u, and range λ. Each bond in the cumulant expansion contributes a factor of $\beta \mathcal{V}$, while each node contributes $N \int d^3 \vec{q}/V$. Dimensional arguments then suggest that a diagram with b bonds and s nodes is proportional to $(\beta u)^b (N\lambda^3/V)^s$. At a given order in βu, the factor of $n\lambda^3$ thus discriminates between weakly and strongly interacting systems. For $n\lambda^3 \ll 1$ (*dilute gas, or short-range interactions*) the diagrams with the fewest nodes should be included. For $n\lambda^3 \gg 1$ (*dense gas, or long-range interactions*) the ring diagrams, which have the most nodes, dominate.

This is precisely in the form of the *van der Waals equation*

$$\left[P + \frac{u_0\Omega}{2}\left(\frac{N}{V}\right)^2\right]\left[V - \frac{N\Omega}{2}\right] = Nk_BT, \tag{5.45}$$

and we can identify the *van der Waals parameters*, $a = u_0\Omega/2$ and $b = \Omega/2$.

Physical interpretation of the van der Waals equation: historically, van der Waals suggested Eq. (5.45) on the basis of experimental results for the equation of state of various gases, toward the end of the nineteenth century. At that time the microscopic interactions between gas particles were not known, and van der Waals postulated the necessity of an attractive interaction between gas atoms based on the observed decreases in pressure. It was only later on that such interactions were observed directly, and then attributed to the induced dipole–dipole forces by London. The physical justification of the correction terms is as follows.

(a) There is a correction to the gas volume V due to the exclusions. At first sight, it may appear surprising that the *excluded volume b* in Eq. (5.45) is one half of the volume that is excluded around each particle. This is because this factor measures a *joint* excluded volume involving all particles in phase space. In fact, the contribution of coordinates to the partition function of the hard-core gas can be estimated at *low densities*, from

$$\mathcal{S}_N = \int{}'\frac{\prod_i d^3\vec{q}_i}{N!} = \frac{1}{N!}V(V-\Omega)(V-2\Omega)\cdots(V-(N-1)\Omega)$$

$$\approx \frac{1}{N!}\left(V - \frac{N\Omega}{2}\right)^N. \tag{5.46}$$

The above result is obtained by adding particles one at a time, and noting that the available volume for the $(m+1)$th particle is $(V-m\Omega)$. At low densities, the overall effect is a reduction of the volume available to each particle by approximately $\Omega/2$. Of course, the above result is only approximate since the effects of excluded volume involving more than two particles are not correctly taken into account. The relatively simple form of Eq. (5.46) is only exact for spatial dimensions $d=1$, and at infinity. As proved in the problems, the exact excluded volume at $d=1$ is in fact Ω.

(b) The decrease in P, due to attractive interactions, is somewhat harder to quantify. In Section 3.6, the gas pressure was related to the impacts of particles on a wall via

$$P = \overline{(nv_x)(2mv_x)}\Big|_{v_x<0} = nm\overline{v_x^2}, \tag{5.47}$$

where the first term is the number of collisions per unit time and area, while the second is the momentum imparted by each particle. For the ideal gas, the usual equation of state is recovered by noting that the average kinetic energy is $m\overline{v_x^2}/2 = k_BT/2$. Attractive interactions lead to a reduction in pressure given by

$$\delta P = \delta n\left(m\overline{v_x^2}\right) + n\delta\left(m\overline{v_x^2}\right). \tag{5.48}$$

While the pressure of a gas is a bulk equilibrium property, in Eq. (5.48) it is related to behavior of the particles close to a surface. The following interpretations of the result depend on distinct assumptions about the relaxation of particles at the surface.

In a canonical ensemble, the gas density is *reduced* at the walls. This is because the particles in the middle of the box experience an attractive potential U from all sides, while at the edge only an attractive energy of $U/2$ is available from half of the space. The resulting change in density is approximately

$$\delta n \approx n \left(e^{-\beta U/2} - e^{-\beta U} \right) \approx \beta n U/2. \tag{5.49}$$

Integrating the interaction of one particle in the bulk with the rest gives

$$U = \int d^3\vec{r}\, \mathcal{V}_{\text{attr.}}(r)\, n = -n\Omega u_0. \tag{5.50}$$

The change in density thus gives the pressure correction of $\delta P = -n^2\Omega u_0/2$ calculated in Eq. (5.45). There is no correction to the kinetic energy of the particles impinging on the wall, since in the canonical formulation the probabilities for momentum and location of the particles are independent variables. The probability distribution for momentum is *uniform* in space, giving the average kinetic energy of $k_B T/2$ for each direction of motion.

The range of decreased density close to the wall in the above description may be quite small; of the order of the range of the interaction between particles. If this length is smaller than the mean free path between collisions, we may well question the assumption that the surface particles can be described by an equilibrium ensemble. An alternative explanation can then be formulated in which particles close to the surface simply follow deterministic Hamiltonian equations in an effective potential, without undergoing any collisions. In this formulation, the impinging particles lose kinetic energy in approaching the wall, since they have to climb out of the potential well set up by the attractions of the bulk particles. The reduction in kinetic energy is given by

$$\delta \frac{mv_x^2}{2} = \frac{1}{2} \int d^3\vec{r}\, \mathcal{V}_{\text{attr.}}(r)\, n = -\frac{1}{2} n\Omega u_0. \tag{5.51}$$

The reduced velocities lead to an *increase* in the surface density in this case, as the slower particles spend a longer time τ in the vicinity of the wall! The relative change in density is given by

$$\frac{\delta n}{n} = \frac{\delta \tau}{\tau} = -\frac{\delta v_x}{v_x} = -\frac{1}{2}\frac{\delta v_x^2}{v_x^2}, \qquad \Longrightarrow \qquad \delta n = -\beta n U/2. \tag{5.52}$$

The increase in density is precisely the opposite of the result of Eq. (5.49) in the equilibrium formulation. However, with the decrease in kinetic energy calculated in Eq. (5.51), it again leads to the correct reduction in pressure.

5.4 Breakdown of the van der Waals equation

As discussed in Section 1.9, mechanical stability of a gas requires the positivity of the isothermal compressibility, $\kappa_T = -V^{-1} \, \partial V/\partial P|_T$. This condition can be obtained by examining density fluctuations in a grand canonical ensemble. The probability of finding N particles in a volume V is given by

$$p(N, V) = \frac{e^{\beta \mu N} Z(T, N, V)}{Q}. \tag{5.53}$$

Since for a gas $\ln Q = -\beta \mathcal{G} = PV/k_B T$, the mean and variance of N simplify to

$$\begin{cases} \langle N \rangle_c = N = \dfrac{\partial(\ln Q)}{\partial(\beta\mu)} = V \left. \dfrac{\partial P}{\partial \mu} \right|_{T,V}, \\[2ex] \langle N^2 \rangle_c = \dfrac{\partial^2(\ln Q)}{\partial(\beta\mu)^2} = \dfrac{\partial \langle N \rangle_c}{\partial(\beta\mu)} = k_B T \left. \dfrac{\partial N}{\partial \mu} \right|_{T,V}. \end{cases} \tag{5.54}$$

Dividing the two equations, and using the chain rule, results in

$$\frac{\langle N^2 \rangle_c}{N} = \frac{k_B T}{V} \left. \frac{\partial N}{\partial P} \right|_{T,V} = -\frac{k_B T}{V} \left. \frac{\partial N}{\partial V} \right|_{P,T} \left. \frac{\partial V}{\partial P} \right|_{N,T} = n k_B T \kappa_T. \tag{5.55}$$

The positivity of κ_T is thus tied to that of the variance of N. A stable value of N corresponds to a *maximum* of the probability $p(N, V)$, that is, a positive compressibility. A negative κ_T actually corresponds to a *minimum* in $p(N, V)$, implying that the system is least likely to be found at such densities. Fluctuations in density will then occur spontaneously and change the density to a stable value.

Any approximate equation of state, such as the van der Waals equation, must at least satisfy the stability requirements. However, the van der Waals isotherms contain a portion with $-\partial P/\partial V|_T < 0$, for temperatures less than a critical value T_c. The negative compressibility implies an instability toward forming domains of lower and higher density, that is, phase separation. The attractive interactions in real gases do indeed result in a liquid–gas phase separation at low temperatures. The isotherms then include a flat portion, $\partial P/\partial V|_T = 0$, at the coexistence of the two phases. Can the (unstable) van der Waals isotherms be used to construct the phase diagram of a real gas?

One way of doing so is by the following *Maxwell construction*: the variations of the chemical potential $\mu(T, P)$ along an isotherm are obtained by integrating Eq. (5.54), as

$$d\mu = \frac{V}{N} dP, \implies \mu(T, P) = \mu(T, P_A) + \int_{P_A}^{P} dP' \frac{V(T, P')}{N}. \tag{5.56}$$

Since the van der Waals isotherms for $T < T_c$ are non-monotonic, there is a range of pressures that correspond to three different values, $\{\mu_\alpha\}$, of the chemical potential. The possibility of several values of μ at a given temperature and pressure indicates phase coexistence. In equilibrium, the number of particles in each phase N_α adjusts so as to minimize the Gibbs free energy $G = \sum_\alpha \mu_\alpha N_\alpha$.

Fig. 5.3 Isotherms of the van der Waals equation have an unphysical portion at low temperature. By contrast, a real gas (inset) has a coexistence region (indicated in gray) with flat isotherms.

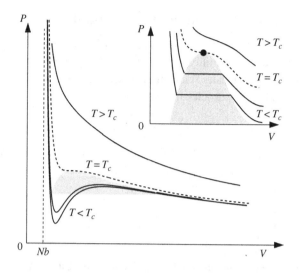

Clearly, the phase with lowest μ_α will acquire all the particles. A phase transition occurs when two branches of allowed chemical potentials intersect. From Eq. (5.56), the critical pressure P_c for this intersection is obtained from the condition

$$\oint_{P_c}^{P_c} dP' V(T, P') = 0. \tag{5.57}$$

A geometrical interpretation of the above result is that P_c corresponds to a pressure that encloses equal areas of the non-monotonic isotherm on each side. The Maxwell construction approach to phase separation is somewhat unsatisfactory, as it relies on integrating a clearly unphysical portion of the van der Waals isotherm. A better approach that makes the approximations involved more apparent is presented in the next section.

Fig. 5.4 Using the Maxwell equal-area construction, the unstable portion of the isotherm from C to D is replaced with the "tie-line" BE. The regions BCO and EDO have equal area. The isothermal portions BC and DE are metastable.

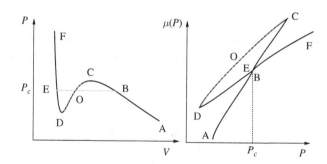

5.5 Mean-field theory of condensation

In principle, all properties of the interacting system, including phase separation, are contained within the thermodynamic potentials that can be obtained by evaluating $Z(T, N)$ or $\mathcal{Q}(T, \mu)$. Phase transitions, however, are characterized by discontinuities in various state functions and must correspond to the appearance of singularities in the partition functions. At first glance, it is somewhat surprising that any singular behavior should emerge from computing such well-behaved integrals (for short-ranged interactions) as

$$Z(T, N, V) = \int \frac{\prod_{i=1}^{N} \mathrm{d}^3\vec{p}_i \mathrm{d}^3\vec{q}_i}{N! h^{3N}} \exp\left[-\beta \sum_{i=1}^{N} \frac{p_i^2}{2m} - \beta \sum_{i<j} \mathcal{V}(\vec{q}_i - \vec{q}_j)\right]. \qquad (5.58)$$

Instead of evaluating the integrals perturbatively, we shall now set up a reasonable approximation scheme. The contributions of the hard core and attractive portions of the potential are again treated separately, and the partition function approximated by

$$Z(T, N, V) \approx \frac{1}{N!} \frac{1}{\lambda^{3N}} \underbrace{V(V - \Omega) \cdots (V - (N-1)\Omega)}_{\text{Excluded volume effects}} \exp(-\beta \bar{U}). \qquad (5.59)$$

Here $\bar{U}$ represents an average attraction energy, obtained by assuming a *uniform density* $n = N/V$, as

$$\bar{U} = \frac{1}{2} \sum_{i,j} \mathcal{V}_{\text{attr.}}(\vec{q}_i - \vec{q}_j) = \frac{1}{2} \int \mathrm{d}^3\vec{r}_1 \mathrm{d}^3\vec{r}_2\, n(\vec{r}_1) n(\vec{r}_2) \mathcal{V}_{\text{attr.}}(\vec{r}_1 - \vec{r}_2)$$
$$\approx \frac{n^2}{2} V \int \mathrm{d}^3\vec{r}\, \mathcal{V}_{\text{attr.}}(\vec{r}) \equiv -\frac{N^2}{2V} u. \qquad (5.60)$$

The key approximation is replacing the variable local density $n(\vec{r})$ with the uniform density n. The parameter u describes the net effect of the attractive interactions. Substituting into Eq. (5.59) leads to the following approximation for the partition function:

$$Z(T, N, V) \approx \frac{(V - N\Omega/2)^N}{N! \lambda^{3N}} \exp\left[\frac{\beta u N^2}{2V}\right]. \qquad (5.61)$$

From the resulting free energy,

$$F = -k_B T \ln Z = -N k_B T \ln(V - N\Omega/2) + N k_B T \ln(N/e) + 3 N k_B T \ln \lambda - \frac{u N^2}{2V}, \qquad (5.62)$$

we obtain the expression for the pressure in the canonical ensemble as

$$P_{\text{can}} = -\left.\frac{\partial F}{\partial V}\right|_{T,N} = \frac{N k_B T}{V - N\Omega/2} - \frac{u N^2}{2V^2}. \qquad (5.63)$$

Remarkably, the uniform density approximation reproduces the van der Waals equation of state. However, the self-consistency of this approximation can now be checked. As long as κ_T is positive, Eq. (5.55) implies that the variance of density vanishes for large volumes as $\langle n^2 \rangle_c = k_B T n^2 \kappa_T / V$. But κ_T diverges at T_c, and at lower temperatures its negativity implies an instability toward density fluctuations as discussed in the previous section. When condensation occurs, there is phase separation into high (liquid) and low (gas) density states, and the uniform density assumption becomes manifestly incorrect. This difficulty is circumvented in the *grand canonical ensemble*. Once the chemical potential is fixed, the number of particles (and hence density) in this ensemble is automatically adjusted to that of the appropriate phase.

As the assumption of a uniform density is correct for both the liquid and gas phases, we can use the approximations of Eqs. (5.60) and (5.61) to estimate the grand partition function

$$\mathcal{Q}(T, \mu, V) = \sum_{N=0}^{\infty} e^{\beta \mu N} Z(T, N, V) \approx \sum_{N=0}^{\infty} \exp\left[N \ln\left(\frac{V}{N} - \frac{\Omega}{2} \right) + \frac{\beta u N^2}{2V} + \Delta N \right],$$
(5.64)

where $\Delta = 1 + \beta \mu - \ln(\lambda^3)$. As in any sum over exponentials in N, the above expression is dominated by a particular value of particle number (hence density), and given by

$$\mathcal{Q}(T, \mu, V) \approx \exp\left\{ \max\left[N\Delta + N \ln\left(\frac{V}{N} - \frac{\Omega}{2} \right) + \frac{\beta u N^2}{2V} \right]_N \right\}.$$
(5.65)

Hence, the grand canonical expression for the gas pressure is obtained from

$$\beta P_{\text{g.c.}} = \frac{\ln \mathcal{Q}}{V} = \max[\Psi(n)]_n,$$
(5.66)

where

$$\Psi(n) = n\Delta + n \ln\left(n^{-1} - \frac{\Omega}{2} \right) + \frac{\beta u}{2} n^2.$$
(5.67)

The possible values of density are obtained from $d\Psi/dn|_{n_\alpha} = 0$, and satisfy

$$\Delta = -\ln\left(n_\alpha^{-1} - \frac{\Omega}{2} \right) + \frac{1}{1 - n_\alpha \Omega / 2} - \beta u n_\alpha.$$
(5.68)

The above equation in fact admits multiple solutions n_α for the density. Substituting the resulting Δ into Eq. (5.66) leads after some manipulation to

$$P_{\text{g.c.}} = \max\left[\frac{n_\alpha k_B T}{1 - n_\alpha \Omega / 2} - \frac{u}{2} n_\alpha^2 \right]_\alpha = \max[P_{\text{can}}(n_\alpha)]_\alpha,$$
(5.69)

that is, the grand canonical and canonical values of pressure are identical at a particular density. However, if Eq. (5.68) admits multiple solutions for the density at a particular chemical potential, the correct density is uniquely determined as the one that maximizes the canonical expression for pressure (or for $\psi(n)$).

The mechanism for the liquid–gas phase transition is therefore the following. The sum in Eq. (5.64) is dominated by two large terms at the liquid and gas densities. At a particular chemical potential, the stable phase is determined by

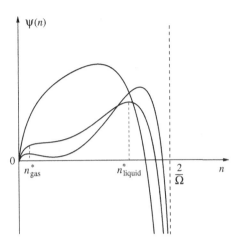

Fig. 5.5 The appropriate uniform density is obtained from the maximum of $\psi(n)$.

the larger of the two terms. The phase transition occurs when the dominant term changes upon varying the temperature. In mathematical form,

$$\ln \mathcal{Q} = \lim_{V \to \infty} \ln \left[e^{\beta V P_{\text{liquid}}} + e^{\beta V P_{\text{gas}}} \right] = \begin{cases} \beta V P_{\text{gas}} & \text{for } T > T^* \\ \beta V P_{\text{liquid}} & \text{for } T < T^* \end{cases}. \tag{5.70}$$

The origin of the singularity in density can thus be traced to the thermodynamic limit of $V \to \infty$. There are no phase transitions in finite systems!

5.6 Variational methods

Perturbative methods provide a systematic way of incorporating the effect of interactions but are impractical for the study of strongly interacting systems. While the first few terms in the virial series slightly modify the behavior of the gas, an infinite number of terms have to be summed to obtain the condensation transition. An alternative, *but approximate*, method for dealing with strongly interacting systems is the use of variational methods.

Suppose that in an appropriate ensemble we need to calculate $Z = \text{tr}\left(e^{-\beta \mathcal{H}}\right)$. In the canonical formulation, Z is the partition function corresponding to the Hamiltonian $\mathcal{H}$ at temperature $k_B T = 1/\beta$, and for a classical system "tr" refers to the integral over the phase space of N particles. However, the method is more general and can be applied to Gibbs or grand partition functions with the appropriate modification of the exponential factor; also in the context of quantum systems where "tr" is a sum over all allowed quantum microstates. Let us assume that calculating Z is very difficult for the (interacting) Hamiltonian $\mathcal{H}$, but that there is another Hamiltonian $\mathcal{H}_0$, *acting on the same set of degrees of freedom*, for which the calculations are easier. We then introduce a Hamiltonian $\mathcal{H}(\lambda) = \mathcal{H}_0 + \lambda(\mathcal{H} - \mathcal{H}_0)$, and a corresponding partition function

$$Z(\lambda) = \text{tr}\{\exp[-\beta\mathcal{H}_0 - \lambda\beta(\mathcal{H} - \mathcal{H}_0)]\}, \qquad (5.71)$$

which interpolates between the two as λ changes from zero to one. It is then easy to prove the convexity condition

$$\frac{d^2 \ln Z(\lambda)}{d\lambda^2} = \beta^2 \left\langle (\mathcal{H} - \mathcal{H}_0)^2 \right\rangle_c \geq 0, \qquad (5.72)$$

where $\langle\rangle$ is an expectation value with the appropriately normalized probability.

Fig. 5.6 A convex curve is always above a tangent line to the curve at any point.

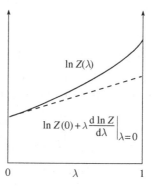

From the convexity of the function, it immediately follows that

$$\ln Z(\lambda) \geq \ln Z(0) + \lambda \left.\frac{d\ln Z}{d\lambda}\right|_{\lambda=0}. \qquad (5.73)$$

But it is easy to see that $d\ln Z/d\lambda|_{\lambda=0} = \beta\langle\mathcal{H}_0 - \mathcal{H}\rangle^0$, where the superscript indicates expectation values with respect to $\mathcal{H}_0$. Setting $\lambda = 1$, we obtain

$$\ln Z \geq \ln Z(0) + \beta\langle\mathcal{H}_0\rangle^0 - \beta\langle\mathcal{H}\rangle^0. \qquad (5.74)$$

Equation (5.74), known as the *Gibbs inequality*, is the basis for most variational estimates. Typically, the "simpler" Hamiltonian $\mathcal{H}_0$ (and hence the right-hand side of Eq. (5.74)) includes several parameters $\{n_\alpha\}$. The best estimate for $\ln Z$ is obtained by finding the maximum of the right-hand side with respect to these parameters. It can now be checked that the approximate evaluation of the grand partition function Q in the preceding section is equivalent to a variational treatment with $\mathcal{H}_0$ corresponding to a gas of hard-core particles of density n, for which (after replacing the sum by its dominant term)

$$\ln Q_0 = \beta\mu N + \ln Z = V\left[n\left(1 + \beta\mu - \ln(\lambda^3)\right) + n\ln\left(n^{-1} - \frac{\Omega}{2}\right)\right]. \qquad (5.75)$$

The difference $\mathcal{H} - \mathcal{H}_0$ contains the attractive portion of the two-body interactions. In the regions of phase space not excluded by the hard-core interactions, the gas in $\mathcal{H}_0$ is described by a uniform density n. Hence,

$$\beta \langle \mathcal{H}_0 - \mathcal{H} \rangle^0 = \beta V \frac{n^2}{2} u, \tag{5.76}$$

and

$$\beta P = \frac{\ln Q}{V} \geq \left[n \left(1 + \beta\mu - \ln(\lambda^3)\right) + n \ln \left(n^{-1} - \frac{\Omega}{2} \right) \right] + \frac{1}{2} \beta u n^2, \tag{5.77}$$

which is the same as Eq. (5.67). The density n is now a parameter on the right-hand side of Eq. (5.77). Obtaining the best variational estimate by maximizing with respect to n is then equivalent to Eq. (5.66).

5.7 Corresponding states

We now have a good perturbative understanding of the behavior of a dilute interacting gas at high temperatures. At lower temperatures, attractive interactions lead to condensation into the liquid state. The qualitative behavior of the phase diagram is the same for most simple gases. There is a line of transitions in the coordinates (P, T) (corresponding to the coexistence of liquid and gas in the (V, T) plane) that terminates at a so-called *critical point.* It is thus possible to transform a liquid to a gas without encountering any singularities. Since the ideal gas law is universal, that is, independent of material, we may hope that there is also a generalized universal equation of state (presumably more complicated) that describes interacting gases, including liquid/gas condensation phenomena. This hope motivated the search for a *law of corresponding states*, obtained by appropriate rescalings of state functions. The most natural choice of scales for pressure, volume, and temperature are those of the critical point, (P_c, V_c, T_c).

The van der Waals equation is an example of a generalized equation of state. Its critical point is found by setting $\partial P / \partial V |_T$ and $\partial^2 P / \partial V^2 |_T$ to zero. The former is the limit of the flat coexistence portion of liquid/gas isotherms; the latter follows from the stability requirement $\kappa_T > 0$ (see the discussion in Section 1.9). The coordinates of the critical point are thus obtained from solving the following coupled equations:

$$\begin{cases} P = \dfrac{k_B T}{v - b} - \dfrac{a}{v^2} \\[2mm] \left. \dfrac{\partial P}{\partial v} \right|_T = -\dfrac{k_B T}{(v - b)^2} + \dfrac{2a}{v^3} = 0 \;, \\[2mm] \left. \dfrac{\partial^2 P}{\partial v^2} \right|_T = \dfrac{2k_B T}{(v - b)^3} - \dfrac{6a}{v^4} = 0 \end{cases} \tag{5.78}$$

where $v = V/N$ is the volume per particle. The solution to these equations is

$$
\left\{
\begin{aligned}
P_c &= \frac{a}{27b^2} \\[4pt]
v_c &= 3b \\[4pt]
k_B T_c &= \frac{8a}{27b}
\end{aligned}
\right. \qquad . \tag{5.79}
$$

Naturally, the critical point depends on the microscopic Hamiltonian (e.g., on the two-body interaction) via the parameters a and b. However, we can scale out such dependencies by measuring P, T, and v in units of P_c, T_c, and v_c. Setting $P_r = P/P_c$, $v_r = v/v_c$, and $T_r = T/T_c$, a reduced version of the van der Waals equation is obtained as

$$
P_r = \frac{8}{3} \frac{T_r}{v_r - 1/3} - \frac{3}{v_r^2}. \tag{5.80}
$$

We have thus constructed a *universal* (material-independent) equation of state. Since the original van der Waals equation depends only on *two* parameters, Eqs. (5.79) predict a universal dimensionless ratio,

$$
\frac{P_c v_c}{k_B T_c} = \frac{3}{8} = 0.375. \tag{5.81}
$$

Experimentally, this ratio is found to be in the range of 0.28 to 0.33. The van der Waals equation is thus not a good candidate for the putative universal equation of state.

We can attempt to construct a generalized equation *empirically* by using *three* independent critical coordinates, and finding $P_r \equiv p_r(v_r, T_r)$ from the collapse of experimental data. Such an approach has some limited success in describing similar gases, for example, the sequence of noble gases Ne, Xe, Kr, etc. However, different types of gases (e.g., diatomic, ionic, etc.) show rather different behaviors. This is not particularly surprising given the differences in the underlying Hamiltonians. We know rigorously from the virial expansion that perturbative treatment of the interacting gas does depend on the details of the microscopic potential. Thus the hope of finding a universal equation of state for all liquids and gases has no theoretical or experimental justification; each case has to be studied separately starting from its microscopic Hamiltonian. It is thus quite surprising that the collapse of experimental data does in fact work very well in the vicinity of the critical point, as described in the next section.

5.8 Critical point behavior

To account for the universal behavior of gases close to their critical point, let us examine the isotherms in the vicinity of (P_c, v_c, T_c). For $T \geq T_c$, a Taylor expansion of $P(T, v)$ in the vicinity of v_c, for any T, gives

$$P(T, v) = P(T, v_c) + \left.\frac{\partial P}{\partial v}\right|_T (v - v_c) + \frac{1}{2} \left.\frac{\partial^2 P}{\partial v^2}\right|_T (v - v_c)^2 + \frac{1}{6} \left.\frac{\partial^3 P}{\partial v^3}\right|_T (v - v_c)^3 + \cdots$$

(5.82)

Since $\partial P/\partial v|_T$ and $\partial^2 P/\partial v^2|_T$ are both zero at T_c, the expansion of the derivatives around the critical point gives

$$P(T, v_c) = P_c + \alpha(T - T_c) + \mathcal{O}\left[(T - T_c)^2\right],$$

$$\left.\frac{\partial P}{\partial v}\right|_{T,v_c} = -a(T - T_c) + \mathcal{O}\left[(T - T_c)^2\right],$$

$$\left.\frac{\partial^2 P}{\partial v^2}\right|_{T,v_c} = b(T - T_c) + \mathcal{O}\left[(T - T_c)^2\right],$$

$$\left.\frac{\partial^3 P}{\partial v^3}\right|_{T_c,v_c} = -c + \mathcal{O}\left[(T - T_c)\right],$$

(5.83)

where a, b, and c are material-dependent constants. The stability condition, $\delta P \delta v \le 0$, requires $a > 0$ (for $T > T_c$) and $c > 0$ (for $T = T_c$), but provides no information on the sign of b. If an analytical expansion is possible, the isotherms of any gas in the vicinity of its critical point must have the general form,

$$P(T, v) = P_c + \alpha(T - T_c) - a(T - T_c)(v - v_c) + \frac{b}{2}(T - T_c)(v - v_c)^2 - \frac{c}{6}(v - v_c)^3 + \cdots$$

(5.84)

Note that the third and fifth terms are of comparable magnitude for $(T - T_c) \sim (v - v_c)^2$. The fourth (and higher-order) terms can then be neglected when this condition is satisfied.

The analytic expansion of Eq. (5.84) results in the following predictions for behavior close to the critical point:

(a) The gas compressibility diverges on approaching the critical point from the high-temperature side, along the critical isochore ($v = v_c$), as

$$\lim_{T \to T_c^+} \kappa(T, v_c) = -\frac{1}{v_c} \left.\frac{\partial P}{\partial v}\right|_T^{-1} = \frac{1}{v_c a(T - T_c)}.$$

(5.85)

(b) The critical isotherm ($T = T_c$) behaves as

$$P = P_c - \frac{c}{6}(v - v_c)^3 + \cdots$$

(5.86)

(c) Equation (5.84) is manifestly inapplicable to $T < T_c$. However, we can try to extract information for the low-temperature side by applying the Maxwell construction to the unstable isotherms (see the problem section). Actually, dimensional considerations are sufficient to show that on approaching T_c from the low-temperature side, the specific volumes of the coexisting liquid and gas phases approach each other as

$$\lim_{T \to T_c^-} (v_{\text{gas}} - v_{\text{liquid}}) \propto (T_c - T)^{1/2}.$$

(5.87)

The liquid–gas transition for $T < T_c$ is accompanied by a discontinuity in density, and the release of latent heat L. This kind of transition is usually referred to as *first order* or discontinuous. The difference between the two phases disappears as the sequence of first-order transitions terminates at the critical point. The singular behavior at this point is attributed to a *second-order* or continuous transition. Equation (5.84) follows from no more than the constraints of mechanical stability, and the assumption of analytical isotherms. Although there are some unknown coefficients, Eqs. (5.86)–(5.87) predict universal forms for the singularities close to the critical point and can be compared to experimental data. The experimental results do indeed confirm the general expectations of singular behavior, and there is universality in the results for different gases that can be summarized as:

(a) The compressibility diverges on approaching the critical point as

$$\lim_{T \to T_c^+} \kappa(T, v_c) \propto (T - T_c)^{-\gamma}, \quad \text{with} \quad \gamma \approx 1.3. \tag{5.88}$$

(b) The critical isotherm behaves as

$$(P - P_c) \propto (v - v_c)^\delta, \quad \text{with} \quad \delta \approx 5.0. \tag{5.89}$$

(c) The difference between liquid and gas densities vanishes close to the critical point as

$$\lim_{T \to T_c^-} (\rho_{\text{liquid}} - \rho_{\text{gas}}) \propto \lim_{T \to T_c^-} (v_{\text{gas}} - v_{\text{liquid}}) \propto (T_c - T)^\beta, \quad \text{with} \quad \beta \approx 0.3. \tag{5.90}$$

These results clearly indicate that the assumption of analyticity of the isotherms, leading to Eq. (5.84), is not correct. The exponents δ, γ, and β appearing in Eqs. (5.89)–(5.90) are known as *critical exponents*. Understanding the origin, universality, and numerical values of these exponents is a fascinating subject covered by the modern theory of critical phenomena, which is explored in a companion volume.

Problems for chapter 5

1. *Debye–Hückel theory and ring diagrams*: the virial expansion gives the gas pressure as an *analytic* expansion in the density $n = N/V$. Long-range interactions can result in *non-analytic* corrections to the ideal gas equation of state. A classic example is the Coulomb interaction in plasmas, whose treatment by Debye–Hückel theory is equivalent to summing all the *ring diagrams* in a cumulant expansion.

   For simplicity consider a gas of N electrons moving in a uniform background of positive charge density Ne/V to ensure overall charge neutrality. The Coulomb interaction takes the form

   $$\mathcal{U}_Q = \sum_{i<j} \mathcal{V}(\vec{q}_i - \vec{q}_j), \quad \text{with} \quad \mathcal{V}(\vec{q}) = \frac{e^2}{4\pi|\vec{q}|} - c.$$

   The constant c results from the background and ensures that the first-order correction vanishes, that is, $\int \mathrm{d}^3\vec{q}\, \mathcal{V}(\vec{q}) = 0$.

(a) Show that the Fourier transform of $\mathcal{V}(\vec{q})$ takes the form

$$\tilde{\mathcal{V}}(\vec{\omega}) = \begin{cases} e^2/\omega^2 & \text{for } \vec{\omega} \neq 0 \\ 0 & \text{for } \vec{\omega} = 0 \end{cases}.$$

(b) In the cumulant expansion for $\langle \mathcal{U}_Q^\ell \rangle_c^0$, we shall retain only the diagrams forming a ring, which are proportional to

$$R_\ell = \int \frac{d^3\vec{q}_1}{V} \cdots \frac{d^3\vec{q}_\ell}{V} \, \mathcal{V}(\vec{q}_1 - \vec{q}_2)\mathcal{V}(\vec{q}_2 - \vec{q}_3) \cdots \mathcal{V}(\vec{q}_\ell - \vec{q}_1).$$

Use properties of Fourier transforms to show that

$$R_\ell = \frac{1}{V^{\ell-1}} \int \frac{d^3\vec{\omega}}{(2\pi)^3} \, \tilde{\mathcal{V}}(\vec{\omega})^\ell.$$

(c) Show that the number of ring graphs generated in $\langle \mathcal{U}_Q^\ell \rangle_c^0$ is

$$S_\ell = \frac{N!}{(N-\ell)!} \times \frac{(\ell-1)!}{2} \approx \frac{(\ell-1)!}{2} N^\ell.$$

(d) Show that the contribution of the ring diagrams can be summed as

$$\ln Z_{\text{rings}} = \ln Z_0 + \sum_{\ell=2}^\infty \frac{(-\beta)^\ell}{\ell!} S_\ell R_\ell$$

$$\approx \ln Z_0 + \frac{V}{2} \int_0^\infty \frac{4\pi\omega^2 d\omega}{(2\pi)^3} \left[\left(\frac{\kappa}{\omega}\right)^2 - \ln\left(1 + \frac{\kappa^2}{\omega^2}\right) \right],$$

where $\kappa = \sqrt{\beta e^2 N/V}$ is the inverse Debye screening length.
(*Hint.* Use $\ln(1+x) = -\sum_{\ell=1}^\infty (-x)^\ell/\ell$.)

(e) The integral in the previous part can be simplified by changing variables to $x = \kappa/\omega$, and performing integration by parts. Show that the final result is

$$\ln Z_{\text{rings}} = \ln Z_0 + \frac{V}{12\pi} \kappa^3.$$

(f) Calculate the correction to pressure from the above ring diagrams.
(g) We can introduce an effective potential $\overline{V}(\vec{q} - \vec{q}')$ between two particles by integrating over the coordinates of all the other particles. This is equivalent to an expectation value that can be calculated perturbatively in a cumulant expansion. If we include only the loopless diagrams (the analog of the rings) between the particles, we have

$$\overline{V}(\vec{q} - \vec{q}') = V(\vec{q} - \vec{q}') + \sum_{\ell=1}^\infty (-\beta N)^\ell \int \frac{d^3\vec{q}_1}{V} \cdots \frac{d^3\vec{q}_\ell}{V} \, \mathcal{V}(\vec{q} - \vec{q}_1)\mathcal{V}(\vec{q}_1 - \vec{q}_2) \cdots$$

$$\mathcal{V}(\vec{q}_\ell - \vec{q}').$$

Show that this sum leads to the screened Coulomb interaction $\overline{V}(\vec{q}) = e^2 \exp(-\kappa|\vec{q}|)/(4\pi|\vec{q}|)$.

* * * * * * *

2. *Virial coefficients*: consider a gas of particles in d-dimensional space interacting through a pairwise central potential, $\mathcal{V}(r)$, where

$$\mathcal{V}(r) = \begin{cases} +\infty & \text{for } 0 < r < a, \\ -\varepsilon & \text{for } a < r < b, \\ 0 & \text{for } b < r < \infty. \end{cases}$$

(a) Calculate the second virial coefficient $B_2(T)$, and comment on its high- and low-temperature behaviors.

(b) Calculate the first correction to isothermal compressibility

$$\kappa_T = -\frac{1}{V} \left.\frac{\partial V}{\partial P}\right|_{T,N}.$$

(c) In the high-temperature limit, reorganize the equation of state into the van der Waals form, and identify the van der Waals parameters.

(d) For $b = a$ (a hard sphere), and $d = 1$, calculate the third virial coefficient $B_3(T)$.

****

3. *Dieterici's equation*: a gas obeys Dieterici's equation of state:

$$P(v - b) = k_B T \exp\left(-\frac{a}{k_B T v}\right),$$

where $v = V/N$.

(a) Find the ratio $Pv/k_B T$ at the critical point.

(b) Calculate the isothermal compressibility κ_T for $v = v_c$ as a function of $T - T_c$.

(c) On the critical isotherm expand the pressure to the lowest non-zero order in $(v - v_c)$.

****

4. *Two-dimensional Coulomb gas*: consider a classical mixture of N positive and N negative charged particles in a *two-dimensional* box of area $A = L \times L$. The Hamiltonian is

$$\mathcal{H} = \sum_{i=1}^{2N} \frac{\vec{p}_i^{\,2}}{2m} - \sum_{i<j}^{2N} c_i c_j \ln |\vec{q}_i - \vec{q}_j|,$$

where $c_i = +c_0$ for $i = 1, \cdots, N$, and $c_i = -c_0$ for $i = N+1, \cdots, 2N$, denote the charges of the particles; $\{\vec{q}_i\}$ and $\{\vec{p}_i\}$ their coordinates and momenta, respectively.

(a) Note that in the interaction term each pair appears only once, and there is no self-interaction $i = j$. How many pairs have repulsive interactions, and how many have attractive interactions?

(b) Write down the expression for the partition function $Z(N, T, A)$ in terms of integrals over $\{\vec{q}_i\}$ and $\{\vec{p}_i\}$. Perform the integrals over the momenta, and rewrite the contribution of the coordinates as a product involving powers of $\{\vec{q}_i\}$, using the identity $e^{\ln x} = x$.

(c) Although it is not possible to perform the integrals over $\{\vec{q}_i\}$ exactly, the dependence of Z on A can be obtained by the simple rescaling of coordinates, $\vec{q}_i{}' = \vec{q}_i/L$. Use the results in parts (a) and (b) to show that $Z \propto A^{2N-\beta c_0^2 N/2}$.

(d) Calculate the two-dimensional pressure of this gas, and comment on its behavior at high and low temperatures.

(e) The unphysical behavior at low temperatures is avoided by adding a hard core that prevents the coordinates of any two particles from coming closer than a distance a. The appearance of two length scales, a and L, makes the scaling analysis of part (c) questionable. By examining the partition function for $N = 1$, obtain an estimate for the temperature T_c at which the short distance scale a becomes important in calculating the partition function, invalidating the result of the previous part. What are the phases of this system at low and high temperatures?

$$********$$

5. *Exact solutions for a one-dimensional gas*: in statistical mechanics, there are very few systems of interacting particles that can be solved *exactly*. Such exact solutions are very important as they provide a check for the reliability of various approximations. A one-dimensional gas with short-range interactions is one such solvable case.

(a) Show that for a potential with a hard core that screens the interactions from further neighbors, the Hamiltonian for N particles can be written as

$$\mathcal{H} = \sum_{i=1}^{N} \frac{p_i^2}{2m} + \sum_{i=2}^{N} \mathcal{V}(x_i - x_{i-1}).$$

The (indistinguishable) particles are labeled with coordinates $\{x_i\}$ such that

$$0 \leq x_1 \leq x_2 \leq \cdots \leq x_N \leq L,$$

where L is the length of the box confining the particles.

(b) Write the expression for the partition function $Z(T, N, L)$. Change variables to $\delta_1 = x_1, \delta_2 = x_2 - x_1, \cdots, \delta_N = x_N - x_{N-1}$, and carefully indicate the allowed ranges of integration and the constraints.

(c) Consider the Gibbs partition function obtained from the Laplace transformation

$$\mathcal{Z}(T, N, P) = \int_0^\infty dL \exp(-\beta P L) Z(T, N, L),$$

and by extremizing the integrand find the standard formula for P in the canonical ensemble.

(d) Change variables from L to $\delta_{N+1} = L - \sum_{i=1}^{N} \delta_i$, and find the expression for $\mathcal{Z}(T, N, P)$ as a product over one-dimensional integrals over each δ_i.

(e) At a fixed pressure P, find expressions for the mean length $L(T, N, P)$, and the density $n = N/L(T, N, P)$ (involving ratios of integrals that should be easy to interpret).

Since the expression for $n(T, P)$ in part (e) is continuous and non-singular for any choice of potential, there is in fact no condensation transition for the one-dimensional gas. By contrast, the approximate van der Waals equation (or the mean-field treatment) incorrectly predicts such a transition.

(f) For a hard-sphere gas, with minimum separation a between particles, calculate the equation of state $P(T, n)$. Compare the excluded volume factor with the approximate result obtained in earlier problems, and also obtain the general virial coefficient $B_\ell(T)$.

****

6. *The Manning transition*: when ionic polymer (polelectrolytes) such as DNA are immersed in water, the negatively charged *counter-ions* go into solution, leaving behind a positively charged polymer. Because of the electrostatic repulsion of the charges left behind, the polymer stretches out into a cylinder of radius a, as illustrated in the figure. While thermal fluctuations tend to make the ions wander about in the solvent, electrostatic attractions favor their return and condensation on the polymer. If the number of counter-ions is N, they interact with the N positive charges left behind on the rod through the potential $U(r) = -2(Ne/L)\ln(r/L)$, where r is the radial coordinate in a cylindrical geometry. If we ignore the Coulomb repulsion between counter-ions, they can be described by the classical Hamiltonian

$$\mathcal{H} = \sum_{i=1}^{N}\left[\frac{p_i^2}{2m} + 2e^2 n \ln\left(\frac{r}{L}\right)\right],$$

where $n = N/L$.

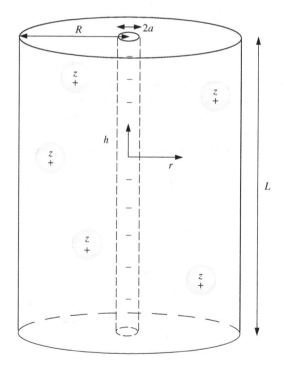

(a) For a cylindrical container of radius R, calculate the canonical partition function Z in terms of temperature T, density n, and radii R and a.

(b) Calculate the probability distribution function $p(r)$ for the radial position of a counter-ion, and its first moment $\langle r \rangle$, the average radial position of a counter-ion.

(c) The behavior of the results calculated above in the limit $R \gg a$ is very different at high and low temperatures. Identify the transition temperature, and characterize the nature of the two phases. In particular, how does $\langle r \rangle$ depend on R and a in each case?

(d) Calculate the pressure exerted by the counter-ions on the wall of the container, at $r = R$, in the limit $R \gg a$, at all temperatures.

(e) The character of the transition examined in part (d) is modified if the Coulomb interactions between counter-ions are taken into account. An approximate approach to the interacting problem is to allow a fraction N_1 of counter-ions to condense along the polymer rod, while the remaining $N_2 = N - N_1$ fluctuate in the solvent. The free counter-ions are again treated as non-interacting particles, governed by the Hamiltonian

$$\mathcal{H} = \sum_{i=1}^{N} \left[\frac{p_i^2}{2m} + 2e^2 n_2 \ln \left(\frac{r}{L} \right) \right],$$

where $n_2 = N_2/L$. *Guess* the equilibrium number of non-interacting ions, N_2^*, and justify your guess by discussing the response of the system to slight deviations from N_2^*. (This is a qualitative question for which no new calculations are needed.)

****

7. *Hard rods*: a collection of N asymmetric molecules in two dimensions may be modeled as a gas of rods, each of length $2l$ and lying in a plane. A rod can move by translation of its center of mass and rotation about the latter, as long as it does not encounter another rod. Without treating the hard-core interaction exactly, we can incorporate it approximately by assuming that the rotational motion of each rod is restricted (by the other rods) to an angle θ, which in turn introduces an excluded volume $\Omega(\theta)$ (associated with each rod). The value of θ is then calculated self-consistently by maximizing the entropy at a given density $n = N/V$, where V is the total accessible area.

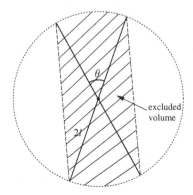

(a) Write down the entropy of such a collection of rods in terms of N, n, Ω, and $A(\theta)$, the phase space volume associated with the rotational freedom of a *single* rod. (You may ignore the momentum contributions throughout, and consider the large N limit.)

(b) Extremizing the entropy as a function of θ, relate the density to Ω, A, and their derivatives Ω', A'; express your result in the form $n = f(\Omega, A, \Omega', A')$.

(c) Express the excluded volume Ω in terms of θ and sketch f as a function of $\theta \in [0, \pi]$, assuming $A \propto \theta$.

(d) Describe the equilibrium state at high densities. Can you identify a phase transition as the density is decreased? Draw the corresponding critical density n_c on your sketch. What is the critical angle θ_c at the transition? You don't need to calculate θ_c explicitly, but give an (implicit) relation defining it. What value does θ adopt at $n < n_c$?

****

8. *Surfactant condensation*: N surfactant molecules are added to the surface of water over an area A. They are subject to a Hamiltonian

$$\mathcal{H} = \sum_{i=1}^{N} \frac{\vec{p}_i^{\,2}}{2m} + \frac{1}{2} \sum_{i,j} \mathcal{V}(\vec{r}_i - \vec{r}_j),$$

where $\vec{r}_i$ and $\vec{p}_i$ are two-dimensional vectors indicating the position and momentum of particle i. (This simple form ignores the couplings to the fluid itself. The actual kinetic and potential energies are more complicated.)

(a) Write down the expression for the partition function $Z(N, T, A)$ in terms of integrals over $\vec{r}_i$ and $\vec{p}_i$, and perform the integrals over the momenta.

The interparticle potential $\mathcal{V}(\vec{r})$ is infinite for separations $|\vec{r}| < a$, and attractive for $|\vec{r}| > a$ such that $\int_a^\infty 2\pi r dr \mathcal{V}(r) = -u_0$.

(b) Estimate the total non-excluded area available in the positional phase space of the system of N particles.

(c) Estimate the total *potential* energy of the system, within a *uniform density approximation* $n = N/A$. Using this potential energy for all configurations allowed in the previous part, write down an approximation for Z.

(d) The surface tension of water without surfactants is σ_0, approximately independent of temperature. Calculate the surface tension $\sigma(n, T)$ in the presence of surfactants.

(e) Show that below a certain temperature, T_c, the expression for σ is manifestly incorrect. What do you think happens at low temperatures?

(f) Compute the heat capacities, C_A, and write down an expression for C_σ without explicit evaluation, due to the surfactants.

****

9. *Critical point behavior*: the pressure P of a gas is related to its density $n = N/V$, and temperature T by the truncated expansion

$$P = k_B T n - \frac{b}{2} n^2 + \frac{c}{6} n^3,$$

where b and c are assumed to be *positive* temperature-independent constants.

(a) Locate the critical temperature T_c below which this equation must be invalid, and the corresponding density n_c and pressure P_c of the critical point. Hence find the ratio $k_B T_c n_c / P_c$.

(b) Calculate the isothermal compressibility $\kappa_T = -\frac{1}{V} \frac{\partial V}{\partial P}\big|_T$, and sketch its behavior as a function of T for $n = n_c$.

(c) On the critical isotherm give an expression for $(P - P_c)$ as a function of $(n - n_c)$.

(d) The instability in the isotherms for $T < T_c$ is avoided by phase separation into a liquid of density n_+ and a gas of density n_-. For temperatures close to T_c, these densities behave as $n_\pm \approx n_c (1 \pm \delta)$. Using a Maxwell construction, or otherwise, find an implicit equation for $\delta(T)$, and indicate its behavior for $(T_c - T) \to 0$. (*Hint.* Along an isotherm, variations of chemical potential obey $d\mu = dP/n$.)

****

10. *The binary alloy:* a binary alloy (as in β brass) consists of N_A atoms of type A, and N_B atoms of type B. The atoms form a simple cubic lattice, each interacting only with its six nearest neighbors. Assume an attractive energy of $-J$ ($J > 0$) between like neighbors $A - A$ and $B - B$, but a repulsive energy of $+J$ for an $A - B$ pair.

(a) What is the minimum energy configuration, or the state of the system at zero temperature?

(b) Estimate the total interaction energy assuming that the atoms are randomly distributed among the N sites; that is, each site is occupied independently with probabilities $p_A = N_A/N$ and $p_B = N_B/N$.

(c) Estimate the mixing entropy of the alloy with the same approximation. Assume $N_A, N_B \gg 1$.

(d) Using the above, obtain a free energy function $F(x)$, where $x = (N_A - N_B)/N$. Expand $F(x)$ to the fourth order in x, and show that the requirement of convexity of F breaks down below a critical temperature T_c. For the remainder of this problem use the expansion obtained in (d) in place of the full function $F(x)$.

(e) Sketch $F(x)$ for $T > T_c$, $T = T_c$, and $T < T_c$. For $T < T_c$ there is a range of compositions $x < |x_{sp}(T)|$, where $F(x)$ is not convex and hence the composition is locally unstable. Find $x_{sp}(T)$.

(f) The alloy globally minimizes its free energy by separating into A-rich and B-rich phases of compositions $\pm x_{eq}(T)$, where $x_{eq}(T)$ minimizes the function $F(x)$. Find $x_{eq}(T)$.

(g) In the (T, x) plane sketch the phase separation boundary $\pm x_{eq}(T)$, and the so-called spinodal line $\pm x_{sp}(T)$. (The spinodal line indicates onset of metastability and hysteresis effects.)

****

6

Quantum statistical mechanics

There are limitations to the applicability of classical statistical mechanics. The need to include quantum mechanical effects becomes especially apparent at low temperatures. In this chapter we shall first demonstrate the failure of the classical results in the contexts of heat capacities of molecular gases and solids, and the ultraviolet catastrophe in black-body radiation. We shall then reformulate statistical mechanics using quantum concepts.

6.1 Dilute polyatomic gases

Consider a dilute gas of polyatomic molecules. The Hamiltonian for *each* molecule of n atoms is

$$\mathcal{H}_1 = \sum_{i=1}^{n} \frac{\vec{p}_i^{\,2}}{2m} + \mathcal{V}(\vec{q}_1, \ldots, \vec{q}_n), \tag{6.1}$$

where the potential energy $\mathcal{V}$ contains all the information on molecular bonds. For simplicity, we have assumed that all atoms in the molecule have the same mass. If the masses are different, the Hamiltonian can be brought into the above form by rescaling the coordinates $\vec{q}_i$ by $\sqrt{m_i/m}$ (and the momenta by $\sqrt{m/m_i}$), where m_i is the mass of the ith atom. Ignoring the interactions *between* molecules, the partition function of a dilute gas is

$$Z(N) = \frac{Z_1^N}{N!} = \frac{1}{N!} \left\{ \int \prod_{i=1}^{n} \frac{\mathrm{d}^3\vec{p}_i \, \mathrm{d}^3\vec{q}_i}{h^3} \, \exp\left[-\beta \sum_{i=1}^{n} \frac{\vec{p}_i^{\,2}}{2m} - \beta \mathcal{V}(\vec{q}_1, \ldots, \vec{q}_n) \right] \right\}^N. \tag{6.2}$$

The chemical bonds that keep the molecule together are usually quite strong (energies of the order of electron volts). At typical accessible temperatures, which are much smaller than the corresponding dissociation temperatures ($\approx 10^4\,\mathrm{K}$), the molecule has a well-defined shape and only undergoes small deformations. The contribution of these deformations to the one-particle partition function Z_1 can be computed as follows:

(a) The first step is to find the equilibrium positions, $(\vec{q}_1^{\,*}, \ldots, \vec{q}_n^{\,*})$, by minimizing the potential $\mathcal{V}$.

(b) The energy cost of small deformations about equilibrium is then obtained by setting $\vec{q}_i = \vec{q}_i^* + \vec{u}_i$, and making an expansion in powers of $\vec{u}$,

$$\mathcal{V} = \mathcal{V}^* + \frac{1}{2} \sum_{i,j=1}^{n} \sum_{\alpha,\beta=1}^{3} \frac{\partial^2 \mathcal{V}}{\partial q_{i,\alpha} \partial q_{j,\beta}} u_{i,\alpha} u_{j,\beta} + \mathcal{O}(u^3). \tag{6.3}$$

(Here $i, j = 1, \cdots, n$ identify the atoms, and $\alpha, \beta = 1, 2, 3$ label a particular component.) Since the expansion is around a stable equilibrium configuration, the first derivatives are absent in Eq. (6.3), and the matrix of second derivatives is *positive definite*, that is, it has only non-negative eigenvalues.

(c) The *normal modes* of the molecule are obtained by diagonalizing the $3n \times 3n$ matrix $\partial^2 \mathcal{V} / \partial q_{i,\alpha} \partial q_{j,\beta}$. The resulting $3n$ eigenvalues, $\{K_s\}$, indicate the *stiffness* of each mode. We can change variables from the original deformations $\{\vec{u}_i\}$ to the amplitudes $\{\tilde{u}_s\}$ of the eigenmodes. The corresponding conjugate momenta are $\{\tilde{p}_s = m\dot{\tilde{u}}_s\}$. Since the transformation from $\{\vec{u}_i\}$ to $\{\tilde{u}_s\}$ is *unitary* (preserving the length of a vector), $\sum_i \vec{p}_i^{\,2} = \sum_s \tilde{p}_s^2$, and the quadratic part of the resulting deformation Hamiltonian is

$$\mathcal{H}_1 = \mathcal{V}^* + \sum_{s=1}^{3n} \left[\frac{1}{2m} \tilde{p}_s^2 + \frac{K_s}{2} \tilde{u}_s^2 \right]. \tag{6.4}$$

(Such transformations are also *canonical*, preserving the measure of integration in phase space, $\prod_{i,\alpha} du_{i,\alpha} dp_{i,\alpha} = \prod_s d\tilde{u}_s d\tilde{p}_s$.)

The average energy of each molecule is the expectation value of the above Hamiltonian. Since each quadratic degree of freedom classically contributes a factor of $k_B T/2$ to the energy,

$$\langle \mathcal{H}_1 \rangle = \mathcal{V}^* + \frac{3n+m}{2} k_B T. \tag{6.5}$$

Only modes with a finite stiffness can store potential energy, and m is defined as the number of such modes with non-zero K_s. The following symmetries of the potential force some eigenvalues to be zero:

(a) *Translation symmetry*: since $\mathcal{V}(\vec{q}_1 + \vec{c}, \cdots, \vec{q}_n + \vec{c}) = \mathcal{V}(\vec{q}_1, \cdots, \vec{q}_n)$, no energy is stored in the center of mass coordinate $\vec{Q} = \sum_\alpha \vec{q}_\alpha / n$, that is, $\mathcal{V}(\vec{Q}) = \mathcal{V}(\vec{Q} + \vec{c})$, and the corresponding three values of K_{trans} are zero.

(b) *Rotation symmetry*: there is also no potential energy associated with rotations of the molecule, and $K_{\text{rot}} = 0$ for the corresponding stiffnesses. The number of rotational modes, $0 \leq r \leq 3$, depends on the shape of the molecule. Generally $r = 3$, except for a single atom that has $r = 0$ (no rotations are possible), and any rod-shaped molecule, which has $r = 2$ as a rotation parallel to its axis does not result in a new configuration.

The remaining $m = 3n - 3 - r$ eigenvectors of the matrix have non-zero stiffness, and correspond to the *vibrational* normal modes. The energy per molecule, from Eq. (6.5), is thus

$$\langle \mathcal{H}_1 \rangle = \frac{6n - 3 - r}{2} k_B T. \tag{6.6}$$

The corresponding heat capacities,

$$C_V = \frac{(6n-3-r)}{2}k_B, \quad \text{and} \quad C_P = C_V + k_B = \frac{(6n-1-r)}{2}k_B, \tag{6.7}$$

are temperature-independent. The ratio $\gamma = C_P/C_V$ is easily measured in adiabatic processes. Values of γ, expected on the basis of the above argument, are listed below for a number of different molecules.

Monatomic	He	$n=1$	$r=0$	$\gamma = 5/3$
Diatomic	O_2 or CO	$n=2$	$r=2$	$\gamma = 9/7$
Linear triatomic	O–C–O	$n=3$	$r=2$	$\gamma = 15/13$
Planar triatomic	H$/^{O}\backslash$H	$n=3$	$r=3$	$\gamma = 14/12 = 7/6$
Tetra-atomic	NH_3	$n=4$	$r=3$	$\gamma = 20/18 = 10/9$

Fig. 6.1 Temperature dependence of the heat capacity for a typical gas of diatomic molecules.

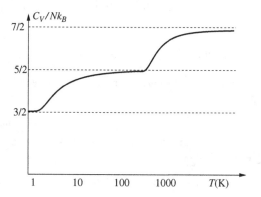

Measurements of the heat capacity of dilute gases do not agree with the above predictions. For example, the value $C_V/k_B = 7/2$, for a diatomic gas such as oxygen, is only observed at temperatures higher than a few thousand Kelvin. At room temperatures, a lower value of 5/2 is observed, while at even lower temperatures of around 10 K, it is further reduced to 3/2. The low-temperature value is similar to that of a monatomic gas, and suggests that no energy is stored in the rotational and vibrational degrees of freedom. These observations can be explained if the allowed energy levels are *quantized*, as follows.

Vibrational modes: a diatomic molecule has one vibrational mode with stiffness $K \equiv m\omega^2$, where ω is the frequency of oscillations. The classical partition function for this mode is

$$\begin{aligned}
Z_{\text{vib}}^c &= \int \frac{dp\,dq}{h} \exp\left[-\beta\left(\frac{p^2}{2m} + \frac{m\omega^2 q^2}{2}\right)\right] \\
&= \frac{1}{h}\sqrt{\left(\frac{2\pi m}{\beta}\right)\left(\frac{2\pi}{\beta m\omega^2}\right)} = \frac{2\pi}{h\beta\omega} = \frac{k_B T}{\hbar\omega},
\end{aligned} \tag{6.8}$$

where $\hbar = h/2\pi$. The corresponding energy stored in this mode,

$$\langle \mathcal{H}_{\text{vib}} \rangle^c = -\frac{\partial \ln Z}{\partial \beta} = \frac{\partial \ln(\beta \hbar \omega)}{\partial \beta} = \frac{1}{\beta} = k_B T, \tag{6.9}$$

comes from $k_B T/2$ per kinetic and potential degrees of freedom. In quantum mechanics, the allowed values of energy are quantized such that

$$\mathcal{H}_{\text{vib}}^q = \hbar \omega \left(n + \frac{1}{2} \right), \tag{6.10}$$

with $n = 0, 1, 2, \cdots$. Assuming that the probability of each discrete level is proportional to its Boltzmann weight (as will be justified later on), there is a normalization factor

$$Z_{\text{vib}}^q = \sum_{n=0}^{\infty} e^{-\beta \hbar \omega (n+1/2)} = \frac{e^{-\beta \hbar \omega/2}}{1 - e^{-\beta \hbar \omega}}. \tag{6.11}$$

The high-temperature limit,

$$\lim_{\beta \to 0} Z_{\text{vib}}^q = \frac{1}{\beta \hbar \omega} = \frac{k_B T}{\hbar \omega},$$

coincides with Eq. (6.8) (due in part to the choice of h as the measure of classical phase space).

The expectation value of vibrational energy is

$$E_{\text{vib}}^q = -\frac{\partial \ln Z}{\partial \beta} = \frac{\hbar \omega}{2} + \frac{\partial}{\partial \beta} \ln \left(1 - e^{-\beta \hbar \omega} \right) = \frac{\hbar \omega}{2} + \hbar \omega \frac{e^{-\beta \hbar \omega}}{1 - e^{-\beta \hbar \omega}}. \tag{6.12}$$

The first term is the energy due to *quantum fluctuations* that are present even in the zero-temperature ground state. The second term describes the additional energy due to thermal fluctuations. The resulting heat capacity,

$$C_{\text{vib}}^q = \frac{dE_{\text{vib}}^q}{dT} = k_B \left(\frac{\hbar \omega}{k_B T} \right)^2 \frac{e^{-\beta \hbar \omega}}{(1 - e^{-\beta \hbar \omega})^2}, \tag{6.13}$$

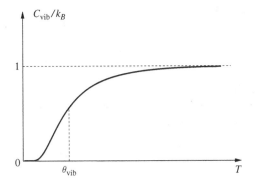

Fig. 6.2 The contribution of (quantized) vibrations to the heat capacity.

achieves the classical value of k_B only at temperatures $T \gg \theta_{\text{vib}}$, where $\theta_{\text{vib}} = \hbar \omega / k_B$ is a characteristic temperature associated with the quanta of vibrational energy. For $T \ll \theta_{\text{vib}}$, C^q_{vib} goes to zero as $\exp(-\theta_{\text{vib}}/T)$. Typical values of θ_{vib} are in the range of 10^3 to 10^4 K, explaining why the classical value of heat capacity is observed only at higher temperatures.

Rotational modes: to account for the low-temperature anomaly in the heat capacity of diatomic molecules, we have to study the quantization of the rotational degrees of freedom. Classically, the orientation of a diatomic molecule is specified by two angles θ and ϕ, and its Lagrangian (equal to the kinetic energy) is

$$\mathcal{L} = \frac{I}{2} \left(\dot{\theta}^2 + \sin^2 \theta \, \dot{\phi}^2 \right), \tag{6.14}$$

where I is the moment of inertia. In terms of the conjugate momenta,

$$p_\theta = \frac{\partial \mathcal{L}}{\mathrm{d}\dot{\theta}} = I \dot{\theta}, \quad p_\phi = \frac{\partial \mathcal{L}}{\mathrm{d}\dot{\phi}} = I \sin^2 \theta \, \dot{\phi}, \tag{6.15}$$

the Hamiltonian for rotations is

$$\mathcal{H}_{\text{rot}} = \frac{1}{2I} \left(p_\theta^2 + \frac{p_\phi^2}{\sin^2 \theta} \right) \equiv \frac{\vec{L}^2}{2I}, \tag{6.16}$$

where $\vec{L}$ is the angular momentum. From the classical partition function,

$$Z^c_{\text{rot}} = \frac{1}{h^2} \int_0^\pi \mathrm{d}\theta \int_0^{2\pi} \mathrm{d}\phi \int_{-\infty}^\infty \mathrm{d}p_\theta \, \mathrm{d}p_\phi \exp\left[-\frac{\beta}{2I} \left(p_\theta^2 + \frac{p_\phi^2}{\sin^2 \theta} \right) \right]$$

$$= \left(\frac{2\pi I}{\beta} \right) \left(\frac{4\pi}{h^2} \right) = \frac{2Ik_B T}{\hbar^2}, \tag{6.17}$$

the stored energy is

$$\langle E_{\text{rot}} \rangle^c = -\frac{\partial \ln Z}{\partial \beta} = \frac{\partial}{\partial \beta} \ln \left(\frac{\beta \hbar^2}{2I} \right) = k_B T, \tag{6.18}$$

as expected for two degrees of freedom. In quantum mechanics, the allowed values of angular momentum are quantized to $\vec{L}^2 = \hbar^2 \ell(\ell+1)$ with $\ell = 0, 1, 2, \cdots$, and each state has a degeneracy of $2\ell + 1$ (along a selected direction, $L_z = -\ell, \cdots, +\ell$). A partition function is now obtained for these levels as

$$Z^q_{\text{rot}} = \sum_{\ell=0}^\infty \exp\left[-\frac{\beta \hbar^2 \ell(\ell+1)}{2I} \right] (2\ell+1) = \sum_{\ell=0}^\infty \exp\left[-\frac{\theta_{\text{rot}} \ell(\ell+1)}{T} \right] (2\ell+1), \tag{6.19}$$

where $\theta_{\text{rot}} = \hbar^2/(2Ik_B)$ is a characteristic temperature associated with quanta of rotational energy. While the sum cannot be analytically evaluated in general, we can study its high- and low-temperature limits.

(a) For $T \gg \theta_{\text{rot}}$, the terms in Eq. (6.19) vary slowly, and the sum can be replaced by the integral

$$\lim_{T \to \infty} Z^q_{\text{rot}} = \int_0^\infty \mathrm{d}x (2x+1) \, \exp\left[-\frac{\theta_{\text{rot}} x(x+1)}{T} \right]$$

$$= \int_0^\infty \mathrm{d}y \, e^{-\theta_{\text{rot}} y/T} = \frac{T}{\theta_{\text{rot}}} = Z^c_{\text{rot}}, \tag{6.20}$$

that is, the classical result of Eq. (6.17) is recovered.

(b) For $T \ll \theta_{\text{rot}}$, the first few terms dominate the sum, and

$$\lim_{T \to \infty} Z_{\text{rot}}^q = 1 + 3\text{e}^{-2\theta_{\text{rot}}/T} + \mathcal{O}(\text{e}^{-6\theta_{\text{rot}}/T}),\tag{6.21}$$

leading to an energy

$$E_{\text{rot}}^q = -\frac{\partial \ln Z}{\partial \beta} \approx -\frac{\partial}{\partial \beta} \ln\left[1 + 3\text{e}^{-2\theta_{\text{rot}}/T}\right] \approx 6k_B \theta_{\text{rot}} \, \text{e}^{-2\theta_{\text{rot}}/T}.\tag{6.22}$$

The resulting heat capacity vanishes at low temperatures as

$$C_{\text{rot}} = \frac{\text{d}E_{\text{rot}}^q}{\text{d}T} = 3k_B \left(\frac{2\theta_{\text{rot}}}{T}\right)^2 \text{e}^{-2\theta_{\text{rot}}/T} + \cdots .\tag{6.23}$$

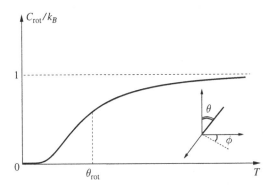

Fig. 6.3 The contribution of (quantized) rotations to the heat capacity of a rod-shaped molecule.

Typical values of θ_{rot} are between 1 and 10 K, explaining the lower temperature shoulder in the heat capacity measurements. At very low temperatures, the only contributions come from the kinetic energy of the center of mass, and the molecule behaves as a monatomic particle. (The heat capacity vanishes at even lower temperatures due to quantum statistics, as will be discussed in the context of identical particles.)

6.2 Vibrations of a solid

Attractive interactions between particles first lead to condensation from a gas to liquid at low temperatures, and finally cause freezing into a solid state at even lower temperatures. For the purpose of discussing its thermodynamics, the solid can be regarded as a very large molecule subject to a Hamiltonian similar to Eq. (6.1), with $n = N \gg 1$ atoms. We can then proceed with the steps outlined in the previous section.

(a) The classical ground state configuration of the solid is obtained by minimizing the potential $\mathcal{V}$. In almost all cases, the minimum energy corresponds to a periodic arrangement of atoms forming a *lattice*. In terms of the three *basis vectors*, $\hat{a}$, $\hat{b}$, and $\hat{c}$, the locations of atoms in a simple crystal are given by

$$\vec{q}^{\,*}(\ell, m, n) = \left[\ell\hat{a} + m\hat{b} + n\hat{c}\right] \equiv \vec{r},\tag{6.24}$$

where $\{\ell, m, n\}$ is a triplet of integers.

(b) At finite temperatures, the atoms may undergo small deformations

$$\vec{q}_{\vec{r}} = \vec{r} + \vec{u}(\vec{r}),$$ (6.25)

with a cost in potential energy of

$$\mathcal{V} = \mathcal{V}^* + \frac{1}{2} \sum_{\substack{\vec{r},\vec{r}' \\ \alpha,\beta}} \frac{\partial^2 \mathcal{V}}{\partial q_{\vec{r},\alpha} \partial q_{\vec{r}',\beta}} u_\alpha(\vec{r})\, u_\beta(\vec{r}') + O(u^3).$$ (6.26)

(c) Finding the normal modes of a crystal is considerably simplified by its *translational symmetry*. In particular, the matrix of second derivatives only depends on the *relative* separation of two points,

$$\frac{\partial^2 \mathcal{V}}{\partial q_{\vec{r},\alpha} \partial q_{\vec{r}',\beta}} = K_{\alpha\beta}(\vec{r} - \vec{r}').$$ (6.27)

It is always possible to take advantage of such a symmetry to at least partially diagonalize the matrix by using a *Fourier basis*,

$$u_\alpha(\vec{r}) = \sum_{\vec{k}}' \frac{e^{i\vec{k}\cdot\vec{r}}}{\sqrt{N}} \tilde{u}_\alpha(\vec{k}).$$ (6.28)

The sum is restricted to *wavevectors* $\vec{k}$ inside a *Brillouin zone*. For example, in a cubic lattice of spacing a, each component of $\vec{k}$ is restricted to the interval $[-\pi/a, \pi/a]$. This is because wavevectors outside this interval carry no additional information as $(k_x + 2\pi m/a)(na) = k_x(na) + 2mn\pi$, and any phase that is a multiple of 2π does not affect the sum in Eq. (6.28).

Fig. 6.4 The Brillouin zone of a square lattice of spacing a.

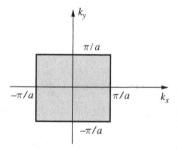

In terms of the Fourier modes, the potential energy of deformations is

$$\mathcal{V} = \mathcal{V}^* + \frac{1}{2N} \sum_{\substack{(\vec{r},\vec{r}'),(\vec{k},\vec{k}') \\ \alpha,\beta}} K_{\alpha\beta}(\vec{r} - \vec{r}')\, e^{i\vec{k}\cdot\vec{r}} \tilde{u}_\alpha(\vec{k})\, e^{i\vec{k}'\cdot\vec{r}'} \tilde{u}_\beta(\vec{k}').$$ (6.29)

We can change variables to relative and center of mass coordinates,

$$\vec{\rho} = \vec{r} - \vec{r}', \quad \text{and} \quad \vec{R} = \frac{\vec{r} + \vec{r}'}{2},$$

by setting

$$\vec{r} = \vec{R} + \frac{\vec{\rho}}{2}, \quad \text{and} \quad \vec{r}' = \vec{R} - \frac{\vec{\rho}}{2}.$$

Equation (6.29) now simplifies to

$$\mathcal{V} = \mathcal{V}^* + \frac{1}{2N} \sum_{\substack{\vec{k},\vec{k}' \\ \alpha,\beta}} \left(\sum_{\vec{R}} e^{i(\vec{k}+\vec{k}')\cdot\vec{R}} \right) \left(\sum_{\vec{\rho}} K_{\alpha\beta}(\vec{\rho}) \, e^{i(\vec{k}-\vec{k}')\cdot\vec{\rho}/2} \tilde{u}_\alpha(\vec{k}) \tilde{u}_\beta(\vec{k}') \right). \quad (6.30)$$

As the sum in the first brackets is $N\delta_{\vec{k}+\vec{k}',0}$,

$$\begin{aligned} \mathcal{V} &= \mathcal{V}^* + \frac{1}{2} \sum_{\vec{k},\alpha,\beta} \left[\sum_{\vec{\rho}} K_{\alpha\beta}(\vec{\rho}) \, e^{i\vec{k}\cdot\vec{\rho}} \right] \tilde{u}_\alpha(\vec{k}) \tilde{u}_\beta(-\vec{k}) \\ &= \mathcal{V}^* + \frac{1}{2} \sum_{\vec{k},\alpha,\beta} \tilde{K}_{\alpha\beta}(\vec{k}) \tilde{u}_\alpha(\vec{k}) \tilde{u}_\beta(\vec{k})^*, \end{aligned} \quad (6.31)$$

where $\tilde{K}_{\alpha\beta}(\vec{k}) = \sum_{\vec{\rho}} K_{\alpha\beta}(\vec{\rho}) \exp(i\vec{k}\cdot\vec{\rho})$, and $\tilde{u}_\beta(\vec{k})^* = \tilde{u}_\beta(-\vec{k})$ is the complex conjugate of $\tilde{u}_\beta(\vec{k})$.

The different Fourier modes are thus decoupled at the quadratic order, and the task of diagonalizing the $3N \times 3N$ matrix of second derivatives is reduced to diagonalizing the 3×3 matrix $\tilde{K}_{\alpha\beta}(\vec{k})$, separately for each $\vec{k}$. The form of $\tilde{K}_{\alpha\beta}$ is further restricted by the point group symmetries of the crystal. The discussion of such constraints is beyond the intent of this section and for simplicity we shall assume that $\tilde{K}_{\alpha\beta}(\vec{k}) = \delta_{\alpha,\beta} \tilde{K}(\vec{k})$ is already diagonal. (For an isotropic material, this implies a specific relation between bulk and shear moduli.)

The kinetic energy of deformations is

$$\sum_{i=1}^{N} \frac{m}{2} \dot{q}_i^2 = \sum_{\vec{k},\alpha} \frac{m}{2} \dot{\tilde{u}}_\alpha(\vec{k}) \dot{\tilde{u}}_\alpha(\vec{k})^* = \sum_{k,\alpha} \frac{1}{2m} \tilde{p}_\alpha(\vec{k}) \tilde{p}_\alpha(\vec{k})^*, \quad (6.32)$$

where

$$\tilde{p}_\alpha(\vec{k}) = \frac{\partial \mathcal{L}}{\partial \dot{\tilde{u}}_\alpha(\vec{k})} = m \dot{\tilde{u}}_\alpha(\vec{k})$$

is the momentum conjugate to $\tilde{u}_\alpha(\vec{k})$. The resulting deformation Hamiltonian,

$$\mathcal{H} = \mathcal{V}^* + \sum_{\vec{k},\alpha} \left[\frac{1}{2m} \left| \tilde{p}_\alpha(\vec{k}) \right|^2 + \frac{\tilde{K}(\vec{k})}{2} \left| \tilde{u}_\alpha(\vec{k}) \right|^2 \right], \quad (6.33)$$

describes $3N$ independent harmonic oscillators of frequencies $\omega_\alpha(\vec{k}) = \sqrt{\tilde{K}(\vec{k})/m}$.

In a classical treatment, each harmonic oscillator of non-zero stiffness contributes $k_B T$ to the internal energy of the solid. At most, six of the $3N$ oscillators are expected to have zero stiffness (corresponding to uniform translations and rotations of the crystal). Thus, up to non-extensive corrections of order $1/N$, the classical internal energy associated with the Hamiltonian (6.33) is $3Nk_BT$, leading to a temperature-independent heat capacity of $3k_B$ per atom. In fact, the measured heat capacity vanishes at low temperatures. We can again relate this observation to the quantization of the energy levels of each oscillator, as discussed in the previous section. Quantizing each harmonic mode separately gives the allowed values of the Hamiltonian as

$$\mathcal{H}^q = \mathcal{V}^* + \sum_{\vec{k},\alpha} \hbar\omega_\alpha(\vec{k}) \left(n_{\vec{k},\alpha} + \frac{1}{2} \right), \tag{6.34}$$

where the set of integers $\{n_{\vec{k},\alpha}\}$ describes the quantum microstate of the oscillators. Since the oscillators are independent, their partition function

$$Z^q = \sum_{\{n_{\vec{k},\alpha}\}} e^{-\beta\mathcal{H}^q} = e^{-\beta E_0} \prod_{\vec{k},\alpha} \sum_{n_{\vec{k},\alpha}} e^{-\beta\hbar\omega_\alpha(\vec{k})n_{\vec{k},\alpha}} = e^{-\beta E_0} \prod_{\vec{k},\alpha} \left[\frac{1}{1-e^{-\beta\hbar\omega_\alpha(\vec{k})}} \right] \tag{6.35}$$

is the product of single oscillator partition functions such as Eq. (6.11). (E_0 includes the ground state energies of all oscillators in addition to $\mathcal{V}^*$.)

The internal energy is

$$E(T) = \langle \mathcal{H}^q \rangle = E_0 + \sum_{\vec{k},\alpha} \hbar\omega_\alpha(\vec{k}) \left\langle n_\alpha(\vec{k}) \right\rangle, \tag{6.36}$$

where the average *occupation numbers* are given by

$$\left\langle n_\alpha(\vec{k}) \right\rangle = \frac{\sum_{n=0}^\infty n e^{-\beta\hbar\omega_\alpha(\vec{k})n}}{\sum_{n=0}^\infty e^{-\beta\hbar\omega_\alpha(\vec{k})n}} = -\frac{\partial}{\partial(\beta\hbar\omega_\alpha(\vec{k}))} \ln\left(\frac{1}{1-e^{-\beta\hbar\omega_\alpha(\vec{k})}} \right)$$
$$= \frac{e^{-\beta\hbar\omega_\alpha(\vec{k})}}{1-e^{-\beta\hbar\omega_\alpha(\vec{k})}} = \frac{1}{e^{\beta\hbar\omega_\alpha(\vec{k})} - 1}. \tag{6.37}$$

As a first attempt at including quantum mechanical effects, we may adopt the *Einstein model* in which all the oscillators are assumed to have the same frequency ω_E. This model corresponds to atoms that are pinned to their ideal location by springs of stiffness $\bar{K} = \partial^2\mathcal{V}/\partial q^2 = m\omega_E^2$. The resulting internal energy,

$$E = E_0 + 3N\frac{\hbar\omega_E \, e^{-\beta\hbar\omega_E}}{1-e^{-\beta\hbar\omega_E}}, \tag{6.38}$$

and heat capacity,

$$C = \frac{dE}{dT} = 3Nk_B \left(\frac{T_E}{T} \right)^2 \frac{e^{-T_E/T}}{(1-e^{-T_E/T})^2}, \tag{6.39}$$

are simply proportional to that of a single oscillator (Eqs. (6.12) and (6.13)). In particular, there is an exponential decay of the heat capacity to zero with a characteristic temperature $T_E = \hbar\omega_E/k_B$. However, the experimentally measured heat capacity decays to zero much more slowly, as T^3.

The discrepancy is resolved through the *Debye model*, which emphasizes that at low temperatures the main contribution to heat capacity is from the

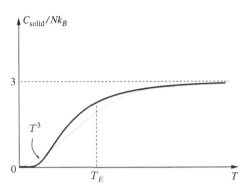

Fig. 6.5 The heat capacity of a solid (gray line) goes to zero more slowly than in the Einstein model (solid line).

oscillators of lowest frequency that are the most easily excited. The lowest energy modes in turn correspond to the smallest wave vectors $k = |\vec{k}|$, or the longest wavelengths $\lambda = 2\pi/k$. Indeed, the modes with $\vec{k} = 0$ simply describe pure translations of the lattice and have zero stiffness. By continuity, we expect $\lim_{\vec{k}\to 0} \tilde{K}(\vec{k}) = 0$, and ignoring considerations of crystal symmetry, the expansion of $\tilde{K}(\vec{k})$ at small wavevectors takes the form

$$\tilde{K}(\vec{k}) = Bk^2 + \mathcal{O}(k^4).$$

The odd terms are absent in the expansion, since $\tilde{K}(\vec{k}) = \tilde{K}(-\vec{k})$ follows from $K(\vec{r} - \vec{r}') = K(\vec{r}' - \vec{r})$ in real space. The corresponding frequencies at long wavelengths are

$$\omega(\vec{k}) = \sqrt{\frac{Bk^2}{m}} = vk, \tag{6.40}$$

where $v = \sqrt{B/m}$ is the *speed of sound* in the crystal. (In a real material, $\tilde{K}_{\alpha\beta}$ is not proportional to $\delta_{\alpha\beta}$, and different *polarizations* of sound have different velocities.)

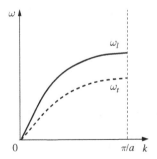

Fig. 6.6 The dispersion of phonons in an isotropic crystal has different branches for longitudinal and transverse waves.

The quanta of vibrational modes are usually referred to as *phonons*. Using the *dispersion relation* in Eq. (6.40), the contribution of low-energy phonons to the internal energy is

$$\langle \mathcal{H}^q \rangle = E_0 + \sum_{\vec{k},\alpha} \frac{\hbar v k}{e^{\beta \hbar v k} - 1}. \tag{6.41}$$

With *periodic boundary conditions* in a box of dimensions $L_x \times L_y \times L_z$, the allowed wavevectors are

$$\vec{k} = \left(\frac{2\pi n_x}{L_x}, \frac{2\pi n_y}{L_y}, \frac{2\pi n_z}{L_z} \right), \tag{6.42}$$

where n_x, n_y, and n_z are integers. In the large size limit, these modes are very densely packed, and the number of modes in a volume element $\mathrm{d}^3 \vec{k}$ is

$$\mathrm{d}\mathcal{N} = \frac{\mathrm{d}k_x}{2\pi/L_x} \frac{\mathrm{d}k_y}{2\pi/L_y} \frac{\mathrm{d}k_z}{2\pi/L_z} = \frac{V}{(2\pi)^3} \mathrm{d}^3 \vec{k} \equiv \rho \mathrm{d}^3 \vec{k}. \tag{6.43}$$

Using the *density of states* ρ, any sum over allowed wavevectors can be replaced by an integral as

$$\lim_{V \to \infty} \sum_{\vec{k}} f(\vec{k}) = \int \mathrm{d}^3 \vec{k} \, \rho f(\vec{k}). \tag{6.44}$$

Hence, Eq. (6.41) can be rewritten as

$$E = E_0 + 3V \int^{\mathrm{B.Z.}} \frac{\mathrm{d}^3 \vec{k}}{(2\pi)^3} \frac{\hbar v k}{e^{\beta \hbar v k} - 1}, \tag{6.45}$$

where the integral is performed over the volume of the Brillouin zone, and the factor of 3 comes from assuming the same sound velocity for the three polarizations.

Due to its dependence on the shape of the Brillouin zone, it is not possible to give a simple closed form expression for the energy in Eq. (6.45). However, we can examine its high- and low-temperature limits. The characteristic *Debye temperature* separating the two limits,

$$T_D = \frac{\hbar v k_{\max}}{k_B} \approx \frac{\hbar v}{k_B} \cdot \frac{\pi}{a}, \tag{6.46}$$

corresponds to the high-frequency modes at the edge of the zone. For $T \gg T_D$, all modes behave classically. The integrand of Eq. (6.45) is just $k_B T$, and since the total number of modes is $3N = 3V \int^{\mathrm{B.Z.}} \mathrm{d}^3 \vec{k}/(2\pi)^3$, the classical results of $E(T) = E_0 + 3N k_B T$ and $C = 3N k_B$ are recovered. For $T \ll T_D$, the factor $\exp(\beta \hbar v k)$ in the denominator of Eq. (6.45) is very large at the Brillouin zone edge. The most important contribution to the integral comes from small k, and the error in extending the integration range to infinity is small. After changing variables to $x = \beta \hbar v |\vec{k}|$, and using $d^3\vec{k} = 4\pi x^2\, dx/(\beta \hbar v)^3$ due to spherical symmetry, Eq. (6.45) gives

$$\lim_{T \ll T_D} E(T) \approx E_0 + \frac{3V}{8\pi^3}\left(\frac{k_B T}{\hbar v}\right)^3 4\pi k_B T \int_0^\infty dx \frac{x^3}{e^x - 1}$$
$$= E_0 + \frac{\pi^2}{10} V \left(\frac{kT}{\hbar v}\right)^3 k_B T. \tag{6.47}$$

(The value of the definite integral, $\pi^4/15 \approx 6.5$, can be found in standard tables.) The resulting heat capacity,

$$C = \frac{dE}{dT} = k_B V \frac{2\pi^2}{5}\left(\frac{k_B T}{\hbar v}\right)^3, \tag{6.48}$$

has the form $C \propto N k_B (T/T_D)^3$ in agreement with observations. The physical interpretation of this result is the following: at temperatures $T \ll T_D$, only a fraction of the phonon modes can be thermally excited. These are the low-frequency phonons with energy quanta $\hbar \omega(\vec{k}) \ll k_B T$. The excited phonons have wavevectors $|\vec{k}| < k^*(T) \approx (k_B T/\hbar v)$. Quite generally, in d space dimensions, the number of these modes is approximately $V k^*(T)^d \sim V(k_B T/\hbar v)^d$. Each excited mode can be treated classically, and contributes roughly $k_B T$ to the internal energy, which thus scales as $E \sim V(k_B T/\hbar v)^d k_B T$. The corresponding heat capacity, $C \sim V k_B (k_B T/\hbar v)^d$, vanishes as T^d.

6.3 Black-body radiation

Phonons correspond to vibrations of a solid medium. There are also (longitudinal) sound modes in liquid and gas states. However, even "empty" vacuum can support fluctuations of the electromagnetic (EM) field, *photons*, which can be thermally excited at finite temperatures. The normal modes of this field are EM waves, characterized by a wavenumber $\vec{k}$, and two possible polarizations α. (Since $\nabla \cdot \vec{E} = 0$ in free space, the electric field must be normal to $\vec{k}$, and only *transverse* modes exist.) With appropriate choice of coordinates, the Hamiltonian for the EM field can be written as a sum of harmonic oscillators,

$$\mathcal{H} = \frac{1}{2} \sum_{\vec{k},\alpha} \left[|\tilde{p}_{\vec{k},\alpha}|^2 + \omega_\alpha(\vec{k})^2 \left|\tilde{u}_\alpha(\vec{k})\right|^2 \right], \tag{6.49}$$

with $\omega_\alpha(\vec{k}) = ck$, where c is the speed of light.

With periodic boundary conditions, the allowed wavevectors in a box of size L are $\vec{k} = 2\pi(n_x, n_y, n_z)/L$, where $\{n_x, n_y, n_z\}$ are integers. However, unlike phonons, there is no Brillouin zone limiting the size of $\vec{k}$, and these integers can be arbitrarily large. The lack of such a restriction leads to the *ultraviolet catastrophe* in a classical treatment: as there is no limit to the wavevector, assigning $k_B T$ per mode leads to an infinite energy stored in the high-frequency modes. (The low frequencies are cut off by the finite size of the box.) It was

indeed to resolve this difficulty that Planck suggested the allowed values of
EM energy must be quantized according to the Hamiltonian

$$\mathcal{H}^q = \sum_{\vec{k},\alpha} \hbar c k \left(n_\alpha(\vec{k}) + \frac{1}{2} \right), \quad \text{with} \quad n_\alpha(\vec{k}) = 0, 1, 2, \cdots . \tag{6.50}$$

As for phonons, the internal energy is calculated from

$$E = \langle \mathcal{H}^q \rangle = \sum_{\vec{k},\alpha} \hbar c k \left(\frac{1}{2} + \frac{e^{-\beta \hbar c k}}{1 - e^{-\beta \hbar c k}} \right) = V E_0 + \frac{2V}{(2\pi)^3} \int d^3\vec{k} \frac{\hbar c k}{e^{\beta \hbar c k} - 1}. \tag{6.51}$$

The zero-point energy is actually infinite, but as only energy differences are
measured, it is usually ignored. The change of variables to $x = \beta \hbar c k$ allows
us to calculate the excitation energy,

$$\begin{aligned}
\frac{E^*}{V} &= \frac{\hbar c}{\pi^2} \left(\frac{k_B T}{\hbar c} \right)^4 \int_0^\infty \frac{dx\, x^3}{e^x - 1} \\
&= \frac{\pi^2}{15} \left(\frac{k_B T}{\hbar c} \right)^3 k_B T.
\end{aligned} \tag{6.52}$$

The EM radiation also exerts a pressure on the walls of the container. From
the partition function of the Hamiltonian (6.50),

$$Z = \sum_{\{n_\alpha(\vec{k})\}} \prod_{\vec{k},\alpha} \exp\left[-\beta \hbar \omega(\vec{k}) \left(n_\alpha(\vec{k}) + \frac{1}{2} \right) \right] = \prod_{\vec{k},\alpha} \frac{e^{-\beta \hbar c k/2}}{1 - e^{-\beta \hbar c k}}, \tag{6.53}$$

the free energy is

$$\begin{aligned}
F &= -k_B T \ln Z = k_B T \sum_{\vec{k},\alpha} \left[\frac{\beta \hbar c k}{2} + \ln\left(1 - e^{-\beta \hbar c k} \right) \right] \\
&= 2V \int \frac{d^3\vec{k}}{(2\pi)^3} \left[\frac{\hbar c k}{2} + k_B T \ln\left(1 - e^{-\beta \hbar c k} \right) \right].
\end{aligned} \tag{6.54}$$

The pressure due to the photon gas is

$$\begin{aligned}
P &= -\left. \frac{\partial F}{\partial V} \right|_T = -\int \frac{d^3\vec{k}}{(2\pi)^3} \left[\hbar c k + 2 k_B T \ln\left(1 - e^{-\beta \hbar c k} \right) \right] \\
&= P_0 - \frac{k_B T}{\pi^2} \int_0^\infty dk\, k^2 \ln\left(1 - e^{-\beta \hbar c k} \right) \quad \text{(integrate by parts)} \\
&= P_0 + \frac{k_B T}{\pi^2} \int_0^\infty dk\, \frac{k^3}{3} \frac{\beta \hbar c e^{-\beta \hbar c k}}{1 - e^{-\beta \hbar c k}} \quad \text{(compare with Eq. (6.51))} \\
&= P_0 + \frac{1}{3} \frac{E}{V}.
\end{aligned} \tag{6.55}$$

Note that there is also an infinite zero-point pressure P_0. Differences in this pressure lead to the *Casimir force* between conducting plates, which is measurable.

The extra pressure of 1/3 times the energy density can be compared with
that of a gas of relativistic particles. (As shown in the problem sets, a dispersion
relation, $\mathcal{E} \propto |\vec{p}|^s$, leads to a pressure, $P = (s/d)(E/V)$ in d dimensions.)

Continuing with the analogy to a gas of particles, if a hole is opened in the container wall, the escaping energy flux per unit area and per unit time is

$$\phi = \langle c_\perp \rangle \frac{E}{V}. \tag{6.56}$$

All photons have speed c, and the average of the component of the velocity perpendicular to the hole can be calculated as

$$\langle c_\perp \rangle = c \times \frac{1}{4\pi} \int_0^{\pi/2} 2\pi \sin\theta \mathrm{d}\theta \cos\theta = \frac{c}{4}, \tag{6.57}$$

resulting in

$$\phi = \frac{1}{4} c \frac{E}{V} = \frac{\pi^2}{60} \frac{k_B^4 T^4}{\hbar^3 c^2}. \tag{6.58}$$

The result, $\phi = \sigma T^4$, is the *Stefan–Boltzmann law* for *black-body radiation*, and

$$\sigma = \frac{\pi^2}{60} \frac{k_B^4}{\hbar^3 c^2} \approx 5.67 \times 10^{-8} \,\mathrm{W\,m^{-2}\,K^{-4}} \tag{6.59}$$

is *Stefan's constant*. Black-body radiation has a characteristic frequency dependence: let $E(T)/V = \int \mathrm{d}k\, \mathcal{E}(k, t)$, where

$$\mathcal{E}(k, T) = \frac{\hbar c}{\pi^2} \frac{k^3}{e^{\beta \hbar c k} - 1} \tag{6.60}$$

is the energy density in wavevector k. The flux of emitted radiation in the interval $[k, k + \mathrm{d}k]$ is $I(k, T)\mathrm{d}k$, where

$$I(k, T) = \frac{c}{4} \mathcal{E}(k, T) = \frac{\hbar c^2}{4\pi^2} \frac{k^3}{e^{\beta \hbar c k} - 1} \rightarrow \begin{cases} c k_B T k^2 / 4\pi & \text{for } k \ll k^*(T) \\ \hbar c^2 k^3 e^{-\beta \hbar c k} / 4\pi^2 & \text{for } k \gg k^*(T). \end{cases} \tag{6.61}$$

The characteristic wavevector $k^*(T) \approx k_B T / \hbar c$ separates quantum and classical regimes. It provides the upper cutoff that eliminates the ultraviolet catastrophe.

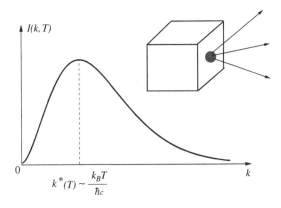

Fig. 6.7 The spectrum of black-body radiation from a system at temperature T.

6.4 Quantum microstates

In the previous sections we indicated several failures of classical statistical mechanics, which were heuristically remedied by assuming quantized energy levels, while still calculating thermodynamic quantities from a partition sum $Z = \sum_n \exp(-\beta E_n)$. This implicitly assumes that the microstates of a quantum system are specified by its discretized energy levels, and governed by a probability distribution similar to a Boltzmann weight. This "analogy" to classical statistical mechanics needs to be justified. Furthermore, quantum mechanics is itself inherently probabilistic, with uncertainties unrelated to those that lead to probabilities in statistical mechanics. Here, we shall construct a quantum formulation of statistical mechanics by closely following the steps that lead to the classical formulation.

Microstates of a classical system of particles are described by the set of coordinates and conjugate momenta $\{\vec{p}_i, \vec{q}_i\}$; that is, by a point in the $6N$-dimensional phase space. In quantum mechanics $\{\vec{q}_i\}$ and $\{\vec{p}_i\}$ are not independent observables. Instead:

- The (micro-)state of a quantum system is completely specified by a *unit vector* $|\Psi\rangle$, which belongs to an infinite-dimensional *Hilbert space*. The vector $|\Psi\rangle$ can be written in terms of its components $\langle n|\Psi\rangle$, which are complex numbers, along a suitable set of orthonormal basis vectors $|n\rangle$. In the convenient notation introduced by Dirac, this decomposition is written as

$$|\Psi\rangle = \sum_n \langle n|\Psi\rangle \, |n\rangle. \tag{6.62}$$

The most familiar basis is that of coordinates $|\{\vec{q}_i\}\rangle$, and $\langle\{\vec{q}_i\}|\Psi\rangle \equiv \Psi(\vec{q}_1, \ldots, \vec{q}_N)$ is the *wave function*. The normalization condition is

$$\langle\Psi|\Psi\rangle = \sum_n \langle\Psi|n\rangle \langle n|\Psi\rangle = 1, \quad \text{where} \quad \langle\Psi|n\rangle \equiv \langle n|\Psi\rangle^*. \tag{6.63}$$

For example, in the coordinate basis, we must require

$$\langle\Psi|\Psi\rangle = \int \prod_{i=1}^{N} \mathrm{d}^d\vec{q}_i \, |\Psi(\vec{q}_1, \ldots, \vec{q}_N)|^2 = 1. \tag{6.64}$$

- Classically, various observables are functions $\mathcal{O}(\{\vec{p}_i, \vec{q}_i\})$, defined in phase space. In quantum mechanics, these functions are replaced by *Hermitian matrices* (operators) in Hilbert space, obtained by substituting operators for $\{\vec{q}_i\}$ and $\{\vec{p}_i\}$ in the classical expression (after proper symmetrization of products, e.g., $pq \to (pq + qp)/2$). These basic operators satisfy the *commutation relations*,

$$[p_j, q_k] \equiv p_j q_k - q_k p_j = \frac{\hbar}{i} \delta_{j,k}. \tag{6.65}$$

For example, in the coordinate basis $|\{\vec{q}_i\}\rangle$, the momentum operators are

$$p_j = \frac{\hbar}{i} \frac{\partial}{\partial q_j}. \tag{6.66}$$

(Note that classical Poisson brackets satisfy $\{p_j, q_k\} = -\delta_{j,k}$. Quite generally, quantum commutation relations are obtained by multiplying the corresponding classical Poisson brackets by $i\hbar$.)

- Unlike in classical mechanics, the value of an operator $\mathcal{O}$ is not uniquely determined for a particular microstate. It is instead a random variable, whose *average* in a state $|\Psi\rangle$ is given by

$$\langle \mathcal{O} \rangle \equiv \langle \Psi | \mathcal{O} | \Psi \rangle \equiv \sum_{m,n} \langle \Psi | m \rangle \langle m | \mathcal{O} | n \rangle \langle n | \Psi \rangle. \tag{6.67}$$

For example,

$$\langle U(\{\vec{q}\}) \rangle = \int \prod_{i=1}^{N} d^3 \vec{q}_i \, \Psi^*(\vec{q}_1, \ldots, \vec{q}_N) U(\{\vec{q}\}) \Psi(\vec{q}_1, \ldots, \vec{q}_N), \quad \text{and}$$

$$\langle K(\{\vec{p}\}) \rangle = \int \prod_{i=1}^{N} d^3 \vec{q}_i \, \Psi^*(\vec{q}_1, \ldots, \vec{q}_N) K\left(\left\{ \frac{\hbar}{i} \frac{d}{d\vec{q}} \right\} \right) \Psi(\vec{q}_1, \ldots, \vec{q}_N).$$

To ensure that the expectation value $\langle \mathcal{O} \rangle$ is real in all states, the operators $\mathcal{O}$ must be *Hermitian*, that is, satisfy

$$\mathcal{O}^\dagger = \mathcal{O}, \quad \text{where} \quad \langle m | \mathcal{O}^\dagger | n \rangle \equiv \langle n | \mathcal{O} | m \rangle^*. \tag{6.68}$$

When replacing $\vec{p}$ and $\vec{q}$ in a classical operator $\mathcal{O}(\{\vec{p}_i, \vec{q}_i\})$ by corresponding matrices, proper symmetrization of various products is necessary to ensure the above hermiticity.

Time evolution of microstates is governed by the Hamiltonian $\mathcal{H}(\{\vec{p}_i, \vec{q}_i\})$. A classical microstate evolves according to Hamilton's equations of motion, while in quantum mechanics the state vector changes in time according to

$$i\hbar \frac{\partial}{\partial t} |\Psi(t)\rangle = \mathcal{H} |\Psi(t)\rangle. \tag{6.69}$$

A convenient basis is one that *diagonalizes* the matrix $\mathcal{H}$. The energy *eigenstates* satisfy $\mathcal{H} |n\rangle = \mathcal{E}_n |n\rangle$, where $\mathcal{E}_n$ are the *eigenenergies*. Substituting $|\Psi(t)\rangle = \sum_n \langle n | \Psi(t) \rangle |n\rangle$ in Eq. (6.69), and taking advantage of the orthonormality condition $\langle m | n \rangle = \delta_{m,n}$, yields

$$i\hbar \frac{d}{dt} \langle n | \Psi(t) \rangle = \mathcal{E}_n \langle n | \Psi(t) \rangle, \quad \Longrightarrow \quad \langle n | \Psi(t) \rangle = \exp\left(-\frac{i \mathcal{E}_n t}{\hbar} \right) \langle n | \Psi(0) \rangle. \tag{6.70}$$

The quantum states at two different times can be related by a *time evolution* operator, as

$$|\Psi(t)\rangle = U(t, t_0) |\Psi(t_0)\rangle, \tag{6.71}$$

which satisfies $i\hbar \partial_t U(t, t_0) = \mathcal{H} U(t, t_0)$, with the boundary condition $U(t_0, t_0) = 1$. *If $\mathcal{H}$ is independent of t*, we can solve these equations to yield

$$U(t, t_0) = \exp\left[-\frac{i}{\hbar} \mathcal{H}(t - t_0) \right]. \tag{6.72}$$

6.5 Quantum macrostates

Macrostates of the system depend on only a few thermodynamic functions. We can form an *ensemble* of a large number $\mathcal{N}$, of microstates μ_α, corresponding to a given macrostate. The different microstates occur with probabilities p_α. (For example, $p_\alpha = 1/\mathcal{N}$ for an unbiased estimate in the absence of any other information.) When we no longer have exact knowledge of the microstate of a system, the system is said to be in a *mixed state*.

Classically, ensemble averages are calculated from

$$\overline{\mathcal{O}(\{\vec{p}_i, \vec{q}_i\})_t} = \sum_\alpha p_\alpha \mathcal{O}(\mu_\alpha(t)) = \int \prod_{i=1}^N \mathrm{d}^3\vec{p}_i \mathrm{d}^3\vec{q}_i \mathcal{O}\left(\{\vec{p}_i, \vec{q}_i\}\right) \rho\left(\{\vec{p}_i, \vec{q}_i\}, t\right), \quad (6.73)$$

where

$$\rho\left(\{\vec{p}_i, \vec{q}_i\}, t\right) = \sum_\alpha p_\alpha \prod_{i=1}^N \delta^3\left(\vec{q}_i - \vec{q}_i(t)_\alpha\right) \delta^3\left(\vec{p}_i - \vec{p}_i(t)_\alpha\right) \quad (6.74)$$

is the *phase space density*.

Similarly, a mixed quantum state is obtained from a set of possible states $\{|\psi_\alpha\rangle\}$, with probabilities $\{p_\alpha\}$. The ensemble average of the quantum mechanical expectation value in Eq. (6.67) is then given by

$$\overline{\langle \mathcal{O} \rangle} = \sum_\alpha p_\alpha \langle \Psi_\alpha | \mathcal{O} | \Psi_\alpha \rangle = \sum_{\alpha, m, n} p_\alpha \langle \Psi_\alpha | m \rangle \langle n | \Psi_\alpha \rangle \langle m | \mathcal{O} | n \rangle$$
$$= \sum_{m,n} \langle n | \rho | m \rangle \langle m | \mathcal{O} | n \rangle = \mathrm{tr}(\rho \mathcal{O}), \quad (6.75)$$

where we have introduced a basis $\{|n\rangle\}$, and defined the *density matrix*

$$\langle n | \rho(t) | m \rangle \equiv \sum_\alpha p_\alpha \langle n | \Psi_\alpha(t) \rangle \langle \Psi_\alpha(t) | m \rangle. \quad (6.76)$$

Classically, the probability (density) $\rho(t)$ is a function defined on phase space. As all operators in phase space, it is replaced by a matrix in quantum mechanics. Stripped of the choice of basis, the density matrix is

$$\rho(t) = \sum_\alpha p_\alpha | \Psi_\alpha(t) \rangle \langle \Psi_\alpha(t) |. \quad (6.77)$$

Clearly, ρ corresponds to a *pure state* if and only if $\rho^2 = \rho$.

The density matrix satisfies the following properties:

(i) *Normalization.* Since each $|\Psi_\alpha\rangle$ is normalized to unity,

$$\langle 1 \rangle = \mathrm{tr}(\rho) = \sum_n \langle n | \rho | n \rangle = \sum_{\alpha, n} p_\alpha |\langle n | \Psi_\alpha \rangle|^2 = \sum_\alpha p_\alpha = 1. \quad (6.78)$$

(ii) *Hermiticity.* The density matrix is *Hermitian*, that is, $\rho^\dagger = \rho$, since

$$\langle m | \rho^\dagger | n \rangle = \langle n | \rho | m \rangle^* = \sum_\alpha p_\alpha \langle \Psi_\alpha | m \rangle \langle n | \Psi_\alpha \rangle = \langle n | \rho | m \rangle, \quad (6.79)$$

ensuring that the averages in Eq. (6.75) are real numbers.

(iii) *Positivity*. For any $|\Phi\rangle$,

$$\langle\Phi|\rho|\Phi\rangle = \sum_\alpha p_\alpha\langle\Phi|\Psi_\alpha\rangle\langle\Psi_\alpha|\Phi\rangle = \sum_\alpha p_\alpha|\langle\Phi|\Psi_\alpha\rangle|^2 \geq 0. \qquad (6.80)$$

Thus, ρ is *positive definite*, implying that all its eigenvalues must be positive.

Liouville's theorem gives the time evolution of the classical density as

$$\frac{d\rho}{dt} = \frac{\partial\rho}{\partial t} - \{\mathcal{H}, \rho\} = 0. \qquad (6.81)$$

It is most convenient to examine the evolution of the quantum density matrix in the basis of energy eigenstates, where according to Eq. (6.70)

$$\begin{aligned}
i\hbar\frac{\partial}{\partial t}\langle n|\rho(t)|m\rangle &= i\hbar\frac{\partial}{\partial t}\sum_\alpha p_\alpha\langle n|\Psi_\alpha(t)\rangle\langle\Psi_\alpha(t)|m\rangle \\
&= \sum_\alpha p_\alpha\left[(\mathcal{E}_n - \mathcal{E}_m)\langle n|\Psi_\alpha\rangle\langle\Psi_\alpha|m\rangle\right] \qquad (6.82) \\
&= \langle n|(\mathcal{H}\rho - \rho\mathcal{H})|m\rangle.
\end{aligned}$$

The final result is a tensorial identity, and hence, *independent* of the choice of basis

$$i\hbar\frac{\partial}{\partial t}\rho = [\mathcal{H}, \rho]. \qquad (6.83)$$

Equilibrium requires *time-independent* averages, and suggests $\partial\rho/\partial t = 0$. This condition is satisfied in both Eqs. (6.81) and (6.83) by choosing $\rho = \rho(\mathcal{H})$. (As discussed in chapter 3, ρ may also depend on conserved quantities $\{L_\alpha\}$ such that $[\mathcal{H}, L_\alpha] = 0$.) Various equilibrium quantum density matrices can now be constructed in analogy with classical statistical mechanics.

Microcanonical ensemble: as the internal energy has a fixed value E, a density matrix that includes this constraint is

$$\rho(E) = \frac{\delta(\mathcal{H} - E)}{\Omega(E)}. \qquad (6.84)$$

In particular, in the basis of energy eigenstates,

$$\langle n|\rho|m\rangle = \sum_\alpha p_\alpha\langle n|\Psi_\alpha\rangle\langle\Psi_\alpha|m\rangle = \begin{cases} \frac{1}{\Omega} & \text{if } \mathcal{E}_n = E, \text{ and } m = n, \\ 0 & \text{if } \mathcal{E}_n \neq E, \text{ or } m \neq n. \end{cases} \qquad (6.85)$$

The first condition states that only eigenstates of the correct energy can appear in the quantum wave function, and that (for $p_\alpha = 1/\mathcal{N}$) such states on average have the same amplitude, $\overline{|\langle n|\Psi\rangle|^2} = 1/\Omega$. This is equivalent to the classical *assumption of equal a priori equilibrium probabilities*. The second (additional) condition states that the Ω eigenstates of energy E are combined in a typical microstate with *independent random phases*. (Note that the normalization condition tr$\rho = 1$ implies that $\Omega(E) = \sum_n \delta(E - E_n)$ is the number of eigenstates of $\mathcal{H}$ with energy E.)

Canonical ensemble: a fixed temperature $T = 1/k_B\beta$ can be achieved by putting the system in contact with a reservoir. By considering the combined system, the canonical density matrix is obtained as

$$\rho(\beta) = \frac{\exp(-\beta\mathcal{H})}{Z(\beta)}. \tag{6.86}$$

The normalization condition $\mathrm{tr}(\rho) = 1$ leads to the quantum partition function

$$Z = \mathrm{tr}\left(e^{-\beta\mathcal{H}}\right) = \sum_n e^{-\beta\mathcal{E}_n}. \tag{6.87}$$

The final sum is over the (discrete) energy levels of $\mathcal{H}$, and justifies the calculations performed in the previous sections.

Grand canonical ensemble: the number of particles N is no longer fixed. Quantum microstates with indefinite particle number span a so-called *Fock space*. The corresponding density matrix is

$$\rho(\beta, \mu) = \frac{e^{-\beta\mathcal{H} + \beta\mu N}}{\mathcal{Q}}, \quad \text{where} \quad \mathcal{Q}(\beta, \mu) = \mathrm{tr}\left(e^{-\beta\mathcal{H} + \beta\mu N}\right) = \sum_{N=0}^{\infty} e^{\beta\mu N} Z_N(\beta).$$

$$\tag{6.88}$$

Example. Consider a single particle in a quantum canonical ensemble in a box of volume V. The energy eigenstates of the Hamiltonian

$$\mathcal{H}_1 = \frac{\vec{p}^2}{2m} = -\frac{\hbar^2}{2m}\nabla^2 \quad \text{(in coordinate basis)}, \tag{6.89}$$

obtained from $\mathcal{H}_1|\vec{k}\rangle = \mathcal{E}(\vec{k})|\vec{k}\rangle$, are

$$\langle \vec{x}|\vec{k}\rangle = \frac{e^{i\vec{k}\cdot\vec{x}}}{\sqrt{V}}, \quad \text{with} \quad \mathcal{E}(\vec{k}) = \frac{\hbar^2 k^2}{2m}. \tag{6.90}$$

With periodic boundary conditions in a box of size L, the allowed values of $\vec{k}$ are $(2\pi/L)(\ell_x, \ell_y, \ell_z)$, where (ℓ_x, ℓ_y, ℓ_z) are integers. A particular vector in this one-particle Hilbert space is specified by its components along each of these basis states (infinite in number). The space of quantum microstates is thus much larger than the corresponding 6-dimensional classical phase space. The partition function for $L \to \infty$,

$$Z_1 = \mathrm{tr}(\rho) = \sum_{\vec{k}} \exp\left(-\frac{\beta\hbar^2 k^2}{2m}\right) = V \int \frac{d^3\vec{k}}{(2\pi)^3} \exp\left(-\frac{\beta\hbar^2 k^2}{2m}\right)$$

$$= \frac{V}{(2\pi)^3}\left(\sqrt{\frac{2\pi m k_B T}{\hbar^2}}\right)^3 = \frac{V}{\lambda^3}, \tag{6.91}$$

coincides with the classical result, and $\lambda = h/\sqrt{2\pi m k_B T}$ (justifying the use of $d^3\vec{p}\, d^3\vec{q}/h^3$ as the correct dimensionless measure of phase space). Elements of the density matrix in a coordinate representation are

$$
\langle \vec{x}'|\rho|\vec{x}\rangle = \sum_{\vec{k}} \langle \vec{x}'|\vec{k}\rangle \frac{e^{-\beta \mathcal{E}(\vec{k})}}{Z_1} \langle \vec{k}|\vec{x}\rangle = \frac{\lambda^3}{V} \int \frac{V d^3\vec{k}}{(2\pi)^3} \frac{e^{-i\vec{k}\cdot(\vec{x}-\vec{x}')}}{V} \exp\left(-\frac{\beta\hbar^2 k^2}{2m}\right)
$$
$$
= \frac{1}{V} \exp\left[-\frac{m(\vec{x}-\vec{x}')^2}{2\beta\hbar^2}\right] = \frac{1}{V} \exp\left[-\frac{\pi(\vec{x}-\vec{x}')^2}{\lambda^2}\right].
$$

(6.92)

The diagonal elements, $\langle \vec{x}|\rho|\vec{x}\rangle = 1/V$, are just the probabilities for finding a particle at $\vec{x}$. The off-diagonal elements have no classical analog. They suggest that the appropriate way to think of the particle is as a wave packet of size $\lambda = h/\sqrt{2\pi m k_B T}$, the *thermal wavelength*. As $T \to \infty$, λ goes to zero, and the classical analysis is valid. As $T \to 0$, λ diverges and quantum mechanical effects take over when λ becomes comparable with the size of the box.

Alternatively, we could have obtained Eq. (6.92), by noting that Eq. (6.86) implies

$$
\frac{\partial}{\partial \beta} Z\rho = -\mathcal{H}Z\rho = \frac{\hbar^2}{2m}\nabla^2 Z\rho.
$$

(6.93)

This is just the diffusion equation (for the free particle), which can be solved subject to the initial condition $\rho(\beta = 0) = 1$ (i.e., $\langle \vec{x}'|\rho(\beta=0)|\vec{x}\rangle = \delta^3(\vec{x} - \vec{x}')/V$) to yield the result of Eq. (6.92).

Problems for chapter 6

1. *One-dimensional chain*: a chain of $N+1$ particles of mass m is connected by N massless springs of spring constant K and relaxed length a. The first and last particles are held fixed at the equilibrium separation of Na. Let us denote the longitudinal displacements of the particles from their equilibrium positions by $\{u_i\}$, with $u_0 = u_N = 0$ since the end particles are fixed. The Hamiltonian governing $\{u_i\}$, and the conjugate momenta $\{p_i\}$, is

$$
\mathcal{H} = \sum_{i=1}^{N-1} \frac{p_i^2}{2m} + \frac{K}{2}\left[u_1^2 + \sum_{i=1}^{N-2}(u_{i+1} - u_i)^2 + u_{N-1}^2\right].
$$

(a) Using the appropriate (sine) Fourier transforms, find the normal modes $\{\tilde{u}_k\}$, and the corresponding frequencies $\{\omega_k\}$.

(b) Express the Hamiltonian in terms of the amplitudes of normal modes $\{\tilde{u}_k\}$, and evaluate the *classical* partition function. (You may integrate the $\{u_i\}$ from $-\infty$ to $+\infty$.)

(c) First evaluate $\langle |\tilde{u}_k|^2\rangle$, and use the result to calculate $\langle u_i^2\rangle$. Plot the resulting squared displacement of each particle as a function of its equilibrium position.

(d) How are the results modified if only the first particle is fixed ($u_0 = 0$), while the other end is free ($u_N \neq 0$)? (Note that this is a much simpler problem as the partition function can be evaluated by changing variables to the $N - 1$ spring extensions.)

****

2. *Black-hole thermodynamics*: according to Bekenstein and Hawking, the entropy of a black hole is proportional to its area A, and given by

$$S = \frac{k_B c^3}{4G\hbar} A.$$

(a) Calculate the escape velocity at a radius R from a mass M using classical mechanics. Find the relationship between the radius and mass of a black hole by setting this escape velocity to the speed of light c. (Relativistic calculations do not modify this result, which was originally obtained by Laplace.)

(b) Does entropy increase or decrease when two black holes collapse into one? What is the entropy change for the Universe (in equivalent number of bits of information) when two solar mass black holes ($M_\odot \approx 2 \times 10^{30}\,\text{kg}$) coalesce?

(c) The internal energy of the black hole is given by the Einstein relation, $E = Mc^2$. Find the temperature of the black hole in terms of its mass.

(d) A "black hole" actually emits thermal radiation due to pair creation processes on its event horizon. Find the rate of energy loss due to such radiation.

(e) Find the amount of time it takes an isolated black hole to evaporate. How long is this time for a black hole of solar mass?

(f) What is the mass of a black hole that is in thermal equilibrium with the current cosmic background radiation at $T = 2.7\,\text{K}$?

(g) Consider a spherical volume of space of radius R. According to the recently formulated *Holographic Principle* there is a maximum to the amount of entropy that this volume of space can have, independent of its contents! What is this maximal entropy?

****

3. *Quantum harmonic oscillator*: consider a single harmonic oscillator with the Hamiltonian

$$\mathcal{H} = \frac{p^2}{2m} + \frac{m\omega^2 q^2}{2}, \quad \text{with} \quad p = \frac{\hbar}{i}\frac{d}{dq}.$$

(a) Find the partition function Z, at a temperature T, and calculate the energy $\langle \mathcal{H} \rangle$.

(b) Write down the formal expression for the canonical density matrix ρ in terms of the eigenstates ($\{|n\rangle\}$), and energy levels ($\{\epsilon_n\}$) of $\mathcal{H}$.

(c) Show that for a general operator $A(x)$,

$$\frac{\partial}{\partial x} \exp[A(x)] \neq \frac{\partial A}{\partial x} \exp[A(x)], \quad \text{unless} \quad \left[A, \frac{\partial A}{\partial x}\right] = 0,$$

while in all cases

$$\frac{\partial}{\partial x} \text{tr}\{\exp[A(x)]\} = \text{tr}\left\{\frac{\partial A}{\partial x}\exp[A(x)]\right\}.$$

(d) Note that the partition function calculated in part (a) does not depend on the mass m, that is, $\partial Z / \partial m = 0$. Use this information, along with the result in part (c), to show that

$$\left\langle \frac{p^2}{2m} \right\rangle = \left\langle \frac{m\omega^2 q^2}{2} \right\rangle .$$

(e) Using the results in parts (d) and (a), or otherwise, calculate $\langle q^2 \rangle$. How are the results in Problem 1 modified at low temperatures by inclusion of quantum mechanical effects?

(f) In a coordinate representation, calculate $\langle q' | \rho | q \rangle$ in the high-temperature limit. One approach is to use the result

$$\exp(\beta A) \exp(\beta B) = \exp \left[\beta(A+B) + \beta^2 [A, B]/2 + \mathcal{O}(\beta^3) \right] .$$

(g) At low temperatures, ρ is dominated by low-energy states. Use the ground state wave function to evaluate the limiting behavior of $\langle q' | \rho | q \rangle$ as $T \to 0$.

(h) Calculate the exact expression for $\langle q' | \rho | q \rangle$.

$$********$$

4. *Relativistic Coulomb gas*: consider a *quantum* system of N positive, and N negative charged relativistic particles in a box of volume $V = L^3$. The Hamiltonian is

$$\mathcal{H} = \sum_{i=1}^{2N} c |\vec{p}_i| + \sum_{i<j}^{2N} \frac{e_i e_j}{|\vec{r}_i - \vec{r}_j|} ,$$

where $e_i = +e_0$ for $i = 1, \cdots, N$, and $e_i = -e_0$ for $i = N+1, \cdots, 2N$, denote the charges of the particles; $\{\vec{r}_i\}$ and $\{\vec{p}_i\}$ their coordinates and momenta, respectively. While this is too complicated a system to solve, we can nonetheless obtain some exact results.

(a) Write down the Schrödinger equation for the eigenvalues $\varepsilon_n(L)$, and (in coordinate space) eigenfunctions $\Psi_n(\{\vec{r}_i\})$. State the constraints imposed on $\Psi_n(\{\vec{r}_i\})$ if the particles are bosons or fermions.

(b) By a change of scale $\vec{r}_i{}' = \vec{r}_i / L$, show that the eigenvalues satisfy a scaling relation $\varepsilon_n(L) = \varepsilon_n(1)/L$.

(c) Using the formal expression for the partition function $Z(N, V, T)$, in terms of the eigenvalues $\{\varepsilon_n(L)\}$, show that Z does not depend on T and V separately, but only on a specific scaling combination of them.

(d) Relate the energy E, and pressure P of the gas to variations of the partition function. Prove the exact result $E = 3PV$.

(e) The Coulomb interaction between charges in d-dimensional space falls off with separation as $e_i e_j / |\vec{r}_i - \vec{r}_j|^{d-2}$. (In $d = 2$ there is a logarithmic interaction.) In what dimension d can you construct an exact relation between E and P for *non-relativistic* particles (kinetic energy $\sum_i \vec{p}_i^2 / 2m$)? What is the corresponding exact relation between energy and pressure?

(f) Why are the above "exact" scaling laws not expected to hold in dense (liquid or solid) Coulomb mixtures?

****

5. *The virial theorem* is a consequence of the invariance of the phase space for a system of N (classical or quantum) particles under canonical transformations, such as a change of scale. In the following, consider N particles with coordinates $\{\vec{q}_i\}$, and conjugate momenta $\{\vec{p}_i\}$ (with $i = 1, \cdots, N$), and subject to a Hamiltonian $\mathcal{H}\left(\{\vec{p}_i\}, \{\vec{q}_i\}\right)$.

(a) *Classical version*: write down the expression for the classical partition function, $Z \equiv Z[\mathcal{H}]$. Show that it is invariant under the rescaling $\vec{q}_1 \to \lambda\vec{q}_1$, $\vec{p}_1 \to \vec{p}_1/\lambda$ of a pair of conjugate variables, that is, $Z[\mathcal{H}_\lambda]$ is independent of λ, where $\mathcal{H}_\lambda$ is the Hamiltonian obtained after the above rescaling.

(b) *Quantum mechanical version*: write down the expression for the quantum partition function. Show that it is also invariant under the rescalings $\vec{q}_1 \to \lambda\vec{q}_1$, $\vec{p}_1 \to \vec{p}_1/\lambda$, where $\vec{p}_i$ and $\vec{q}_i$ are now quantum mechanical operators. (*Hint.* Start with the time-independent Schrödinger equation.)

(c) Now assume a Hamiltonian of the form

$$\mathcal{H} = \sum_i \frac{\vec{p}_i^{\,2}}{2m} + V\left(\{\vec{q}_i\}\right).$$

Use the result that $Z[\mathcal{H}_\lambda]$ is independent of λ to prove the *virial* relation

$$\left\langle \frac{\vec{p}_1^{\,2}}{m} \right\rangle = \left\langle \frac{\partial V}{\partial \vec{q}_1} \cdot \vec{q}_1 \right\rangle,$$

where the brackets denote thermal averages.

(d) The above relation is sometimes used to estimate the mass of distant galaxies. The stars on the outer boundary of the G-8.333 galaxy have been measured to move with velocity $v \approx 200\,\mathrm{km\,s^{-1}}$. Give a numerical estimate of the ratio of the G-8.333's mass to its size.

****

6. *Electron spin*: the Hamiltonian for an electron in a magnetic field $\vec{B}$ is

$$\mathcal{H} = -\mu_B \vec{\sigma}.\vec{B}, \quad \text{where} \quad \sigma_x = \begin{pmatrix} 0 & 1 \\ 1 & 0 \end{pmatrix}, \quad \sigma_y = \begin{pmatrix} 0 & -i \\ i & 0 \end{pmatrix}, \quad \text{and} \quad \sigma_z = \begin{pmatrix} 1 & 0 \\ 0 & -1 \end{pmatrix}$$

are the Pauli spin operators, and μ_B is the Bohr magneton.

(a) In the quantum canonical ensemble evaluate the density matrix if $\vec{B}$ is along the z direction.

(b) Repeat the calculation assuming that $\vec{B}$ points along the x direction.

(c) Calculate the average energy in each of the above cases.

****

7. *Quantum rotor*: consider a rotor in two dimensions with

$$\mathcal{H} = -\frac{\hbar^2}{2I}\frac{d^2}{d\theta^2}, \quad \text{and} \quad 0 \le \theta < 2\pi.$$

(a) Find the eigenstates and energy levels of the system.

(b) Write the expression for the density matrix $\langle \theta' | \rho | \theta \rangle$ in a canonical ensemble of temperature T, and evaluate its low- and high-temperature limits.

****

8. *Quantum mechanical entropy*: a quantum mechanical system (defined by a Hamiltonian $\mathcal{H}$), at temperature T, is described by a density matrix $\rho(t)$, which has an associated entropy $S(t) = -\text{tr}\rho(t)\ln\rho(t)$.

(a) Write down the time evolution equation for the density matrix, and calculate dS/dt.

(b) Using the method of Lagrange multipliers, find the density operator ρ_{max} that maximizes the functional $S[\rho]$, subject to the constraint of fixed average energy $\langle \mathcal{H} \rangle = \text{tr}\rho\mathcal{H} = E$.

(c) Show that the solution to part (b) is stationary, that is, $\partial \rho_{max}/\partial t = 0$.

****

9. *Ortho/para-hydrogen*: hydrogen molecules can exist in ortho and para states.

(a) The two electrons of H_2 in para-hydrogen form a singlet (antisymmetric) state. The orbital angular momentum can thus only take even values; that is,

$$\mathcal{H}_p = \frac{\hbar^2}{2I}\ell(\ell+1),$$

where $\ell = 0, 2, 4, \cdots$. Calculate the rotational partition function of para-hydrogen, and evaluate its low- and high-temperature limits.

(b) In ortho-hydrogen the electrons are in a triply degenerate symmetric state, hence

$$\mathcal{H}_o = \frac{\hbar^2}{2I}\ell(\ell+1),$$

with $\ell = 1, 3, 5, \cdots$. Calculate the rotational partition function of ortho-hydrogen, and evaluate its low- and high-temperature limits.

(c) For an equilibrium gas of N hydrogen molecules calculate the partition function. (*Hint.* Sum over contributions from mixtures of N_p para- and $N_o = N - N_p$ ortho-hydrogen particles. Ignore vibrational degrees of freedom.)

(d) Write down the expression for the rotational contribution to the internal energy $\langle E_{\text{rot.}} \rangle$, and comment on its low- and high-temperature limits.
Actually, due to small transition rates between ortho- and para-hydrogen, in most circumstances the mixture is not in equilibrium.

****

10. *van Leeuwen's theorem*: Consider a gas of charged particles subject to a general Hamiltonian of the form

$$\mathcal{H} = \sum_{i=1}^{N} \frac{\vec{p}_i^2}{2m} + U(\vec{q}_1, \cdots, \vec{q}_N).$$

In an external magnetic field, $\vec{B}$, the canonical momenta, $\vec{p}_n$, are replaced with $\vec{p}_n - e\vec{A}$, where $\vec{A}$ is the vector potential, $\vec{B} = \vec{\nabla} \times \vec{A}$. Show that if quantum effects are ignored, the thermodynamics of the problem is independent of $\vec{B}$.

****

7

Ideal quantum gases

7.1 Hilbert space of identical particles

In chapter 4, we discussed the Gibbs paradox for the mixing entropy of gases of identical particles. This difficulty was overcome by postulating that the phase space for N identical particles must be divided by $N!$, the number of permutations. This is not quite satisfactory as the classical equations of motion implicitly treat the particles as distinct. In quantum mechanics, by contrast, the identity of particles appears at the level of allowed states in Hilbert space. For example, the probability of finding two identical particles at positions $\vec{x}_1$ and $\vec{x}_2$, obtained from the expectation value of position operators, is given by $|\Psi(\vec{x}_1, \vec{x}_2)|^2$, where $\Psi(\vec{x}_1, \vec{x}_2)$ is the state vector in the coordinate representation (i.e., the wave function). Since the exchange of particles 1 and 2 leads to the same configuration, we must have $|\Psi(\vec{x}_1, \vec{x}_2)|^2 = |\Psi(\vec{x}_2, \vec{x}_1)|^2$. For a single-valued function, this leads to two possibilities[1]

$$|\psi(1, 2)\rangle = +|\psi(2, 1)\rangle, \qquad \text{or} \qquad |\psi(1, 2)\rangle = -|\psi(2, 1)\rangle. \tag{7.1}$$

The Hilbert space used to describe identical particles is thus restricted to obey certain symmetries.

For a system of N identical particles, there are $N!$ permutations P, forming a group S_N. There are several ways for representing a permutation; for example, $P(1\ 2\ 3\ 4) = (3\ 2\ 4\ 1)$ for $N = 4$ can alternatively be indicated by

$$P = \begin{pmatrix} 1 & 2 & 3 & 4 \\ 3 & 2 & 4 & 1 \end{pmatrix}$$

---

[1] A complex phase is ruled out by considering the action of the operator E_{12}, which exchanges the labels 1 and 2. Since $E_{12}^2 = I$, the identity operator, its squared eigenvalue must be unity, ruling out $e^{i\phi}$ unless $\phi = 0$ or π. This constraint is circumvented by *anyons*, which are described by multivalued wave functions with so-called fractional statistics.

Any permutation can be obtained from a sequence of two particle exchanges. For example, the above permutation is obtained by the exchanges (1,3) and (1,4) performed in sequence. The *parity* of a permutation is defined as

$$(-1)^P \equiv \begin{cases} +1 & \text{if } P \text{ involves an } even \text{ number of exchanges, e.g. } (1\ 2\ 3) \to (2\ 3\ 1) \\ -1 & \text{if } P \text{ involves an } odd \text{ number of exchanges, e.g. } (1\ 2\ 3) \to (2\ 1\ 3) \end{cases}.$$

(Note that if lines are drawn connecting the initial and final locations of each integer in the above representation of P, the parity is (-1) raised to the number of intersections of these lines; four in the above example.)

The action of permutations on an N-particle quantum state leads to a *representation* of the permutation group in Hilbert space. Requiring the wave function to be single-valued, and to give equal probabilities under particle exchange, restricts the representation to be either fully symmetric or anti-symmetric. This allows for two types of identical particles in nature:

(1) **Bosons** correspond to the fully symmetric representation such that

$$P|\psi(1,\cdots,N)\rangle = +|\psi(1,\cdots,N)\rangle.$$

(2) **Fermions** correspond to the fully anti-symmetric representation such that

$$P|\psi(1,\cdots,N)\rangle = (-1)^P|\psi(1,\cdots,N)\rangle.$$

Of course, the Hamiltonian for identical particles must itself be symmetric, that is, $P\mathcal{H} = \mathcal{H}$. However, for a given $\mathcal{H}$, there are many eigenstates with different symmetries under permutations. To select the correct set of eigenstates, in quantum mechanics the statistics of the particles (bosons or fermions) is specified independently. For example, consider N non-interacting particles in a box of volume V, with a Hamiltonian

$$\mathcal{H} = \sum_{\alpha=1}^{N} \mathcal{H}_\alpha = \sum_{\alpha=1}^{N} \left(-\frac{\hbar^2}{2m} \nabla_\alpha^2 \right). \tag{7.2}$$

Each $\mathcal{H}_\alpha$ can be separately diagonalized, with plane wave states $\{|\vec{k}\rangle\}$ and corresponding energies $\mathcal{E}(\vec{k}) = \hbar^2 k^2 / 2m$. Using sums and products of these one-particle states, we can construct the following N-particle states:

(1) The *product* Hilbert space is obtained by simple multiplication of the one-body states, that is,

$$|\vec{k}_1,\cdots,\vec{k}_N\rangle_\otimes \equiv |\vec{k}_1\rangle \cdots |\vec{k}_N\rangle. \tag{7.3}$$

In the coordinate representation,

$$\langle \vec{x}_1, \cdots, \vec{x}_N | \vec{k}_1, \cdots, \vec{k}_N \rangle_\otimes = \frac{1}{V^{N/2}} \exp \left(i \sum_{\alpha=1}^N \vec{k}_\alpha \cdot \vec{x}_\alpha \right), \tag{7.4}$$

and

$$\mathcal{H} | \vec{k}_1, \cdots, \vec{k}_N \rangle_\otimes = \left(\sum_{\alpha=1}^N \frac{\hbar^2}{2m} k_\alpha^2 \right) | \vec{k}_1, \cdots, \vec{k}_N \rangle_\otimes. \tag{7.5}$$

But the product states do not satisfy the symmetry requirements for identical particles, and we must find the appropriate *subspaces* of correct symmetry.

(2) The *fermionic* subspace is constructed as

$$| \vec{k}_1, \cdots, \vec{k}_N \rangle_- = \frac{1}{\sqrt{N_-}} \sum_P (-1)^P P | \vec{k}_1, \cdots, \vec{k}_N \rangle_\otimes, \tag{7.6}$$

where the sum is over all $N!$ permutations. If any one-particle label $\vec{k}$ appears more than once in the above list, the result is zero and there is no anti-symmetrized state. Anti-symmetrization is possible only when all the N values of $\vec{k}_\alpha$ are different. In this case, there are $N!$ terms in the above sum, and $N_- = N!$ is necessary to ensure normalization. For example, a two-particle anti-symmetrized state is

$$| \vec{k}_1, \vec{k}_2 \rangle_- = \frac{(| \vec{k}_1, \vec{k}_2 \rangle - | \vec{k}_2, \vec{k}_1 \rangle)}{\sqrt{2}}.$$

(If not otherwise indicated, $| \vec{k}_1, \cdots, \vec{k}_N \rangle$ refers to the product state.)

(3) Similarly, the *bosonic* subspace is constructed as

$$| \vec{k}_1, \cdots, \vec{k}_N \rangle_+ = \frac{1}{\sqrt{N_+}} \sum_P P | \vec{k}_1, \cdots, \vec{k}_N \rangle_\otimes. \tag{7.7}$$

In this case, there are no restrictions on the allowed values of $\vec{k}$. A particular one-particle state may be repeated $n_{\vec{k}}$ times in the list, with $\sum_{\vec{k}} n_{\vec{k}} = N$. As we shall prove shortly, proper normalization requires $N_+ = N! \prod_{\vec{k}} n_{\vec{k}}!$. For example, a correctly normalized three-particle bosonic state is constructed from two one-particle states $|\alpha\rangle$, and one one-particle state $|\beta\rangle$ as ($n_\alpha = 2$, $n_\beta = 1$, and $N_+ = 3!2!1! = 12$)

$$|\alpha\alpha\beta\rangle_+ = \frac{|\alpha\rangle|\alpha\rangle|\beta\rangle + |\alpha\rangle|\beta\rangle|\alpha\rangle + |\beta\rangle|\alpha\rangle|\alpha\rangle + |\alpha\rangle|\alpha\rangle|\beta\rangle + |\beta\rangle|\alpha\rangle|\alpha\rangle + |\alpha\rangle|\beta\rangle|\alpha\rangle}{\sqrt{12}}$$

$$= \frac{1}{\sqrt{3}} \left(|\alpha\rangle|\alpha\rangle|\beta\rangle + |\alpha\rangle|\beta\rangle|\alpha\rangle + |\beta\rangle|\alpha\rangle|\alpha\rangle \right).$$

It is convenient to discuss bosons and fermions simultaneously by defining

$$|\{\vec{k}\}\rangle_\eta = \frac{1}{\sqrt{N_\eta}} \sum_P \eta^P P |\{\vec{k}\}\rangle, \quad \text{with } \eta = \begin{cases} +1 & \text{for bosons} \\ -1 & \text{for fermions} \end{cases}, \tag{7.8}$$

with $(+1)^P \equiv 1$, and $(-1)^P$ used to indicate the parity of P as before. Each state is *uniquely* specified by a set of *occupation numbers* $\{n_{\vec{k}}\}$, such that $\sum_{\vec{k}} n_{\vec{k}} = N$, and

(1) For fermions, $|\{\vec{k}\}\rangle_- = 0$, unless $n_{\vec{k}} = 0$ or 1, and $N_- = N! \prod_k n_{\vec{k}}! = N!$.
(2) For bosons, any $\vec{k}$ may be repeated $n_{\vec{k}}$ times, and the normalization is calculated from

$$_+\langle\{\vec{k}\}|\{\vec{k}\}\rangle_+ = \frac{1}{N_+} \sum_{P,P'} \langle P\{\vec{k}\}|P'\{\vec{k}\}\rangle = \frac{N!}{N_+} \sum_P \langle\{\vec{k}\}|P\{\vec{k}\}\rangle$$

$$= \frac{N! \prod_{\vec{k}} n_{\vec{k}}!}{N_+} = 1, \quad \Rightarrow \quad N_+ = N! \prod_k n_{\vec{k}}!. \tag{7.9}$$

(The term $\langle\{\vec{k}\}|P\{\vec{k}\}\rangle$ is zero unless the permuted $\vec{k}$'s are identical to the original set, which happens $\prod_{\vec{k}} n_{\vec{k}}!$ times.)

7.2 Canonical formulation

Using the states constructed in the previous section, we can calculate the canonical density matrix for non-interacting identical particles. In the coordinate representation we have

$$\langle\{\vec{x}'\}|\rho|\{\vec{x}\}\rangle_\eta = \sum_{\{\vec{k}\}}' \sum_{P,P'} \eta^P \eta^{P'} \langle\{\vec{x}'\}|P'\{\vec{k}\}\rangle \rho(\{\vec{k}\}) \langle P\{\vec{k}\}|\{\vec{x}\}\rangle \frac{1}{N_\eta}, \tag{7.10}$$

where $\rho(\{\vec{k}\}) = \exp\left[-\beta\left(\sum_{\alpha=1}^N \hbar^2 k_\alpha^2/2m\right)\right]/Z_N$. The sum $\sum_{\{\vec{k}_1,\vec{k}_2,\cdots,\vec{k}_N\}}'$ is restricted to ensure that each identical particle state appears once and only once. In both the bosonic and fermionic subspaces the set of occupation numbers $\{n_{\vec{k}}\}$ uniquely identifies a state. We can, however, remove this restriction from Eq. (7.10) if we divide by the resulting over-counting factor (for bosons) of $N!/(\prod_{\vec{k}} n_{\vec{k}}!)$, that is,

$$\sum_{\{\vec{k}\}}' = \sum_{\{\vec{k}\}} \frac{\prod_{\vec{k}} n_{\vec{k}}!}{N!}.$$

(Note that for fermions, the $(-1)^P$ factors cancel out the contributions from cases where any $n_{\vec{k}}$ is larger than one.) Therefore,

$$\langle\{\vec{x}'\}|\rho|\{\vec{x}\}\rangle = \sum_{\{\vec{k}\}} \frac{\prod_{\vec{k}} n_{\vec{k}}!}{N!} \cdot \frac{1}{N! \prod_{\vec{k}} n_{\vec{k}}!} \cdot$$

$$\sum_{P,P'} \frac{\eta^P \eta^{P'}}{Z_N} \exp\left(-\beta \sum_{\alpha=1}^N \frac{\hbar^2 k_\alpha^2}{2m}\right) \langle\{\vec{x}'\}|P'\{\vec{k}\}\rangle\langle P\{\vec{k}\}|\{\vec{x}\}\rangle. \tag{7.11}$$

In the limit of large volume, the sums over $\{\vec{k}\}$ can be replaced by integrals, and using the plane wave representation of wave functions, we have

$$\langle\{\vec{x}'\}|\rho|\{\vec{x}\}\rangle = \frac{1}{Z_N(N!)^2} \sum_{P,P'} \eta^P \eta^{P'} \int \prod_{\alpha=1}^N \frac{V \, d^3\vec{k}_\alpha}{(2\pi)^3} \exp\left(-\frac{\beta\hbar^2 k_\alpha^2}{2m}\right)$$

$$\times \left\{\frac{\exp\left[-i\sum_{\alpha=1}^N (\vec{k}_{P\alpha}\cdot\vec{x}_\alpha - \vec{k}_{P'\alpha}\cdot\vec{x}'_\alpha)\right]}{V^N}\right\}. \tag{7.12}$$

We can order the sum in the exponent by focusing on a particular $\vec{k}$-vector. Since $\sum_\alpha f(P\alpha)g(\alpha) = \sum_\beta f(\beta)g(P^{-1}\beta)$, where $\beta = P\alpha$ and $\alpha = P^{-1}\beta$, we obtain

$$\langle\{\vec{x}'\}|\rho|\{\vec{x}\}\rangle = \frac{1}{Z_N(N!)^2} \sum_{P,P'} \eta^P \eta^{P'} \prod_{\alpha=1}^{N} \left[\int \frac{d^3\vec{k}_\alpha}{(2\pi)^3} e^{-i\vec{k}_\alpha \cdot \left(\vec{x}_{P^{-1}\alpha} - \vec{x}'_{P'^{-1}\alpha}\right) - \beta\hbar^2 k_\alpha^2/2m} \right].$$
$$(7.13)$$

The Gaussian integrals in the square brackets are equal to

$$\frac{1}{\lambda^3} \exp\left[-\frac{\pi}{\lambda^2} \left(\vec{x}_{P^{-1}\alpha} - \vec{x}'_{P'^{-1}\alpha}\right)^2 \right].$$

Setting $\beta = P^{-1}\alpha$ in Eq. (7.13) gives

$$\langle\{\vec{x}'\}|\rho|\{\vec{x}\}\rangle = \frac{1}{Z_N\lambda^{3N}(N!)^2} \sum_{P,P'} \eta^P \eta^{P'} \exp\left[-\frac{\pi}{\lambda^2} \sum_{\beta=1}^{N} \left(\vec{x}_\beta - \vec{x}'_{P'^{-1}P\beta}\right)^2 \right]. \quad (7.14)$$

Finally, we set $Q = P'^{-1}P$, and use the results $\eta^P = \eta^{P^{-1}}$, and $\eta^Q = \eta^{P'^{-1}P} = \eta^{P'}\eta^P$, to get (after performing $\sum_P = N!$)

$$\langle\{\vec{x}'\}|\rho|\{\vec{x}\}\rangle = \frac{1}{Z_N\lambda^{3N}N!} \sum_Q \eta^Q \exp\left[-\frac{\pi}{\lambda^2} \sum_{\beta=1}^{N} \left(\vec{x}_\beta - \vec{x}'_{Q\beta}\right)^2 \right]. \quad (7.15)$$

The canonical partition function, Z_N, is obtained from the normalization condition

$$\mathrm{tr}(\rho) = 1, \quad \Longrightarrow \quad \int \prod_{\alpha=1}^{N} d^3\vec{x}_\alpha \langle\{\vec{x}\}|\rho|\{\vec{x}\}\rangle = 1,$$

as

$$Z_N = \frac{1}{N!\lambda^{3N}} \int \prod_{\alpha=1}^{N} d^3\vec{x}_\alpha \sum_Q \eta^Q \exp\left[-\frac{\pi}{\lambda^2} \sum_{\beta=1}^{N} \left(\vec{x}_\beta - \vec{x}_{Q\beta}\right)^2 \right]. \quad (7.16)$$

The quantum partition function thus involves a sum over $N!$ possible permutations. The classical result $Z_N = \left(V/\lambda^3\right)^N /N!$ is obtained from the term corresponding to no particle exchange, $Q \equiv 1$. The division by $N!$ finally justifies the factor that was (somewhat artificially) introduced in classical statistical mechanics to deal with the phase space of identical particles. However, this classical result is only valid at very high temperatures and is modified by the quantum corrections coming from the remaining permutations. As any permutation involves a product of factors $\exp[-\pi(\vec{x}_1 - \vec{x}_2)^2/\lambda^2]$, its contributions vanish as $\lambda \to 0$ for $T \to \infty$.

The lowest order correction comes from the simplest permutation, which is the exchange of two particles. The exchange of particles 1 and 2 is accompanied by a factor of $\eta \exp[-2\pi(\vec{x}_1 - \vec{x}_2)^2/\lambda^2]$. As each of the possible $N(N-1)/2$ pairwise exchanges gives the same contribution to Z_N, we get

$$Z_N = \frac{1}{N!\lambda^{3N}} \int \prod_{\alpha=1}^{N} d^3\vec{x}_\alpha \left\{ 1 + \frac{N(N-1)}{2} \eta \exp\left[-\frac{2\pi}{\lambda^2}(\vec{x}_1 - \vec{x}_2)^2 \right] + \cdots \right\}. \quad (7.17)$$

For any $\alpha \geq 3$, $\int \mathrm{d}^3\vec{x}_\alpha = V$; in the remaining two integrations we can use the relative, $\vec{r}_{12} = \vec{x}_2 - \vec{x}_1$, and center of mass coordinates to get

$$
Z_N = \frac{1}{N!\lambda^{3N}} V^N \left[1 + \frac{N(N-1)}{2V} \eta \int \mathrm{d}^3\vec{r}_{12} \mathrm{e}^{-2\pi\vec{r}_{12}^2/\lambda^2} + \cdots \right]
$$
$$
= \frac{1}{N!} \left(\frac{V}{\lambda^3} \right)^N \left[1 + \frac{N(N-1)}{2V} \cdot \left(\sqrt{\frac{2\pi\lambda^2}{4\pi}} \right)^3 \eta + \cdots \right]. \tag{7.18}
$$

From the corresponding free energy,

$$
F = -k_B T \ln Z_N = -N k_B T \ln \left[\frac{\mathrm{e}}{\lambda^3} \cdot \frac{V}{N} \right] - \frac{k_B T N^2}{2V} \cdot \frac{\lambda^3}{2^{3/2}} \eta + \cdots, \tag{7.19}
$$

the gas pressure is computed as

$$
P = -\left. \frac{\partial F}{\partial V} \right|_T = \frac{N k_B T}{V} - \frac{N^2 k_B T}{V^2} \cdot \frac{\lambda^3}{2^{5/2}} \eta + \cdots = n k_B T \left[1 - \frac{\eta\lambda^3}{2^{5/2}} n + \cdots \right]. \tag{7.20}
$$

Note that the first quantum correction is equivalent to a second virial coefficient of

$$
B_2 = -\frac{\eta\lambda^3}{2^{5/2}}. \tag{7.21}
$$

The resulting correction to pressure is negative for bosons, and positive for fermions. In the classical formulation, a second virial coefficient was obtained from a two-body interaction. The classical potential $\mathcal{V}(\vec{r})$ that leads to a virial series for the partition function as in Eq. (7.17) is obtained from

$$
f(\vec{r}) = \mathrm{e}^{-\beta\mathcal{V}(\vec{r})} - 1 = \eta\mathrm{e}^{-2\pi\vec{r}^2/\lambda^2}, \quad \Longrightarrow
$$
$$
\mathcal{V}(\vec{r}) = -k_B T \ln \left[1 + \eta\mathrm{e}^{-2\pi\vec{r}^2/\lambda^2} \right] \approx -k_B T \eta\mathrm{e}^{-2\pi\vec{r}^2/\lambda^2}. \tag{7.22}
$$

(The final approximation corresponds to high temperatures, where only the first correction is important.) Thus the effects of quantum statistics at high temperatures are approximately equivalent to introducing an interaction between particles. The interaction is attractive for bosons, repulsive for fermions, and operates over distances of the order of the thermal wavelength λ.

Fig. 7.1 The effective classical potential that mimicks quantum correlations at high temperatures is attractive for bosons, and repulsive for fermions.

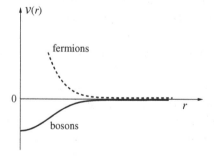

7.3 Grand canonical formulation

Calculating the partition function by performing all the sums in Eq. (7.16) is a formidable task. Alternatively, we can compute Z_N in the energy basis as

$$Z_N = \text{tr}\left(e^{-\beta\mathcal{H}}\right) = \sum_{\{\vec{k}_\alpha\}}' \exp\left[-\beta\sum_{\alpha=1}^N \mathcal{E}(\vec{k}_\alpha)\right] = \sum_{\{n_{\vec{k}}\}}' \exp\left[-\beta\sum_{\vec{k}} \mathcal{E}(\vec{k})n(\vec{k})\right]. \quad (7.23)$$

These sums are still difficult to perform due to the restrictions of symmetry on the allowed values of $\vec{k}$ or $\{n_{\vec{k}}\}$: the occupation numbers $\{n_{\vec{k}}\}$ are restricted to $\sum_{\vec{k}} n_{\vec{k}} = N$, and $n_{\vec{k}} = 0, 1, 2, \cdots$ for bosons, while $n_{\vec{k}} = 0$ or 1 for fermions. As before, the first constraint can be removed by looking at the grand partition function,

$$\mathcal{Q}_\eta(T, \mu) = \sum_{N=0}^\infty e^{\beta\mu N} \sum_{\{n_{\vec{k}}\}}' \exp\left[-\beta\sum_{\vec{k}} \mathcal{E}(\vec{k})n_{\vec{k}}\right]$$
$$= \sum_{\{n_{\vec{k}}\}} \prod_{\vec{k}}^\eta \exp\left[-\beta(\mathcal{E}(\vec{k}) - \mu)n_{\vec{k}}\right]. \quad (7.24)$$

The sums over $\{n_{\vec{k}}\}$ can now be performed independently for each $\vec{k}$, subject to the restrictions on occupation numbers imposed by particle symmetry.

- For *fermions*, $n_{\vec{k}} = 0$ or 1, and

$$\mathcal{Q}_- = \prod_{\vec{k}}\left[1 + \exp\left(\beta\mu - \beta\mathcal{E}(\vec{k})\right)\right]. \quad (7.25)$$

- For *bosons*, $n_{\vec{k}} = 0, 1, 2, \cdots$, and summing the geometric series gives

$$\mathcal{Q}_+ = \prod_{\vec{k}}\left[1 - \exp\left(\beta\mu - \beta\mathcal{E}(\vec{k})\right)\right]^{-1}. \quad (7.26)$$

The results for both cases can be presented simultaneously as

$$\ln\mathcal{Q}_\eta = -\eta\sum_{\vec{k}} \ln\left[1 - \eta\exp\left(\beta\mu - \beta\mathcal{E}(\vec{k})\right)\right], \quad (7.27)$$

with $\eta = -1$ for fermions, and $\eta = +1$ for bosons.

In the grand canonical formulation, different one-particle states are occupied independently, with a joint probability

$$p_\eta\left(\{n(\vec{k})\}\right) = \frac{1}{\mathcal{Q}_\eta} \prod_{\vec{k}} \exp\left[-\beta(\mathcal{E}(\vec{k}) - \mu)n_{\vec{k}}\right]. \quad (7.28)$$

The *average occupation number* of a state of energy $\mathcal{E}(\vec{k})$ is given by

$$\langle n_{\vec{k}}\rangle_\eta = -\frac{\partial\ln\mathcal{Q}_\eta}{\partial(\beta\mathcal{E}(\vec{k}))} = \frac{1}{z^{-1}e^{\beta\mathcal{E}(\vec{k})} - \eta}, \quad (7.29)$$

where $z = \exp(\beta\mu)$. The average values of the particle number and internal energy are then given by

$$
\begin{cases}
N_\eta = \sum_{\vec{k}} \langle n_{\vec{k}} \rangle_\eta = \sum_{\vec{k}} \dfrac{1}{z^{-1} e^{\beta\mathcal{E}(\vec{k})} - \eta} \\[3mm]
E_\eta = \sum_{\vec{k}} \mathcal{E}(\vec{k}) \langle n_{\vec{k}} \rangle_\eta = \sum_{\vec{k}} \dfrac{\mathcal{E}(\vec{k})}{z^{-1} e^{\beta\mathcal{E}(\vec{k})} - \eta}
\end{cases}
\tag{7.30}
$$

7.4 Non-relativistic gas

Quantum particles are further characterized by a *spin s*. In the absence of a magnetic field, different spin states have the same energy, and a *spin degeneracy* factor, $g = 2s + 1$, multiplies Eqs. (7.27)–(7.30). In particular, for a non-relativistic gas in three dimensions ($\mathcal{E}(\vec{k}) = \hbar^2 k^2 / 2m$, and $\sum_{\vec{k}} \to V \int d^3\vec{k}/(2\pi)^3$) these equations reduce to

$$
\begin{cases}
\beta P_\eta = \dfrac{\ln \mathcal{Q}_\eta}{V} = -\eta g \displaystyle\int \dfrac{d^3\vec{k}}{(2\pi)^3} \ln\left[1 - \eta z \exp\left(-\dfrac{\beta\hbar^2 k^2}{2m} \right) \right], \\[4mm]
n_\eta \equiv \dfrac{N_\eta}{V} = g \displaystyle\int \dfrac{d^3\vec{k}}{(2\pi)^3} \dfrac{1}{z^{-1} \exp\left(\dfrac{\beta\hbar^2 k^2}{2m} \right) - \eta}, \\[4mm]
\varepsilon_\eta \equiv \dfrac{E_\eta}{V} = g \displaystyle\int \dfrac{d^3\vec{k}}{(2\pi)^3} \dfrac{\hbar^2 k^2}{2m} \dfrac{1}{z^{-1} \exp\left(\dfrac{\beta\hbar^2 k^2}{2m} \right) - \eta}.
\end{cases}
\tag{7.31}
$$

To simplify these equations, we change variables to $x = \beta\hbar^2 k^2/(2m)$, so that

$$
k = \frac{\sqrt{2mk_B T}}{\hbar} x^{1/2} = \frac{2\pi^{1/2}}{\lambda} x^{1/2}, \qquad \Longrightarrow \qquad dk = \frac{\pi^{1/2}}{\lambda} x^{-1/2} dx.
$$

Substituting into Eqs. (7.31) gives

$$
\begin{cases}
\beta P_\eta = -\eta \dfrac{g}{2\pi^2} \dfrac{4\pi^{3/2}}{\lambda^3} \displaystyle\int_0^\infty dx\, x^{1/2} \ln\left(1 - \eta z e^{-x} \right) \\[4mm]
\qquad = \dfrac{g}{\lambda^3} \dfrac{4}{3\sqrt{\pi}} \displaystyle\int_0^\infty \dfrac{dx\, x^{3/2}}{z^{-1} e^x - \eta} \quad \text{(integration by parts)}, \\[4mm]
n_\eta = \dfrac{g}{\lambda^3} \dfrac{2}{\sqrt{\pi}} \displaystyle\int_0^\infty \dfrac{dx\, x^{1/2}}{z^{-1} e^x - \eta}, \\[4mm]
\beta\varepsilon_\eta = \dfrac{g}{\lambda^3} \dfrac{2}{\sqrt{\pi}} \displaystyle\int_0^\infty \dfrac{dx\, x^{3/2}}{z^{-1} e^x - \eta}.
\end{cases}
\tag{7.32}
$$

We now define two sets of functions by

$$f_m^\eta(z) = \frac{1}{(m-1)!} \int_0^\infty \frac{\mathrm{d}x\, x^{m-1}}{z^{-1}\mathrm{e}^x - \eta}. \tag{7.33}$$

For non-integer arguments, the function $m! \equiv \Gamma(m+1)$ is defined by the integral $\int_0^\infty \mathrm{d}x\, x^m\, \mathrm{e}^{-x}$. In particular, from this definition it follows that $(1/2)! = \sqrt{\pi}/2$, and $(3/2)! = (3/2)\sqrt{\pi}/2$. Equations (7.32) now take the simple forms

$$\begin{cases} \beta P_\eta = \dfrac{g}{\lambda^3} f_{5/2}^\eta(z), \\[2mm] \quad n_\eta = \dfrac{g}{\lambda^3} f_{3/2}^\eta(z), \\[2mm] \quad \varepsilon_\eta = \dfrac{3}{2} P_\eta. \end{cases} \tag{7.34}$$

These results completely describe the thermodynamics of ideal quantum gases as a function of z. To find the equation of state explicitly in the form $P_\eta(n_\eta, T)$ requires solving for z in terms of density. To this end, we need to understand the behavior of the functions $f_m^\eta(z)$.

The high-temperature, low-density (non-degenerate) limit will be examined first. In this limit, z is small and

$$\begin{aligned} f_m^\eta(z) &= \frac{1}{(m-1)!} \int_0^\infty \frac{\mathrm{d}x\, x^{m-1}}{z^{-1}\mathrm{e}^x - \eta} = \frac{1}{(m-1)!} \int_0^\infty \mathrm{d}x\, x^{m-1}\, (z\mathrm{e}^{-x})\, (1 - \eta z\mathrm{e}^{-x})^{-1} \\[2mm] &= \frac{1}{(m-1)!} \int_0^\infty \mathrm{d}x\, x^{m-1} \sum_{\alpha=1}^\infty (z\mathrm{e}^{-x})^\alpha\, \eta^{\alpha+1} \\[2mm] &= \sum_{\alpha=1}^\infty \eta^{\alpha+1} z^\alpha \frac{1}{(m-1)!} \int_0^\infty \mathrm{d}x\, x^{m-1} \mathrm{e}^{-\alpha x} \\[2mm] &= \sum_{\alpha=1}^\infty \eta^{\alpha+1} \frac{z^\alpha}{\alpha^m} = z + \eta \frac{z^2}{2^m} + \frac{z^3}{3^m} + \eta \frac{z^4}{4^m} + \cdots . \end{aligned} \tag{7.35}$$

We thus find (self-consistently) that $f_m^\eta(z)$, and hence $n_\eta(z)$ and $P_\eta(z)$, are indeed small as $z \to 0$. Equations (7.34) in this limit give

$$\begin{cases} \dfrac{n_\eta \lambda^3}{g} = f_{3/2}^\eta(z) = z + \eta \dfrac{z^2}{2^{3/2}} + \dfrac{z^3}{3^{3/2}} + \eta \dfrac{z^4}{4^{3/2}} + \cdots , \\[3mm] \dfrac{\beta P_\eta \lambda^3}{g} = f_{5/2}^\eta(z) = z + \eta \dfrac{z^2}{2^{5/2}} + \dfrac{z^3}{3^{5/2}} + \eta \dfrac{z^4}{4^{5/2}} + \cdots . \end{cases} \tag{7.36}$$

The first of the above equations can be solved perturbatively, by the recursive procedure of substituting the solution up to a lower order, as

$$\begin{aligned} z &= \frac{n_\eta \lambda^3}{g} - \eta \frac{z^2}{2^{3/2}} - \frac{z^3}{3^{3/2}} - \cdots \\[2mm] &= \left(\frac{n_\eta \lambda^3}{g}\right) - \frac{\eta}{2^{3/2}} \left(\frac{n_\eta \lambda^3}{g}\right)^2 - \cdots \\[2mm] &= \left(\frac{n_\eta \lambda^3}{g}\right) - \frac{\eta}{2^{3/2}} \left(\frac{n_\eta \lambda^3}{g}\right)^2 + \left(\frac{1}{4} - \frac{1}{3^{3/2}}\right) \left(\frac{n_\eta \lambda^3}{g}\right)^3 - \cdots . \end{aligned} \tag{7.37}$$

Substituting this solution into the second leads to

$$\frac{\beta P_\eta \lambda^3}{g} = \left(\frac{n_\eta \lambda^3}{g}\right) - \frac{\eta}{2^{3/2}}\left(\frac{n_\eta \lambda^3}{g}\right)^2 + \left(\frac{1}{4} - \frac{1}{3^{3/2}}\right)\left(\frac{n_\eta \lambda^3}{g}\right)^3$$
$$+ \frac{\eta}{2^{5/2}}\left(\frac{n_\eta \lambda^3}{g}\right)^2 - \frac{1}{8}\left(\frac{n_\eta \lambda^3}{g}\right)^3 + \frac{1}{3^{5/2}}\left(\frac{n_\eta \lambda^3}{g}\right)^3 + \cdots.$$

The pressure of the quantum gas can thus be expressed in the form of a virial expansion, as

$$P_\eta = n_\eta k_B T \left[1 - \frac{\eta}{2^{5/2}}\left(\frac{n_\eta \lambda^3}{g}\right) + \left(\frac{1}{8} - \frac{2}{3^{5/2}}\right)\left(\frac{n_\eta \lambda^3}{g}\right)^2 + \cdots\right]. \tag{7.38}$$

The second virial coefficient $B_2 = -\eta\lambda^3/(2^{5/2}g)$ agrees with Eq. (7.21) computed in the canonical ensemble for $g = 1$. The natural (dimensionless) expansion parameter is $n_\eta \lambda^3/g$, and quantum mechanical effects become important when $n_\eta \lambda^3 \geq g$; in the so-called *quantum degenerate limit*. The behavior of fermi and bose gases is very different in this degenerate limit of low temperatures and high densities, and the two cases will be discussed separately in the following sections.

7.5 The degenerate fermi gas

At zero temperature, the *fermi occupation number*,

$$\langle n_{\vec{k}}\rangle_- = \frac{1}{e^{\beta\left(\mathcal{E}(k)-\mu\right)} + 1}, \tag{7.39}$$

is one for $\mathcal{E}(\vec{k}) < \mu$, and zero otherwise. The limiting value of μ at zero temperature is called the *fermi energy*, $\mathcal{E}_F$, and all one-particle states of energy less than $\mathcal{E}_F$ are occupied, forming a *fermi sea*. For the ideal gas with $\mathcal{E}(\vec{k}) = \hbar^2 k^2/(2m)$, there is a corresponding *fermi wavenumber* k_F, calculated from

$$N = \sum_{|\vec{k}|\leq k_F}(2s+1) = gV\int^{k<k_F}\frac{d^3\vec{k}}{(2\pi)^3} = g\frac{V}{6\pi^2}k_F^3. \tag{7.40}$$

In terms of the density $n = N/V$,

$$k_F = \left(\frac{6\pi^2 n}{g}\right)^{1/3}, \qquad \Longrightarrow \qquad \mathcal{E}_F(n) = \frac{\hbar^2 k_F^2}{2m} = \frac{\hbar^2}{2m}\left(\frac{6\pi^2 n}{g}\right)^{2/3}. \tag{7.41}$$

Note that while in a classical treatment the ideal gas has a large density of states at $T = 0$ (from $\Omega_{\text{Classical}} \propto V^N/N!$), the quantum fermi gas has a unique ground state with $\Omega = 1$. Once the one-particle momenta are specified (all $\vec{k}$ for $|\vec{k}| < k_F$), there is only one anti-symmetrized state, as constructed in Eq. (7.6).

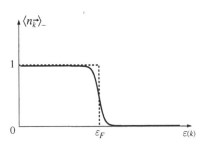

Fig. 7.2 The fermi
distribution for
occupation numbers, at
zero (dotted), and finite
(solid) temperatures.

To see how the fermi sea is modified at small temperatures, we need the behavior of $f_m^-(z)$ for large z, which, after integration by parts, is

$$f_m^-(z) = \frac{1}{m!} \int_0^\infty dx\, x^m \frac{d}{dx}\left(\frac{-1}{z^{-1}e^x + 1}\right).$$

Since the fermi occupation number changes abruptly from one to zero, its derivative in the above equation is sharply peaked. We can expand around this peak by setting $x = \ln z + t$, and extending the range of integration to $-\infty < t < +\infty$, as

$$
\begin{aligned}
f_m^-(z) &\approx \frac{1}{m!} \int_{-\infty}^\infty dt\, (\ln z + t)^m \frac{d}{dt}\left(\frac{-1}{e^t + 1}\right) \\
&= \frac{1}{m!} \int_{-\infty}^\infty dt \sum_{\alpha=0}^\infty \binom{m}{\alpha} t^\alpha (\ln z)^{m-\alpha} \frac{d}{dt}\left(\frac{-1}{e^t + 1}\right) \\
&= \frac{(\ln z)^m}{m!} \sum_{\alpha=0}^\infty \frac{m!}{\alpha!(m-\alpha)!} (\ln z)^{-\alpha} \int_{-\infty}^\infty dt\, t^\alpha \frac{d}{dt}\left(\frac{-1}{e^t + 1}\right).
\end{aligned}
$$ (7.42)

Using the (anti-)symmetry of the integrand under $t \to -t$, and undoing the integration by parts, yields

$$\frac{1}{\alpha!}\int_{-\infty}^\infty dt\, t^\alpha \frac{d}{dt}\left(\frac{-1}{e^t+1}\right) = \begin{cases} 0 & \text{for } \alpha \text{ odd,} \\[2ex] \dfrac{2}{(\alpha-1)!}\displaystyle\int_0^\infty dt\, \dfrac{t^{\alpha-1}}{e^t+1} = 2f_\alpha^-(1) & \text{for } \alpha \text{ even.} \end{cases}$$

Inserting the above into Eq. (7.42), and using tabulated values for the integrals $f_\alpha^-(1)$, leads to the *Sommerfeld expansion*,

$$
\begin{aligned}
\lim_{z\to\infty} f_m^-(z) &= \frac{(\ln z)^m}{m!} \sum_{\alpha=0}^{\text{even}} 2f_\alpha^-(1) \frac{m!}{(m-\alpha)!} (\ln z)^{-\alpha} \\
&= \frac{(\ln z)^m}{m!}\left[1 + \frac{\pi^2}{6}\frac{m(m-1)}{(\ln z)^2} + \frac{7\pi^4}{360}\frac{m(m-1)(m-2)(m-3)}{(\ln z)^4} + \cdots\right].
\end{aligned}
$$ (7.43)

In the degenerate limit, the density and chemical potential are related by

$$\frac{n\lambda^3}{g} = f_{3/2}^-(z) = \frac{(\ln z)^{3/2}}{(3/2)!}\left[1 + \frac{\pi^2}{6}\frac{3}{2}\frac{1}{2}(\ln z)^{-2} + \cdots\right] \gg 1.$$ (7.44)

Fig. 7.3 The function
$f_{3/2}^-(z)$ determines the
chemical potential μ.

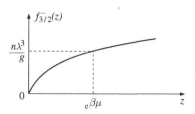

The lowest-order result reproduces the expression in Eq. (7.40) for the fermi energy,

$$\lim_{T\to 0}\ln z = \left[\frac{3}{4\sqrt{\pi}}\,\frac{n\lambda^3}{g}\right]^{2/3} = \frac{\beta\hbar^2}{2m}\left(\frac{6\pi^2 n}{g}\right)^{2/3} = \beta\mathcal{E}_F.$$

Inserting the zero-temperature limit into Eq. (7.44) gives the first-order correction,

$$\ln z = \beta\mathcal{E}_F\left[1 + \frac{\pi^2}{8}\left(\frac{k_B T}{\mathcal{E}_F}\right)^2 + \cdots\right]^{-2/3} = \beta\mathcal{E}_F\left[1 - \frac{\pi^2}{12}\left(\frac{k_B T}{\mathcal{E}_F}\right)^2 + \cdots\right]. \quad (7.45)$$

Fig. 7.4 Temperature
dependence of the
chemical potential for a
fermi gas.

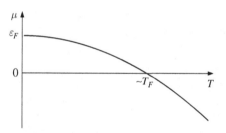

The appropriate dimensionless expansion parameter is $(k_B T/\mathcal{E}_F)$. Note that the fermion chemical potential $\mu = k_B T \ln z$ is positive at low temperatures, and negative at high temperatures (from Eq. (7.37)). It changes sign at a temperature of the order of $\mathcal{E}_F/k_B$, set by the fermi energy.

The low-temperature expansion for the pressure is

$$\beta P = \frac{g}{\lambda^3}f_{5/2}^-(z) = \frac{g}{\lambda^3}\frac{(\ln z)^{5/2}}{(5/2)!}\left[1 + \frac{\pi^2}{6}\frac{5}{2}\frac{3}{2}(\ln z)^{-2} + \cdots\right]$$

$$= \frac{g}{\lambda^3}\frac{8(\beta\mathcal{E}_F)^{5/2}}{15\sqrt{\pi}}\left[1 - \frac{5}{2}\frac{\pi^2}{12}\left(\frac{k_B T}{\mathcal{E}_F}\right)^2 + \cdots\right]\left[1 + \frac{5\pi^2}{8}\left(\frac{k_B T}{\mathcal{E}_F}\right)^2 + \cdots\right] \quad (7.46)$$

$$= \beta P_F\left[1 + \frac{5}{12}\pi^2\left(\frac{k_B T}{\mathcal{E}_F}\right)^2 + \cdots\right],$$

where $P_F = (2/5)n\mathcal{E}_F$ is the *fermi pressure*. Unlike its classical counterpart, the fermi gas at zero temperature has finite pressure and internal energy.

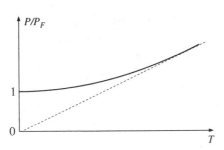

Fig. 7.5 Temperature dependence of the pressure of a fermi gas.

The low-temperature expansion for the internal energy is obtained easily from Eq. (7.46) using

$$\frac{E}{V} = \frac{3}{2}P = \frac{3}{5}nk_BT_F\left[1 + \frac{5}{12}\pi^2\left(\frac{T}{T_F}\right)^2 + \cdots\right], \tag{7.47}$$

where we have introduced the *fermi temperature* $T_F = \mathcal{E}_F/k_B$. Equation (7.47) leads to a low-temperature heat capacity,

$$C_V = \frac{dE}{dT} = \frac{\pi^2}{2}Nk_B\left(\frac{T}{T_F}\right) + \mathcal{O}\left(\frac{T}{T_F}\right)^3. \tag{7.48}$$

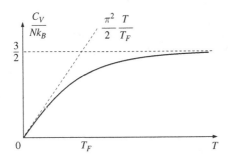

Fig. 7.6 Temperature dependence of the heat capacity of a fermi gas.

The linear scaling of the heat capacity as $T \to 0$ is a general feature of a fermi gas, valid in all dimensions. It has the following simple physical interpretation: the probability of occupying single-particle states, Eq. (7.39), is very close to a step function at small temperatures. Only particles within a distance of approximately k_BT of the fermi energy can be thermally excited. This represents only a small fraction, T/T_F, of the total number of fermions. Each excited particle gains an energy of the order of k_BT, leading to a change in the internal energy of approximately $k_BTN(T/T_F)$. Hence the heat capacity is given by $C_V = dE/dT \sim Nk_BT/T_F$. This conclusion is also valid for an interacting fermi gas.

The fact that only a small number, $N(T/T_F)$, of fermions are excited at small temperatures accounts for many interesting properties of fermi gases. For example, the magnetic susceptibility of a classical gas of N non-interacting particles of magnetic moment μ_B follows the Curie law, $\chi \propto N\mu_B^2/(k_B T)$. Since quantum mechanically, only a fraction of spins contributes at low-temperatures, the low-temperature susceptibility saturates to a (Pauli) value of $\chi \propto N\mu_B^2/(k_B T_F)$. (See the problems for details of this calculation.)

7.6 The degenerate bose gas

The average *bose occupation number*,

$$\langle n_{\vec{k}} \rangle_+ = \frac{1}{\exp\left[\beta\left(\mathcal{E}(\vec{k}) - \mu\right)\right] - 1}, \tag{7.49}$$

must certainly be positive. This requires $\mathcal{E}(\vec{k}) - \mu$ to be positive for all $\vec{k}$, and hence $\mu < \min\left[\mathcal{E}(\vec{k})\right]_{\vec{k}} = 0$ (for $\mathcal{E}(\vec{k}) = \hbar^2 k^2/2m$). At high temperatures (classical limit), μ is large and *negative*, and increases toward zero like $k_B T \ln(n\lambda^3/g)$ as temperature is reduced. In the degenerate quantum limit, μ approaches its limiting value of zero. To see how this limit is achieved, and to find out about the behavior of the degenerate bose gas, we have to examine the limiting behavior of the functions $f_m^+(z)$ in Eqs. (7.34) as $z = \exp(\beta\mu)$ goes to unity.

The functions $f_m^+(z)$ are monotonically increasing with z in the interval $0 \leq z \leq 1$. The maximum value attained at $z = 1$ is given by

$$\zeta_m \equiv f_m^+(1) = \frac{1}{(m-1)!} \int_0^\infty \frac{\mathrm{d}x\, x^{m-1}}{e^x - 1}. \tag{7.50}$$

The integrand has a pole as $x \to 0$, where it behaves as $\int \mathrm{d}x x^{m-2}$. Therefore, ζ_m is finite for $m > 1$ and infinite for $m \leq 1$. A useful recursive property of these functions is (for $m > 1$)

$$
\begin{aligned}
\frac{\mathrm{d}}{\mathrm{d}z} f_m^+(z) &= \int_0^\infty \mathrm{d}x\, \frac{x^{m-1}}{(m-1)!} \frac{\mathrm{d}}{\mathrm{d}z}\left(\frac{1}{z^{-1}e^x - 1}\right) \\
&\qquad \left(\text{use } \frac{\mathrm{d}}{\mathrm{d}z} f\left(z^{-1}e^x\right) = -\frac{e^x}{z^2} f' = -\frac{1}{z}\frac{\mathrm{d}}{\mathrm{d}x} f\left(z^{-1}e^x\right)\right) \\
&= -\frac{1}{z} \int_0^\infty \mathrm{d}x\, \frac{x^{m-1}}{(m-1)!} \frac{\mathrm{d}}{\mathrm{d}x}\left(\frac{1}{z^{-1}e^x - 1}\right) \quad \text{(integrate by parts)} \\
&= \frac{1}{z} \int_0^\infty \mathrm{d}x\, \frac{x^{m-2}}{(m-2)!} \frac{1}{z^{-1}e^x - 1} = \frac{1}{z} f_{m-1}^+(z).
\end{aligned}
\tag{7.51}
$$

Hence, a sufficiently high derivative of $f_m^+(z)$ will be divergent at $z = 1$ for all m.

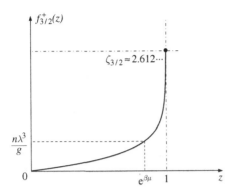

Fig. 7.7 Graphical calculation of the fugacity of a three-dimensional ideal bose gas.

The density of *excited states* for the non-relativistic bose gas in three dimensions is thus bounded by

$$n_\times = \frac{g}{\lambda^3} f_{3/2}^+(z) \leq n^* = \frac{g}{\lambda^3} \zeta_{3/2}. \tag{7.52}$$

At sufficiently high temperatures, such that

$$\frac{n\lambda^3}{g} = \frac{n}{g} \left(\frac{h}{\sqrt{2\pi m k_B T}} \right)^3 \leq \zeta_{3/2} \approx 2.612 \cdots, \tag{7.53}$$

this bound is not relevant, and $n_\times = n$. However, on lowering temperature, the limiting density of excited states is achieved at

$$T_c(n) = \frac{h^2}{2\pi m k_B} \left(\frac{n}{g\zeta_{3/2}} \right)^{2/3}. \tag{7.54}$$

For $T \leq T_c$, z gets stuck to unity ($\mu = 0$). The limiting density of excited states, $n^* = g\zeta_{3/2}/\lambda^3 \propto T^{3/2}$, is then less than the total particle density (and independent of it). The remaining gas particles, with density $n_0 = n - n^*$, occupy the lowest energy state with $\vec{k} = 0$. This phenomenon, of a *macroscopic occupation* of a single one-particle state, is a consequence of quantum statistics, and known as *Bose–Einstein condensation*.

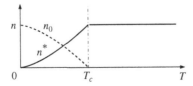

Fig. 7.8 Temperature dependence of the thermally excited bose particle density (solid line). For $T < T_c(n)$, the remaining density occupies the ground state (dashed line).

Fig. 7.9 Temperature
dependence of the
pressure of the bose gas,
at two densities $n_1 < n_2$.

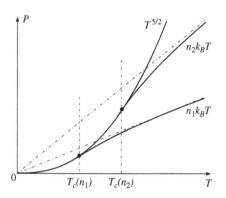

The bose condensate has some unusual properties. The gas pressure for $T < T_c$,

$$\beta P = \frac{g}{\lambda^3} f_{5/2}^+(1) = \frac{g}{\lambda^3} \zeta_{5/2} \approx 1.341 \frac{g}{\lambda^3}, \tag{7.55}$$

vanishes as $T^{5/2}$ and is *independent of density*. This is because only the excited fraction n^* has finite momentum and contributes to the pressure. Alternatively, bose condensation can be achieved at a fixed temperature by increasing density (reducing volume). From Eq. (7.53), the transition occurs at a specific volume

$$v^* = \frac{1}{n^*} = \frac{\lambda^3}{g\zeta_{3/2}}. \tag{7.56}$$

For $v < v^*$, the pressure–volume isotherm is flat, since $\partial P/\partial v \propto \partial P/\partial n = 0$ from Eq. (7.55). The flat portion of isotherms is reminiscent of coexisting liquid and gas phases. We can similarly regard bose condensation as the coexistence of a "normal gas" of specific volume v^*, and a "liquid" of volume 0. The vanishing of the "liquid" volume is an unrealistic feature due to the absence of any interaction potential between the particles.

Bose condensation combines features of discontinuous (first-order), and continuous (second-order) transitions; there is a finite latent heat while the compressibility diverges. The latent heat of the transition can be obtained from the *Clausius–Clapeyron* equation, which gives the change of the transition temperature with pressure as

$$\left.\frac{dT}{dP}\right|_{\text{Coexistence}} = \frac{\Delta V}{\Delta S} = \frac{T_c(v^* - v_0)}{L}. \tag{7.57}$$

Since Eq. (7.55) gives the gas pressure right up to the transition point,

$$\left.\frac{dP}{dT}\right|_{\text{Coexistence}} = \frac{5}{2}\frac{P}{T}. \tag{7.58}$$

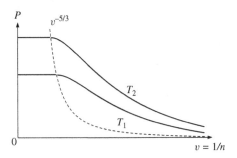

Fig. 7.10 Isotherms of
the bose gas at two
temperatures $T_1 < T_2$.

Using the above equations we find a latent heat,

$$L = T_c v^* \frac{dP}{dT}\Big|_{\text{Coexistence}} = \frac{5}{2} P v^* = \frac{5}{2} \frac{g}{\lambda^3} \zeta_{5/2} k_B T_c \left(\frac{g}{\lambda^3} \zeta_{3/2} \right)^{-1},$$

$$\Longrightarrow \qquad L = \frac{5}{2} \frac{\zeta_{5/2}}{\zeta_{3/2}} k_B T_c \approx 1.28 k_B T_c. \tag{7.59}$$

To find the compressibility $\kappa_T = \partial n / \partial P|_T / n$, take derivatives of Eqs. (7.34), and take advantage of the identity in Eq. (7.51) to get

$$\begin{cases} \dfrac{dP}{dz} = \dfrac{g k_B T}{\lambda^3} \dfrac{1}{z} f_{3/2}^+(z) \\[2mm] \dfrac{dn}{dz} = \dfrac{g}{\lambda^3} \dfrac{1}{z} f_{1/2}^+(z) \end{cases}. \tag{7.60}$$

The ratio of these equations leads to

$$\kappa_T = \frac{f_{1/2}^+(z)}{n k_B T f_{3/2}^+(z)}, \tag{7.61}$$

which diverges at the transition since $\lim_{z \to 1} f_{1/2}^+(z) \to \infty$, that is, the isotherms approach the flat coexistence portion tangentially.

From the expression for energy in the grand canonical ensemble,

$$E = \frac{3}{2} PV = \frac{3}{2} V \frac{g}{\lambda^3} k_B T f_{5/2}^+(z) \propto T^{5/2} f_{5/2}^+(z), \tag{7.62}$$

and using Eq. (7.51), the heat capacity is obtained as

$$C_{V,N} = \frac{dE}{dT}\Big|_{V,N} = \frac{3}{2} V \frac{g}{\lambda^3} k_B T \left[\frac{5}{2T} f_{5/2}^+(z) + \frac{1}{z} f_{3/2}^+(z) \frac{dz}{dT}\Big|_{V,N} \right]. \tag{7.63}$$

The derivative $dz/dT|_{V,N}$ is found from the condition of fixed particle number, using

$$\frac{dN}{dT}\Big|_V = 0 = \frac{g}{\lambda^3} V \left[\frac{3}{2T} f_{3/2}^+(z) + \frac{1}{z} f_{1/2}^+(z) \frac{dz}{dT}\Big|_{V,N} \right]. \tag{7.64}$$

Substituting the solution

$$\frac{T}{z} \frac{dz}{dT}\Big|_{V,N} = -\frac{3}{2} \frac{f_{3/2}^+(z)}{f_{1/2}^+(z)}$$

Fig. 7.11 Heat capacity of the bose gas has a cusp at T_c.

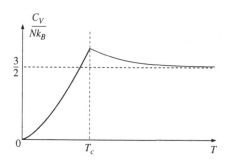

into Eq. (7.63) yields

$$\frac{C_V}{Vk_B} = \frac{3}{2}\frac{g}{\lambda^3}\left[\frac{5}{2}f_{5/2}^+(z) - \frac{3}{2}\frac{f_{3/2}^+(z)^2}{f_{1/2}^+(z)}\right]. \tag{7.65}$$

Expanding the result in powers of z indicates that at high temperatures the heat capacity is larger than the classical value; $C_V/Nk_B = 3/2[1 + n\lambda^3/2^{7/2} + \cdots]$. At low temperatures, $z = 1$ and

$$\frac{C_V}{Nk_B} = \frac{15}{4}\frac{g}{n\lambda^3}\zeta_{5/2} = \frac{15}{4}\frac{\zeta_{5/2}}{\zeta_{3/2}}\left(\frac{T}{T_c}\right)^{3/2}. \tag{7.66}$$

The origin of the $T^{3/2}$ behavior of the heat capacity at low temperatures is easily understood. At $T = 0$ all particles occupy the $\vec{k} = 0$ state. At small but finite temperatures there is occupation of states of finite momentum, up to a value of approximately k_m such that $\hbar^2 k_m^2/2m = k_B T$. Each of these states has an energy proportional to $k_B T$. The excitation energy in d dimensions is thus given by $E_\times \propto V k_m^d k_B T$. The resulting heat capacity is $C_V \propto V k_B T^{d/2}$. The reasoning is similar to that used to calculate the heat capacities of a phonon (or photon) gas. The difference in the power laws simply originates from the difference in the energy spectrum of low-energy excitations ($\mathcal{E}(k) \propto k^2$ in the former and $\mathcal{E}(k) \propto k$ for the latter). In both cases the total number of excitations is not conserved, corresponding to $\mu = 0$. For the bose gas, this lack of conservation only persists up to the transition temperature, at which point all particles are excited out of the reservoir with $\mu = 0$ at $\vec{k} = 0$. C_V is continuous at T_c, reaching a maximum value of approximately $1.92k_B$ per particle, but has a discontinuous derivative at this point.

7.7 Superfluid He⁴

Interesting examples of quantum fluids are provided by the isotopes of helium. The two electrons of He occupy the 1s orbital with opposite spins. The filled orbital makes this noble gas particularly inert. There is still a van der Waals interaction between two He atoms, but the interatomic potential has a shallow minimum of depth 9 K at a separation of roughly 3 Å. The weakness of this

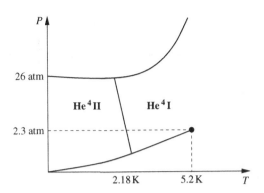

Fig. 7.12 Schematic phase diagram of He^4 with a phase transition between normal and superfluid forms of the liquid.

interaction makes He a universal *wetting agent*. As the He atoms have a stronger attraction for practically all other molecules, they easily spread over the surface of any substance. Due to its light mass, the He atom undergoes large zero-point fluctuations at $T = 0$. These fluctuations are sufficient to *melt* a solid phase, and thus He remains in a liquid state at ordinary pressures. Pressures of over 25 atmospheres are required to sufficiently localize the He atoms to result in a solid phase at $T = 0$. A remarkable property of the quantum liquid at zero temperature is that, unlike its classical counterpart, it has zero entropy.

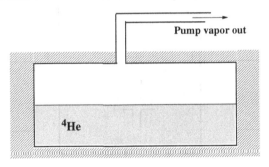

Fig. 7.13 Evaporative cooling of helium by pumping out vapor.

The lighter isotope of He^3 has three nucleons and obeys fermi statistics. The liquid phase is very well described by a fermi gas of the type discussed in Section 7.5. The interactions between atoms do not significantly change the non-interacting picture described earlier. By contrast, the heavier isotope of He^4 is a boson. Helium can be cooled down by a process of evaporation: liquid helium is placed in an isolated container, in equilibrium with the gas phase at a finite vapor density. As the helium gas is pumped out, part of the liquid evaporates to take its place. The evaporation process is accompanied by the release of latent heat, which cools the liquid. The boiling liquid is quite active and turbulent, just as a boiling pot of water. However, when the liquid is cooled down to below 2.2 K, it suddenly becomes quiescent and the turbulence disappears. The liquids on the two sides of this phase transition are usually referred to as HeI and HeII.

Fig. 7.14 Pushing superfluid helium through a (super-)leak can be achieved for $P_1 \rightarrow P_2$, and leads to cooling of the intake chamber.

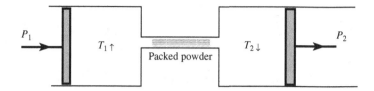

HeII has unusual hydrodynamic properties. It flows through the finest capillaries without any resistance. Consider an experiment that pushes HeII from one container to another through a small tube packed with powder. For ordinary fluids a finite pressure difference between the containers, proportional to viscosity, is necessary to maintain the flow. HeII flows even in the limit of zero pressure difference and acts as if it has zero viscosity. For this reason it is referred to as a *superfluid*. The superflow is accompanied by heating of the container that loses HeII and cooling of the container that accepts it; the *mechano-caloric effect*. Conversely, HeII acts to remove temperature differences by flowing from hot regions. This is demonstrated dramatically by the *fountain effect*, in which the superfluid spontaneously moves up a tube from a heated container.

Fig. 7.15 Helium spurts out of the (resistively) heated chamber in the "fountain effect."

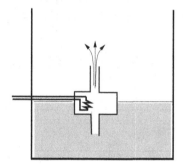

In some other circumstances HeII behaves as a viscous fluid. A classical method for measuring the viscosity of a liquid is via torsional oscillators: a collection of closely spaced disks connected to a shaft is immersed in the fluid and made to oscillate. The period of oscillations is proportional to the moment of inertia, which is modified by the quantity of fluid that is dragged by the oscillator. This experiment was performed on HeII by Andronikashvili, who indeed found a finite viscous drag. Furthermore, the changes in the frequency of the oscillator with temperature indicate that the quantity of fluid that is dragged by the oscillator starts to decrease below the transition temperature. The measured *normal density* vanishes as $T \rightarrow 0$, approximately as T^4.

In 1938, Fritz London suggested that a good starting hypothesis is that the transition to the superfluid state is related to the Bose–Einstein condensation. This hypothesis can account for a number of observations:

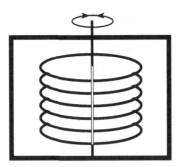

Fig. 7.16 A torsional oscillator can be used to measure the "normal fraction" in a superfluid.

(1) The critical temperature of an ideal bose gas of volume $v = 46.2\,\text{Å}^3$ per particle is obtained from Eq. (7.54) as

$$T_c = \frac{h^2}{2\pi m_{He} k_B} \left(v\, \zeta_{3/2}\right)^{-2/3} \approx 3.14\,\text{K}. \tag{7.67}$$

The actual transition temperature of $T_c \approx 2.18\,\text{K}$ is not far from this value.

(2) The origin of the transition has to be related to quantum statistics as He^3, which is atomically similar, but a fermion, does not have a similar transition. (Actually He^3 does become superfluid but at temperatures of only a few milli-Kelvin. This follows a pairing of He^3 atoms, which changes their statistics.)

(3) A bose condensate can account for the observed thermo-mechanical properties of HeII. The expression for pressure in Eq. (7.55) is only a function of temperature, and not density. As such, variations in pressure are accompanied by changes in temperature. This is also the reason for the absence of boiling activity in HeII. Bubbles nucleate and grow in a boiling liquid at local hot spots. In an ordinary fluid, variations in temperature relax to equilibrium only through the slow process of *heat diffusion*. By contrast, if the local pressure is a function of temperature only, there will be increased pressure at a hot spot. The fluid flows in response to pressure gradients and removes them very rapidly (at the speed of sound in the medium).

(4) Hydrodynamic behavior of HeII can be explained by Tisza's *two-fluid model*, which postulates the coexistence of two components for $T < T_c$:

    (a) A *normal* component of density ρ_n, moving with velocity $\vec{v}_n$, and having a finite entropy density s_n.

    (b) A *superfluid* component of density ρ_s, which flows without viscosity and with no vorticity ($\nabla \times \vec{u}_s = 0$), and has zero entropy density, $s_s = 0$.

In the *super-leak* experiments, it is the superfluid component that passes through, reducing the entropy and hence temperature. In the Andronikashvili experiment, the normal component sticks to the torsional oscillator and gets dragged by it. This experiment thus gives the ratio of ρ_n to ρ_s.

There are, however, many important differences between superfluid helium and the ideal bose condensate:

(1) Interactions certainly play an important role in the liquid state. The Bose–Einstein condensate has infinite compressibility, while HeII has a finite density, related to atomic volume, and is essentially incompressible.

(2) It can be shown that, even at $T = 0$, the ideal bose condensate is not superfluid. This is because the low-energy spectrum, $\mathcal{E}(\vec{k}) = \hbar^2 k^2 / 2m$, admits too many excitations. Any external body moving through such a fluid can easily lose energy to it by exciting these modes, leading to a finite viscosity.

(3) The detailed functional forms of the heat capacity and superfluid density are very different from their counterparts in the ideal bose condensate. The measured heat capacity diverges at the transition with a characteristic shape similar to λ, and vanishes at low temperatures as T^3 (compared with $T^{3/2}$ for the ideal bose gas). The superfluid density obtained in the Andronikashvilli experiment vanishes as $(T_c - T)^{2/3}$ at the transition, while the normal component vanishes as approximately T^4 as $T \to 0$ (compared with $T^{3/2}$ for the condensate density in Eq. (7.52)). Understanding the nature of the singular behavior close to T_c is beyond the scope of this discussion, and will be taken up in the accompanying volume. The behavior close to zero temperature, however, suggests a different spectrum of low-energy excitations.

Based on the shape of the experimentally measured heat capacity, Landau suggested that the spectrum of low-energy excitations is actually similar to that of phonons. This is a consequence of the interactions between the particles. The low-energy excitations of a classical liquid are longitudinal sound waves. (In comparison, a solid admits three such excitations, two transverse and one longitudinal.) The correspondence principle suggests that quantized versions of such modes should also be present in a quantum liquid. As in the case of phonons, a linear spectrum of excitations leads to a heat capacity that vanishes as T^3. The speed of sound waves can be computed from the coefficient of the T^3 dependence (see problems) as $v \approx 240\,\mathrm{m\,s^{-1}}$. A further anomaly in heat capacity can be explained by assuming that the spectrum of excitations bends down and has a minimum in the vicinity of a wavenumber $k_0 \approx 2\,\text{Å}^{-1}$. The excitations in the vicinity of this minimum are referred to as *rotons*, and have an energy

$$\mathcal{E}_{\mathrm{roton}}(\vec{k}) = \Delta + \frac{\hbar^2}{2\mu}(k - k_0)^2, \tag{7.68}$$

with $\Delta \approx 8.6\,\mathrm{K}$, and $\mu \approx 0.16 m_{\mathrm{He}}$. The spectrum postulated by Landau was confirmed directly by neutron scattering measurements in the 1950s.

Problems for chapter 7

1. *Identical particle pair*: let $Z_1(m)$ denote the partition function for a single quantum particle of mass m in a volume V.

   (a) Calculate the partition function of two such particles, if they are bosons, and also if they are (spinless) fermions.

(b) Use the classical approximation $Z_1(m) = V/\lambda^3$ with $\lambda = h/\sqrt{2\pi m k_B T}$. Calculate the corrections to the energy E, and the heat capacity C, due to bose or fermi statistics.

(c) At what temperature does the approximation used above break down?

$$********$$

2. *Generalized ideal gas*: consider a gas of non-interacting identical (spinless) quantum particles with an energy spectrum $\varepsilon = |\vec{p}/\hbar|^s$, contained in a box of "volume" V in d dimensions.

(a) Calculate the grand potential $\mathcal{G}_\eta = -k_B T \ln \mathcal{Q}_\eta$, and the density $n = N/V$, at a chemical potential μ. Express your answers in terms of s, d, and $f_m^\eta(z)$, where $z = e^{\beta\mu}$, and

$$f_m^\eta(z) = \frac{1}{\Gamma(m)} \int_0^\infty \frac{dx\, x^{m-1}}{z^{-1}e^x - \eta}.$$

(*Hint.* Use integration by parts on the expression for $\ln \mathcal{Q}_\eta$.)

(b) Find the ratio PV/E, and compare it with the classical result obtained previously.

(c) *For fermions*, calculate the dependence of E/N, and P, on the density $n = N/V$, at zero temperature. (*Hint.* $f_m(z) \to (\ln z)^m/m!$ as $z \to \infty$.)

(d) *For bosons*, find the dimension $d_\ell(s)$, below which there is no bose condensation. Is there condensation for $s = 2$ at $d = 2$?

$$********$$

3. *Pauli paramagnetism*: calculate the contribution of electron spin to its magnetic susceptibility as follows. Consider non-interacting electrons, each subject to a Hamiltonian

$$\mathcal{H}_1 - \frac{\vec{p}^2}{2m} - \mu_0 \vec{\sigma} \cdot \vec{B},$$

where $\mu_0 = e\hbar/2mc$, and the eigenvalues of $\vec{\sigma} \cdot \vec{B}$ are $\pm B$.
(The orbital effect, $\vec{p} \to \vec{p} - e\vec{A}$, has been ignored.)

(a) Calculate the grand potential $\mathcal{G}_- = -k_B T \ln \mathcal{Q}_-$, at a chemical potential μ.

(b) Calculate the densities $n_+ = N_+/V$, and $n_- = N_-/V$, of electrons pointing parallel and anti-parallel to the field.

(c) Obtain the expression for the magnetization $M = \mu_0(N_+ - N_-)$, and expand the result for small B.

(d) Sketch the zero-field susceptibility $\chi(T) = \partial M/\partial B|_{B=0}$, and indicate its behavior at low and high temperatures.

(e) Estimate the magnitude of χ/N for a typical metal at room temperature.

$$********$$

4. *Freezing of* He^3: at low temperatures He^3 can be converted from liquid to solid by application of pressure. A peculiar feature of its phase boundary is that $(dP/dT)_{melting}$ is negative at temperatures below 0.3 K [$(dP/dT)_m \approx -30$ atm K^{-1} at $T \approx 0.1$ K]. We will use a simple model of liquid and solid phases of He^3 to account for this feature.

(a) In the solid phase, the He^3 atoms form a crystal lattice. Each atom has nuclear spin of 1/2. Ignoring the interaction between spins, what is the entropy per particle s_s, due to the spin degrees of freedom?

(b) Liquid He^3 is modeled as an ideal fermi gas, with a volume of 46 Å^3 per atom. What is its fermi temperature T_F, in degrees Kelvin?

(c) How does the heat capacity of liquid He^3 behave at low temperatures? Write down an expression for C_V in terms of N, T, k_B, T_F, *up to a numerical constant*, that is valid for $T \ll T_F$.

(d) Using the result in (c), calculate the entropy per particle s_ℓ, in the liquid at low temperatures. For $T \ll T_F$, which phase (solid or liquid) has the higher entropy?

(e) By equating chemical potentials, or by any other technique, prove the Clausius–Clapeyron equation $(dP/dT)_{melting} = (s_\ell - s_s)/(v_\ell - v_s)$, where v_ℓ and v_s are the volumes per particle in the liquid and solid phases, respectively.

(f) It is found experimentally that $v_\ell - v_s = 3$ Å^3 per atom. Using this information, plus the results obtained in previous parts, estimate $(dP/dT)_{melting}$ at $T \ll T_F$.

****

5. *Non-interacting fermions*: consider a grand canonical ensemble of non-interacting *fermions* with chemical potential μ. The one-particle states are labeled by a wave vector $\vec{k}$, and have energies $\mathcal{E}(\vec{k})$.

(a) What is the joint probability $P(\{n_{\vec{k}}\})$ of finding a set of occupation numbers $\{n_{\vec{k}}\}$ of the one-particle states?

(b) Express your answer to part (a) in terms of the average occupation numbers $\{\langle n_{\vec{k}}\rangle_-\}$.

(c) A random variable has a set of ℓ discrete outcomes with probabilities p_n, where $n = 1, 2, \cdots, \ell$. What is the entropy of this probability distribution? What is the maximum possible entropy?

(d) Calculate the entropy of the probability distribution for fermion occupation numbers in part (b), and comment on its zero-temperature limit.

(e) Calculate the variance of the total number of particles $\langle N^2 \rangle_c$, and comment on its zero-temperature behavior.

(f) The number fluctuations of a gas are related to its compressibility κ_T, and number density $n = N/V$, by

$$\langle N^2 \rangle_c = Nnk_BT\kappa_T.$$

Give a *numerical estimate* of the compressibility of the fermi gas in a metal at $T = 0$ in units of Å^3 eV^{-1}.

****

6. *Stoner ferromagnetism*: the conduction electrons in a metal can be treated as a gas of fermions of spin 1/2 (with up/down degeneracy), and density $n = N/V$. The Coulomb repulsion favors wave functions that are anti-symmetric in position coordinates, thus keeping the electrons apart. Because of the full (position *and* spin) anti-symmetry of fermionic wave functions, this interaction may be approximated by an effective spin–spin coupling that favors states with parallel spins. In this simple approximation, the net effect is described by an interaction energy

$$U = \alpha \frac{N_+ N_-}{V},$$

where N_+ and $N_- = N - N_+$ are the numbers of electrons with up and down spins, and V is the volume. (The parameter α is related to the scattering length a by $\alpha = 4\pi\hbar^2 a/m$.)

(a) The ground state has two fermi seas filled by the spin-up and spin-down electrons. Express the corresponding fermi wavevectors $k_{F\pm}$ in terms of the densities $n_\pm = N_\pm/V$.

(b) Calculate the kinetic energy density of the ground state as a function of the densities $n_\pm$, and fundamental constants.

(c) Assuming small deviations $n_\pm = n/2 \pm \delta$ from the symmetric state, expand the kinetic energy to *fourth order in* δ.

(d) Express the spin–spin interaction density U/V in terms of n and δ. Find the critical value of α_c, such that for $\alpha > \alpha_c$ the electron gas can lower its total energy by spontaneously developing a magnetization. (This is known as the *Stoner instability*.)

(e) Explain qualitatively, and sketch the behavior of, the spontaneous magnetization as a function of α.

$$********$$

7. *Boson magnetism*: consider a gas of non-interacting spin 1 bosons, each subject to a Hamiltonian

$$\mathcal{H}_1(\vec{p}, s_z) = \frac{\vec{p}^{\,2}}{2m} - \mu_0 s_z B,$$

where $\mu_0 = e\hbar/mc$, and s_z takes *three* possible values of $(-1, 0, +1)$. (The orbital effect, $\vec{p} \to \vec{p} - e\vec{A}$, has been ignored.)

(a) In a grand canonical ensemble of chemical potential μ, what are the average occupation numbers $\left\{ \langle n_+(\vec{k}) \rangle, \langle n_0(\vec{k}) \rangle, \langle n_-(\vec{k}) \rangle \right\}$ of one-particle states of wavenumber $\vec{k} = \vec{p}/\hbar$?

(b) Calculate the average total numbers $\{N_+, N_0, N_-\}$ of bosons with the three possible values of s_z in terms of the functions $f_m^+(z)$.

(c) Write down the expression for the magnetization $M(T, \mu) = \mu_0(N_+ - N_-)$, and by expanding the result for small B find the *zero-field susceptibility* $\chi(T, \mu) = \partial M/\partial B|_{B=0}$.

To find the behavior of $\chi(T, n)$, where $n = N/V$ is the total density, proceed as follows:

(d) For $B = 0$, find the high-temperature expansion for $z(\beta, n) = e^{\beta\mu}$, correct to second order in n. Hence obtain the first correction from quantum statistics to $\chi(T, n)$ at high temperatures.

(e) Find the temperature $T_c(n, B = 0)$ of Bose–Einstein condensation. What happens to $\chi(T, n)$ on approaching $T_c(n)$ from the high-temperature side?

(f) What is the chemical potential μ for $T < T_c(n)$, at a small but finite value of B? Which one-particle state has a macroscopic occupation number?

(g) Using the result in (f), find the spontaneous magnetization

$$\overline{M}(T, n) = \lim_{B \to 0} M(T, n, B).$$

****

8. *Dirac fermions* are non-interacting particles of spin 1/2. The one-particle states come in pairs of positive and negative energies,

$$\mathcal{E}_\pm(\vec{k}) = \pm\sqrt{m^2 c^4 + \hbar^2 k^2 c^2},$$

independent of spin.

(a) For *any* fermionic system of chemical potential μ, show that the probability of finding an occupied state of energy $\mu + \delta$ is the same as that of finding an unoccupied state of energy $\mu - \delta$. (δ is any constant energy.)

(b) At zero temperature all negative energy Dirac states are occupied and all positive energy ones are empty, that is, $\mu(T = 0) = 0$. Using the result in (a) find the chemical potential at finite temperature T.

(c) Show that the mean excitation energy of this system at finite temperature satisfies

$$E(T) - E(0) = 4V \int \frac{d^3\vec{k}}{(2\pi)^3} \frac{\mathcal{E}_+(\vec{k})}{\exp\left(\beta \mathcal{E}_+(\vec{k})\right) + 1} \quad.$$

(d) Evaluate the integral in part (c) for *massless Dirac particles* (i.e., for $m = 0$).

(e) Calculate the heat capacity, C_V, of such massless Dirac particles.

(f) Describe the qualitative dependence of the heat capacity at low temperature if the particles are massive.

****

9. *Numerical estimates*: the following table provides typical values for the fermi energy and fermi temperature for (i) electrons in a typical metal; (ii) nucleons in a heavy nucleus; and (iii) He^3 atoms in liquid He^3 (atomic volume $= 46.2$ Å^3 per atom).

(a) Estimate the ratio of the electron and phonon heat capacities at room temperature for a typical metal.

(b) Compare the thermal wavelength of a neutron at room temperature to the minimum wavelength of a phonon in a typical crystal.

(c) Estimate the degeneracy discriminant, $n\lambda^3$, for hydrogen, helium, and oxygen gases at room temperature and pressure. At what temperatures do quantum mechanical effects become important for these gases?

	$n(1/m^3)$	m (kg)	ε_F (eV)	T_F (K)
Electron	10^{29}	9×10^{-31}	4.4	5×10^4
Nucleons	10^{44}	1.6×10^{-27}	1.0×10^8	1.1×10^{12}
Liquid He^3	2.6×10^{28}	4.6×10^{-27}	10^{-3}	10^1

(d) Experiments on He^4 indicate that at temperatures below 1 K, the heat capacity is given by $C_V = 20.4T^3$ J kg^{-1} K^{-1}. Find the low-energy excitation spectrum, $\mathcal{E}(k)$, of He^4.

(*Hint.* There is only one non-degenerate branch of such excitations.)

****

10. *Solar interior*: according to astrophysical data, the plasma at the center of the Sun has the following properties:

Temperature: $T = 1.6 \times 10^7$ K
Hydrogen density: $\rho_H = 6 \times 10^4$ kg m^{-3}
Helium density: $\rho_{He} = 1 \times 10^5$ kg m^{-3}.

(a) Obtain the thermal wavelengths for electrons, protons, and α-particles (nuclei of He).
(b) Assuming that the gas is ideal, determine whether the electron, proton, or α-particle gases are degenerate in the quantum mechanical sense.
(c) Estimate the total gas pressure due to these gas particles near the center of the Sun.
(d) Estimate the total radiation pressure close to the center of the Sun. Is it matter, or radiation pressure, that prevents the gravitational collapse of the Sun?

****

11. *Bose condensation in d dimensions*: consider a gas of non-interacting (spinless) bosons with an energy spectrum $\epsilon = p^2/2m$, contained in a box of "volume" $V = L^d$ in d dimensions.

(a) Calculate the grand potential $\mathcal{G} = -k_B T \ln \mathcal{Q}$, and the density $n = N/V$, at a chemical potential μ. Express your answers in terms of d and $f_m^+(z)$, where $z = e^{\beta \mu}$, and

$$f_m^+(z) = \frac{1}{\Gamma(m)} \int_0^\infty \frac{x^{m-1}}{z^{-1}e^x - 1} dx.$$

(*Hint.* Use integration by parts on the expression for $\ln \mathcal{Q}$.)

(b) Calculate the ratio PV/E, and compare it with the classical value.
(c) Find the critical temperature, $T_c(n)$, for Bose–Einstein condensation.
(d) Calculate the heat capacity $C(T)$ for $T < T_c(n)$.
(e) Sketch the heat capacity at all temperatures.
(f) Find the ratio, $C_{max}/C(T \to \infty)$, of the maximum heat capacity to its classical limit, and evaluate it in $d = 3$.

(g) How does the above calculated ratio behave as $d \to 2$? In what dimensions are your results valid? Explain.

****

12. *Exciton dissociation in a semiconductor*: shining an intense laser beam on a semiconductor can create a metastable collection of electrons (charge $-e$, and effective mass m_e) and holes (charge $+e$, and effective mass m_h) in the bulk. The oppositely charged particles may pair up (as in a hydrogen atom) to form a gas of *excitons*, or they may dissociate into a plasma. We shall examine a much simplified model of this process.

(a) Calculate the free energy of a gas composed of N_e electrons and N_h holes, at temperature T, treating them as classical non-interacting particles of masses m_e and m_h.

(b) By pairing into an excition, the electron hole pair lowers its energy by ε. (The binding energy of a hydrogen-like exciton is $\varepsilon \approx me^4/(2\hbar^2\epsilon^2)$, where ϵ is the dielectric constant, and $m^{-1} = m_e^{-1} + m_h^{-1}$.) Calculate the free energy of a gas of N_p excitons, treating them as classical non-interacting particles of mass $m = m_e + m_h$.

(c) Calculate the chemical potentials μ_e, μ_h, and μ_p of the electron, hole, and exciton states, respectively.

(d) Express the equilibrium condition between excitons and electrons/holes in terms of their chemical potentials.

(e) At a high temperature T, find the density n_p of excitons, as a function of the total density of excitations $n \approx n_e + n_h$.

****

13. *Freezing of He^4*: at low temperatures He^4 can be converted from liquid to solid by application of pressure. An interesting feature of the phase boundary is that the melting pressure is reduced slightly from its $T = 0$ K value, by approximately 20 N m^{-2} at its minimum at $T = 0.8$ K. We will use a simple model of liquid and solid phases of He^4 to account for this feature.

(a) The important excitations in liquid He^4 at $T < 1$ K are phonons of velocity c. Calculate the contribution of these modes to the heat capacity per particle, C_V^ℓ/N, of the liquid.

(b) Calculate the low-temperature heat capacity per particle, C_V^s/N, of solid He^4 in terms of longitudinal and transverse sound velocities c_L and c_T.

(c) Using the above results calculate the entropy difference $(s_\ell - s_s)$, assuming a single sound velocity $c \approx c_L \approx c_T$, and approximately equal volumes per particle $v_\ell \approx v_s \approx v$. Which phase (solid or liquid) has the higher entropy?

(d) Assuming a small (temperature-independent) volume difference $\delta v = v_\ell - v_s$, calculate the form of the melting curve. To explain the anomaly described at the beginning, which phase (solid or liquid) must have the higher density?

****

14. *Neutron star core*: Professor Rajagopal's group at MIT has proposed that a new phase of QCD matter may exist in the core of neutron stars. This phase can be viewed as a condensate of quarks in which the low-energy excitations are approximately

$$\mathcal{E}(\vec{k})_{\pm} =_{\pm} \hbar^2 \frac{\left(|\vec{k}| - k_F\right)^2}{2M}.$$

The excitations are fermionic, with a degeneracy of $g = 2$ from spin.

(a) At zero temperature all negative energy states are occupied and all positive energy ones are empty, that is, $\mu(T = 0) = 0$. By relating occupation numbers of states of energies $\mu + \delta$ and $\mu - \delta$, or otherwise, find the chemical potential at finite temperatures T.

(b) Assuming a constant density of states near $k = k_F$, that is, setting $d^3k \approx 4\pi k_F^2 dq$ with $q = |\vec{k}| - k_F$, show that the mean excitation energy of this system at finite temperature is

$$E(T) - E(0) \approx 2gV \frac{k_F^2}{\pi^2} \int_0^\infty dq \frac{\mathcal{E}_+(q)}{\exp\left(\beta\mathcal{E}_+(q)\right) + 1}.$$

(c) Give a closed form answer for the excitation energy by evaluating the above integral.

(d) Calculate the heat capacity, C_V, of this system, and comment on its behavior at low temperature.

****

15. *Non-interacting bosons*: consider a grand canonical ensemble of non-interacting *bosons* with chemical potential μ. The one-particle states are labeled by a wave vector $\vec{q}$, and have energies $\mathcal{E}(\vec{q})$.

(a) What is the joint probability $P(\{n_{\vec{q}}\})$ of finding a set of occupation numbers $\{n_{\vec{q}}\}$ of the one-particle states, in terms of the fugacities $z_{\vec{q}} \equiv \exp\left[\beta(\mu - \mathcal{E}(\vec{q}))\right]$?

(b) For a particular $\vec{q}$, calculate the characteristic function $\langle \exp\left[ikn_{\vec{q}}\right]\rangle$.

(c) Using the result of part (b), **or otherwise**, give expressions for the mean and variance of $n_{\vec{q}}$ occupation number $\langle n_{\vec{q}}\rangle$.

(d) Express the variance in part (c) in terms of the mean occupation number $\langle n_{\vec{q}}\rangle$.

(e) Express your answer to part (a) in terms of the occupation numbers $\{\langle n_{\vec{q}}\rangle\}$.

(f) Calculate the entropy of the probability distribution for bosons, in terms of $\{\langle n_{\vec{q}}\rangle\}$, and comment on its zero-temperature limit.

****

16. *Relativistic bose gas in d dimensions*: consider a gas of non-interacting (spinless) bosons with energy $\epsilon = c\left|\vec{p}\right|$, contained in a box of "volume" $V = L^d$ in d dimensions.

(a) Calculate the grand potential $\mathcal{G} = -k_B T \ln \mathcal{Q}$, and the density $n = N/V$, at a chemical potential μ. Express your answers in terms of d and $f_m^+(z)$, where $z = e^{\beta\mu}$, and

$$f_m^+(z) = \frac{1}{(m-1)!} \int_0^\infty \frac{x^{m-1}}{z^{-1}e^x - 1} dx.$$

(*Hint.* Use integration by parts on the expression for $\ln \mathcal{Q}$.)

(b) Calculate the gas pressure P, its energy E, and compare the ratio $E/(PV)$ to the classical value.

(c) Find the critical temperature, $T_c(n)$, for Bose–Einstein condensation, indicating the dimensions where there is a transition.

(d) What is the temperature dependence of the heat capacity $C(T)$ for $T < T_c(n)$?

(e) Evaluate the dimensionless heat capacity $C(T)/(Nk_B)$ at the critical temperature $T = T_c$, and compare its value to the classical (high-temperature) limit.

$$********$$

17. *Graphene* is a single sheet of carbon atoms bonded into a *two-dimensional* hexagonal lattice. It can be obtained by exfoliation (repeated peeling) of graphite. The band structure of graphene is such that the single-particle excitations behave as relativistic *Dirac fermions*, with a spectrum that at low energies can be approximated by

$$\mathcal{E}_{\pm}(\vec{k}) = \pm \hbar v \left| \vec{k} \right|.$$

There is spin degeneracy of $g = 2$, and $v \approx 10^6 \, \mathrm{m \, s^{-1}}$. Experiments on unusual transport properties of graphene were reported in *Nature* **438**, 197 (2005). In this problem, you shall calculate the heat capacity of this material.

(a) If at zero temperature all negative energy states are occupied and all positive energy ones are empty, find the chemical potential $\mu(T)$.

(b) Show that the mean excitation energy of this system at finite temperature satisfies

$$E(T) - E(0) = 4A \int \frac{\mathrm{d}^2 \vec{k}}{(2\pi)^2} \frac{\mathcal{E}_+(\vec{k})}{\exp\left(\beta \mathcal{E}_+(\vec{k})\right) + 1}.$$

(c) Give a closed form answer for the excitation energy by evaluating the above integral.

(d) Calculate the heat capacity, C_V, of such massless Dirac particles.

(e) Explain qualitatively the contribution of phonons (lattice vibrations) to the heat capacity of graphene. The typical sound velocity in graphite is of the order of $2 \times 10^4 \, \mathrm{m \, s^{-1}}$. Is the low temperature heat capacity of graphene controlled by phonon or electron contributions?

$$********$$

Solutions to selected problems from chapter 1

1. *Surface tension*: thermodynamic properties of the interface between two phases are described by a state function called the surface tension S. It is defined in terms of the work required to increase the surface area by an amount dA through $dW = SdA$.

(a) By considering the work done against surface tension in an infinitesimal change in radius, show that the pressure inside a spherical drop of water of radius R is larger than outside pressure by $2S/R$. What is the air pressure inside a soap bubble of radius R?

The work done by a water droplet on the outside world, needed to increase the radius from R to $R + \Delta R$, is

$$\Delta W = (P - P_o) \cdot 4\pi R^2 \cdot \Delta R,$$

where P is the pressure inside the drop and P_o is the atmospheric pressure. In equilibrium, this should be equal to the increase in the surface energy $S\Delta A = S \cdot 8\pi R \cdot \Delta R$, where S is the surface tension, and

$$\Delta W_{\text{total}} = 0, \quad \Longrightarrow \quad \Delta W_{\text{pressure}} = -\Delta W_{\text{surface}},$$

resulting in

$$(P - P_o) \cdot 4\pi R^2 \cdot \Delta R = S \cdot 8\pi R \cdot \Delta R, \quad \Longrightarrow \quad (P - P_o) = \frac{2S}{R}.$$

In a soap bubble, there are two air–soap surfaces with almost equal radii of curvatures, and

$$P_{\text{film}} - P_o = P_{\text{interior}} - P_{\text{film}} = \frac{2S}{R},$$

leading to

$$P_{\text{interior}} - P_o = \frac{4S}{R}.$$

Hence, the air pressure inside the bubble is larger than atmospheric pressure by $4S/R$.

(b) A water droplet condenses on a solid surface. There are three surface tensions involved, S_{aw}, S_{sw}, and S_{sa}, where a, s, and w refer to air, solid, and water, respectively. Calculate the angle of contact, and find the condition for the appearance of a water film (complete wetting).

When steam condenses on a solid surface, water either forms a droplet, or spreads on the surface. There are two ways to consider this problem:

Method 1. Energy associated with the interfaces.

In equilibrium, the total energy associated with the three interfaces should be at a minimum, and therefore

$$dE = S_{aw}\,dA_{aw} + S_{as}\,dA_{as} + S_{ws}\,dA_{ws} = 0.$$

Since the total surface area of the solid is constant,

$$dA_{as} + dA_{ws} = 0.$$

From geometrical considerations (see proof below), we obtain

$$dA_{ws}\cos\theta = dA_{aw}.$$

From these equations, we obtain

$$dE = (S_{aw}\cos\theta - S_{as} + S_{ws})\,dA_{ws} = 0, \quad \Longrightarrow \quad \cos\theta = \frac{S_{as} - S_{ws}}{S_{aw}}.$$

Proof of $dA_{ws}\cos\theta = dA_{aw}$: *consider a droplet that is part of a sphere of radius R, which is cut by the substrate at an angle θ. The areas of the involved surfaces are*

$$A_{ws} = \pi(R\sin\theta)^2, \qquad \text{and} \qquad A_{aw} = 2\pi R^2(1 - \cos\theta).$$

Let us consider a small change in shape, accompanied by changes in R and θ. These variations should preserve the volume of water, that is, constrained by

$$V = \frac{\pi R^3}{3}\left(\cos^3\theta - 3\cos\theta + 2\right).$$

Introducing $x = \cos\theta$, we can rewrite the above results as

$$
\begin{cases}
A_{ws} = \pi R^2\left(1 - x^2\right), \\[2mm]
A_{aw} = 2\pi R^2\left(1 - x\right), \\[2mm]
V = \dfrac{\pi R^3}{3}\left(x^3 - 3x + 2\right).
\end{cases}
$$

The variations of these quantities are then obtained from

$$
\begin{cases}
dA_{ws} = 2\pi R\left[\dfrac{dR}{dx}(1 - x^2) - Rx\right]dx, \\[3mm]
dA_{aw} = 2\pi R\left[2\dfrac{dR}{dx}(1 - x) - R\right]dx, \\[3mm]
dV = \pi R^2\left[\dfrac{dR}{dx}(x^3 - 3x + 2) + R(x^2 - x)\right]dx = 0.
\end{cases}
$$

From the last equation, we conclude

$$\frac{1}{R}\frac{dR}{dx} = -\frac{x^2 - x}{x^3 - 3x + 2} = -\frac{x+1}{(x-1)(x+2)}.$$

Substituting for dR/dx *gives*

$$dA_{ws} = 2\pi R^2 \frac{dx}{x+2}, \qquad \text{and} \qquad dA_{aw} = 2\pi R^2 \frac{x \cdot dx}{x+2},$$

resulting in the required result of

$$dA_{aw} = x \cdot dA_{ws} = dA_{ws}\cos\theta.$$

Method 2. Balancing forces on the contact line.

   Another way to interpret the result is to consider the force balance of the equilibrium surface tension on the contact line. There are four forces acting on the line: (1) the surface tension at the water–gas interface, (2) the surface tension at the solid–water interface, (3) the surface tension at the gas–solid interface, and (4) the force downward by solid–contact line interaction. The last force ensures that the contact line stays on the solid surface, and is downward since the contact line is allowed to move only horizontally without friction. These forces should cancel along both the y and x directions. The latter gives the condition for the contact angle known as Young's equation,

$$\mathcal{S}_{as} = \mathcal{S}_{aw}\cdot\cos\theta + \mathcal{S}_{ws}, \qquad \Longrightarrow \qquad \cos\theta = \frac{\mathcal{S}_{as} - \mathcal{S}_{ws}}{\mathcal{S}_{aw}}.$$

The critical condition for the complete wetting occurs when $\theta = 0$, *or* $\cos\theta = 1$, *that is, for*

$$\cos\theta_C = \frac{\mathcal{S}_{as} - \mathcal{S}_{ws}}{\mathcal{S}_{aw}} = 1.$$

Complete wetting of the substrate thus occurs whenever

$$\mathcal{S}_{aw} \le \mathcal{S}_{as} - \mathcal{S}_{ws}.$$

(c) In the realm of "large" bodies gravity is the dominant force, while at "small" distances surface tension effects are all important. At room temperature, the surface tension of water is $\mathcal{S}_o \approx 7 \times 10^{-2}\,\mathrm{N\,m^{-1}}$. Estimate the typical length scale that separates "large" and "small" behaviors. Give a couple of examples for where this length scale is important.

   Typical length scales at which the surface tension effects become significant are given by the condition that the forces exerted by surface tension and relevant pressures become comparable, or by the condition that the surface energy is comparable to the other energy changes of interest.

Example 1. Size of water drops not much deformed on a non-wetting surface. This is given by equalizing the surface energy and the gravitational energy,

$$S \cdot 4\pi R^2 \approx mgR = \rho VgR = \frac{4\pi}{3}R^4 g,$$

leading to

$$R \approx \sqrt{\frac{3S}{\rho g}} \approx \sqrt{\frac{3 \times 7 \times 10^{-2}\,\mathrm{N/m}}{10^3\,\mathrm{kg/m^3} \times 10\,\mathrm{m/s^2}}} \approx 1.5 \times 10^{-3}\,\mathrm{m} = 1.5\,\mathrm{mm}.$$

Example 2. Swelling of spherical gels in a saturated vapor: osmotic pressure of the gel (about 1 atm) = surface tension of water, gives

$$\pi_{gel} \approx \frac{N}{V}k_B T \approx \frac{2S}{R},$$

where N is the number of counter-ions within the gel. Thus,

$$R \approx \left(\frac{2 \times 7 \times 10^{-2}\,\mathrm{N/m}}{10^5\,\mathrm{N/m^2}}\right) \approx 10^{-6}\,\mathrm{m}.$$

$$* \; * \; * \; * \; * \; * \; * \; *$$

2. *Surfactants*: surfactant molecules such as those in soap or shampoo prefer to spread on the air–water surface rather than dissolve in water. To see this, float a hair on the surface of water and gently touch the water in its vicinity with a piece of soap. (This is also why a piece of soap can power a toy paper boat.)

(a) The air–water surface tension S_o (assumed to be temperature-independent) is reduced roughly by Nk_BT/A, where N is the number of surfactant particles, and A is the area. Explain this result qualitatively.

   Typical surfactant molecules have a hydrophilic (water-soluble) head and a hydrophobic (oil-soluble) tail, and prefer to go to the interface between water and air, or water and oil. Some examples are,

$$CH_3 - (CH_2)_{11} - SO_3^- \cdot Na^+,$$

$$CH_3 - (CH_2)_{11} - N^+(CH_3)_3 \cdot Cl^-,$$

$$CH_3 - (CH_2)_{11} - O - (CH_2 - CH_2 - O)_{12} - H.$$

The surfactant molecules spread over the surface of water and behave as a two-dimensional gas. The gas has a pressure proportional to the density and the absolute temperature, which comes from the two-dimensional degrees of freedom of the molecules. Thus the surfactants lower the free energy of the surface when the surface area is increased.

$$\Delta F_{\mathrm{surfactant}} = \frac{N}{A}k_B T \cdot \Delta A = (S - S_o) \cdot \Delta A, \quad \Longrightarrow \quad S = S_o - \frac{N}{A}k_B T.$$

(Note that surface tension is defined with a sign opposite to that of hydrostatic pressure.)

(b) Place a drop of water on a clean surface. Observe what happens to the air–water-surface contact angle as you gently touch the droplet surface with a small piece of soap, and explain the observation.

As shown in the previous problem, the contact angle satisfies

$$\cos\theta = \frac{\mathcal{S}_{as} - \mathcal{S}_{ws}}{\mathcal{S}_{aw}}.$$

Touching the surface of the droplet with a small piece of soap reduces $\mathcal{S}_{aw}$, hence $\cos\theta$ increases, or equivalently, the angle θ decreases.

(c) More careful observations show that at higher surfactant densities

$$\left.\frac{\partial \mathcal{S}}{\partial A}\right|_T = \frac{Nk_BT}{(A-Nb)^2} - \frac{2a}{A}\left(\frac{N}{A}\right)^2, \quad \text{and} \quad \left.\frac{\partial T}{\partial \mathcal{S}}\right|_A = -\frac{A-Nb}{Nk_B},$$

where a and b are constants. Obtain the expression for $\mathcal{S}(A,T)$ and explain qualitatively the origin of the corrections described by a and b.

When the surfactant molecules are dense their interaction becomes important, resulting in

$$\left.\frac{\partial \mathcal{S}}{\partial A}\right|_T = \frac{Nk_BT}{(A-Nb)^2} - \frac{2a}{A}\left(\frac{N}{A}\right)^2,$$

and

$$\left.\frac{\partial T}{\partial \mathcal{S}}\right|_A = -\frac{A-Nb}{Nk_B}.$$

Integrating the first equation gives

$$\mathcal{S}(A,t) = f(T) - \frac{Nk_BT}{A-Nb} + a\left(\frac{N}{A}\right)^2,$$

where $f(T)$ is a function of only T, while integrating the second equation yields

$$\mathcal{S}(A,T) = g(A) - \frac{Nk_BT}{A-Nb},$$

with $g(A)$ a function of only A. By comparing these two equations we get

$$\mathcal{S}(A,T) = \mathcal{S}_o - \frac{Nk_BT}{A-Nb} + a\left(\frac{N}{A}\right)^2,$$

where $\mathcal{S}_o$ represents the surface tension in the absence of surfactants and is independent of A and T. The equation resembles the van der Waals equation of state for gas–liquid systems. The factor Nb in the second term represents the excluded volume effect due to the finite size of the surfactant molecules. The last term represents the binary interaction between two surfactant molecules. If surfactant molecules attract each other, the coefficient a is positive and the surface tension increases.

(d) Find an expression for $C_{\mathcal{S}} - C_A$ in terms of $\left.\frac{\partial E}{\partial A}\right|_T = \left.\frac{\partial E}{\partial A}\right|_{\mathcal{S}}$, $\mathcal{S}$, $\left.\frac{\partial \mathcal{S}}{\partial A}\right|_T$, and $\left.\frac{\partial T}{\partial \mathcal{S}}\right|_A$.

Taking A and T as independent variables, we obtain

$$đQ = dE - \mathcal{S}\cdot dA, \implies đQ = \left.\frac{\partial E}{\partial A}\right|_T dA + \left.\frac{\partial E}{\partial T}\right|_A dT - \mathcal{S}\cdot dA,$$

and

$$\delta Q = \left(\left.\frac{\partial E}{\partial A}\right|_T - \mathcal{S}\right)dA + \left.\frac{\partial E}{\partial T}\right|_A dT.$$

From the above result, the heat capacities are obtained as

$$
\begin{cases}
C_A \equiv \left.\dfrac{\delta Q}{\delta T}\right|_A = \left.\dfrac{\partial E}{\partial T}\right|_A \\[4mm]
C_{\mathcal{S}} \equiv \left.\dfrac{\delta Q}{\delta T}\right|_{\mathcal{S}} = \left(\left.\dfrac{\partial E}{\partial A}\right|_T - \mathcal{S}\right)\left.\dfrac{\partial A}{\partial T}\right|_{\mathcal{S}} + \left.\dfrac{\partial E}{\partial T}\right|_{\mathcal{S}}
\end{cases},
$$

resulting in

$$
C_{\mathcal{S}} - C_A = \left(\left.\dfrac{\partial E}{\partial A}\right|_T - \mathcal{S}\right)\left.\dfrac{\partial A}{\partial T}\right|_{\mathcal{S}}.
$$

Using the chain rule

$$
\left.\dfrac{\partial T}{\partial \mathcal{S}}\right|_A \cdot \left.\dfrac{\partial \mathcal{S}}{\partial A}\right|_T \cdot \left.\dfrac{\partial A}{\partial T}\right|_{\mathcal{S}} = -1,
$$

we obtain

$$
C_{\mathcal{S}} - C_A = \left(\left.\dfrac{\partial E}{\partial A}\right|_T - \mathcal{S}\right)\cdot\left(\dfrac{-1}{\left.\frac{\partial T}{\partial \mathcal{S}}\right|_A \cdot \left.\frac{\partial \mathcal{S}}{\partial A}\right|_T}\right).
$$

* * * * * * * *

3. *Temperature scales*: prove the equivalence of the ideal gas temperature scale Θ, and the thermodynamic scale T, by performing a Carnot cycle on an ideal gas. The ideal gas satisfies $PV = Nk_B\Theta$, and its internal energy E is a function of Θ only. However, *you may not assume that $E \propto \Theta$.* You may wish to proceed as follows:

(a) Calculate the heat exchanges Q_H and Q_C as a function of Θ_H, Θ_C, and the volume expansion factors.

The ideal gas temperature is defined through the equation of state

$$
\theta = \dfrac{PV}{Nk_B}.
$$

The thermodynamic temperature is defined for a reversible Carnot cycle by

$$
\dfrac{T_{\text{hot}}}{T_{\text{cold}}} = \dfrac{Q_{\text{hot}}}{Q_{\text{cold}}}.
$$

For an ideal gas, the internal energy is a function only of θ, that is, $E = E(\theta)$, and

$$
đQ = dE - đW = \dfrac{dE}{d\theta}\cdot d\theta + PdV.
$$

Consider the Carnot cycle indicated in the figure. For the segment 1 to 2, which undergoes an isothermal expansion, we have

$$
d\theta = 0, \Longrightarrow đQ_{\text{hot}} = PdV, \quad \text{and} \quad P = \dfrac{Nk_B\theta_{\text{hot}}}{V}.
$$

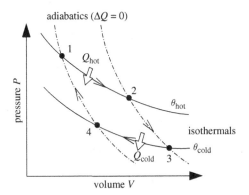

Hence, the heat input of the cycle is related to the expansion factor by

$$Q_{\text{hot}} = \int_{V_1}^{V_2} Nk_B\theta_{\text{hot}} \frac{dV}{V} = Nk_B\theta_{\text{hot}} \ln\left(\frac{V_2}{V_1}\right).$$

A similar calculation along the low-temperature isotherm yields

$$Q_{\text{cold}} = \int_{V_4}^{V_3} Nk_B\theta_{\text{cold}} \frac{dV}{V} = Nk_B\theta_{\text{cold}} \ln\left(\frac{V_3}{V_4}\right),$$

and thus

$$\frac{Q_{\text{hot}}}{Q_{\text{cold}}} = \frac{\theta_{\text{hot}}}{\theta_{\text{cold}}} \frac{\ln(V_2/V_1)}{\ln(V_3/V_4)}.$$

(b) Calculate the volume expansion factor in an adiabatic process as a function of Θ.

Next, we calculate the volume expansion/compression ratios in the adiabatic processes. Along an adiabatic segment

$$đQ = 0, \Longrightarrow 0 = \frac{dE}{d\theta} \cdot d\theta + \frac{Nk_B\theta}{V} \cdot dV, \Longrightarrow \frac{dV}{V} = -\frac{1}{Nk_B\theta} \frac{dE}{d\theta} \cdot d\theta.$$

Integrating the above between the two temperatures, we obtain

$$\begin{cases} \ln\left(\dfrac{V_3}{V_2}\right) = -\dfrac{1}{Nk_B} \displaystyle\int_{\theta_{\text{cold}}}^{\theta_{\text{hot}}} \dfrac{1}{\theta} \dfrac{dE}{d\theta} \cdot d\theta, & \text{and} \\[3mm] \ln\left(\dfrac{V_4}{V_1}\right) = -\dfrac{1}{Nk_B} \displaystyle\int_{\theta_{\text{cold}}}^{\theta_{\text{hot}}} \dfrac{1}{\theta} \dfrac{dE}{d\theta} \cdot d\theta. \end{cases}$$

While we cannot explicitly evaluate the integral (since $E(\theta)$ is arbitrary), we can nonetheless conclude that

$$\frac{V_1}{V_4} = \frac{V_2}{V_3}.$$

(c) Show that $Q_H/Q_C = \Theta_H/\Theta_C$.

Combining the results of parts (a) and (b), we observe that

$$\frac{Q_{\text{hot}}}{Q_{\text{cold}}} = \frac{\theta_{\text{hot}}}{\theta_{\text{cold}}}.$$

Since the thermodynamic temperature scale is defined by

$$\frac{Q_{\text{hot}}}{Q_{\text{cold}}} = \frac{T_{\text{hot}}}{T_{\text{cold}}},$$

we conclude that θ and T are proportional. If we further define $\theta(\text{triple point}_{H_2O}) = T(\text{triple point}_{H_2O}) = 273.16$, θ and T become identical.

$$* * * * * * *$$

4. *Equations of state*: the equation of state constrains the form of internal energy as in the following examples.

(a) Starting from $dE = TdS - PdV$, show that the equation of state $PV = Nk_BT$ in fact implies that E can only depend on T.

Since there is only one form of work, we can choose any two parameters as independent variables. For example, selecting T and V, such that $E = E(T, V)$, and $S = S(T, V)$, we obtain

$$dE = TdS - PdV = T \left.\frac{\partial S}{\partial T}\right|_V dT + T \left.\frac{\partial S}{\partial V}\right|_T dV - PdV,$$

resulting in

$$\left.\frac{\partial E}{\partial V}\right|_T = T \left.\frac{\partial S}{\partial V}\right|_T - P.$$

Using Maxwell's relation[1]

$$\left.\frac{\partial S}{\partial V}\right|_T = \left.\frac{\partial P}{\partial T}\right|_V,$$

we obtain

$$\left.\frac{\partial E}{\partial V}\right|_T = T \left.\frac{\partial P}{\partial T}\right|_V - P.$$

Since $T \left.\frac{\partial P}{\partial T}\right|_V = T\frac{Nk_B}{V} = P$, for an ideal gas, $\left.\frac{\partial E}{\partial V}\right|_T = 0$. Thus E depends only on T, that is, $E = E(T)$.

(b) What is the most general equation of state consistent with an internal energy that depends only on temperature?

If $E = E(T)$,

$$\left.\frac{\partial E}{\partial V}\right|_T = 0, \quad \Longrightarrow \quad T \left.\frac{\partial P}{\partial T}\right|_V = P.$$

The solution for this equation is $P = f(V)T$, where $f(V)$ is any function of only V.

(c) Show that for a van der Waals gas C_V is a function of temperature alone.

The van der Waals equation of state is given by

$$\left[P - a\left(\frac{N}{V}\right)^2\right] \cdot (V - Nb) = Nk_BT,$$

---

[1] $dL = Xdx + Ydy + \cdots, \implies \left.\frac{\partial X}{\partial y}\right|_x = \left.\frac{\partial Y}{\partial x}\right|_y = \frac{\partial^2 L}{\partial x \cdot \partial y}.$

or

$$P = \frac{N k_B T}{(V - Nb)} + a \left(\frac{N}{V} \right)^2 .$$

From these equations, we conclude that

$$C_V \equiv \left. \frac{\partial E}{\partial T} \right|_V , \implies \left. \frac{\partial C_V}{\partial V} \right|_T = \frac{\partial^2 E}{\partial V \partial T} = \frac{\partial}{\partial T} \left\{ T \left. \frac{\partial P}{\partial T} \right|_V - P \right\} = T \left. \frac{\partial^2 P}{\partial T^2} \right|_V = 0.$$

****

5. *The Clausius–Clapeyron equation* describes the variation of boiling point with pressure. It is usually derived from the condition that the chemical potentials of the gas and liquid phases are the same at coexistence.

   From the equations

$$\mu_{\text{liquid}}(P, T) = \mu_{\text{gas}}(P, T),$$

   and

$$\mu_{\text{liquid}}(P + dP, T + dT) = \mu_{\text{gas}}(P + dP, T + dT),$$

   we conclude that along the coexistence line

$$\left. \frac{dP}{dT} \right|_{\text{coX}} = \frac{\left. \frac{\partial \mu_g}{\partial T} \right|_P - \left. \frac{\partial \mu_l}{\partial T} \right|_P}{\left. \frac{\partial \mu_l}{\partial P} \right|_T - \left. \frac{\partial \mu_g}{\partial P} \right|_T} .$$

   The variations of the Gibbs free energy, $G = N\mu(P, T)$, from the extensivity condition are given by

$$V = \left. \frac{\partial G}{\partial P} \right|_T , \quad S = - \left. \frac{\partial G}{\partial T} \right|_P .$$

   In terms of intensive quantities

$$v = \frac{V}{N} = \left. \frac{\partial \mu}{\partial P} \right|_T , \quad s = \frac{S}{N} = - \left. \frac{\partial \mu}{\partial T} \right|_P ,$$

   where s and v are molar entropy and volume, respectively. Thus, the coexistence line satisfies the condition

$$\left. \frac{dP}{dT} \right|_{\text{coX}} = \frac{S_g - S_l}{V_g - V_l} = \frac{s_g - s_l}{v_g - v_l} .$$

For an alternative derivation, consider a Carnot engine using one mole of water. At the source (P, T) the latent heat L is supplied converting water to steam. There is a volume increase V associated with this process. The pressure is adiabatically decreased to $P - dP$. At the sink $(P - dP, T - dT)$ steam is condensed back to water.

(a) Show that the work output of the engine is $W = V\mathrm{d}P + \mathcal{O}(\mathrm{d}P^2)$. Hence obtain the Clausius–Clapeyron equation

$$\left.\frac{\mathrm{d}P}{\mathrm{d}T}\right|_{\text{boiling}} = \frac{L}{TV}. \tag{1}$$

If we approximate the adiabatic processes as taking place at constant volume V (vertical lines in the P − V diagram), we find

$$W = \oint P\mathrm{d}V = PV - (P-\mathrm{d}P)V = V\mathrm{d}P.$$

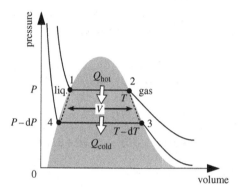

Here, we have neglected the volume of the liquid state, which is much smaller than that of the gas state. As the error is of the order of

$$\left.\frac{\partial V}{\partial P}\right|_S \mathrm{d}P \cdot \mathrm{d}P = O(\mathrm{d}P^2),$$

we have

$$W = V\mathrm{d}P + O(\mathrm{d}P^2).$$

The efficiency of any Carnot cycle is given by

$$\eta = \frac{W}{Q_H} = 1 - \frac{T_C}{T_H},$$

and in the present case,

$$Q_H = L, \quad W = V\mathrm{d}P, \quad T_H = T, \quad T_C = T - \mathrm{d}T.$$

Substituting these values in the universal formula for efficiency, we obtain the Clausius–Clapeyron equation

$$\frac{V\mathrm{d}P}{L} = \frac{\mathrm{d}T}{T}, \quad \text{or} \quad \left.\frac{\mathrm{d}P}{\mathrm{d}T}\right|_{\text{coX}} = \frac{L}{T \cdot V}.$$

(b) What is wrong with the following argument: "The heat Q_H supplied at the source to convert one mole of water to steam is $L(T)$. At the sink $L(T-dT)$ is supplied to condense one mole of steam to water. The difference $dTdL/dT$ must equal the work $W = VdP$, equal to LdT/T from Eq. (1). Hence $dL/dT = L/T$, implying that L is proportional to T!"

The statement "At the sink $L(T-dT)$ is supplied to condense one mole of water" is incorrect. In the $P-V$ diagram shown, the state at "1" corresponds to pure water, "2" corresponds to pure vapor, but the states "3" and "4" have two phases coexisting. In going from the state 3 to 4 less than one mole of steam is converted to water. Part of the steam has already been converted into water during the adiabatic expansion $2 \to 3$, and the remaining portion is converted in the adiabatic compression $4 \to 1$. Thus the actual latent heat should be less than the contribution by one mole of water.

(c) Assume that L is approximately temperature-independent, and that the volume change is dominated by the volume of steam treated as an ideal gas, that is, $V = Nk_BT/P$. Integrate Eq. (1) to obtain $P(T)$.

For an ideal gas

$$V = \frac{Nk_BT}{P}, \implies \left.\frac{dP}{dT}\right|_{coX} = \frac{LP}{Nk_BT^2}, \quad \text{or} \quad \frac{dP}{P} = \frac{L}{Nk_BT^2}dT.$$

Integrating this equation, the boiling temperature is obtained as a function of the pressure P, as

$$P = C \cdot \exp\left(-\frac{L}{k_BT_{\text{Boiling}}}\right).$$

(d) A hurricane works somewhat like the engine described above. Water evaporates at the warm surface of the ocean, steam rises up in the atmosphere, and condenses to water at the higher and cooler altitudes. The Coriolis force converts the upward suction of the air to spiral motion. (Using ice and boiling water, you can create a little storm in a tea cup.) Typical values of warm ocean surface and high altitude temperatures are $80°$ F and $-120°$ F, respectively. The warm water surface layer must be at least 200 feet thick to provide sufficient water vapor, as the hurricane needs to condense about 90 million tons of water vapor per hour to maintain itself. Estimate the maximum possible efficiency, and power output, of such a hurricane. (The latent heat of vaporization of water is about 2.3×10^6 J kg^{-1}.)

For $T_C = -120°$ F $= 189°$ K, and $T_H = 80°$ F $= 300°$ K, the limiting efficiency, as that of a Carnot engine, is

$$\eta_{\text{max}} = \frac{T_H - T_C}{T_H} = 0.37.$$

The output power, equal to (input power) × (efficiency), is

$$\text{power output} = \frac{90 \times 10^6 \text{ tons}}{\text{hr}} \cdot \frac{1\,\text{hr}}{3600\,\text{sec}} \cdot \frac{1000\,\text{kg}}{\text{ton}} \cdot \frac{2.3 \times 10^6\,\text{J}}{\text{kg}} \times 0.37 = 2 \times 10^{13}\,\text{watts}.$$

****

6. *Glass*: liquid quartz, if cooled slowly, crystallizes at a temperature T_m, and releases latent heat L. Under more rapid cooling conditions, the liquid is supercooled and becomes glassy.

(a) As both phases of quartz are almost incompressible, there is no work input, and changes in internal energy satisfy $dE = TdS + \mu dN$. Use the extensivity condition to obtain the expression for μ in terms of E, T, S, and N.

Since in the present context we are considering only chemical work, we can regard entropy as a function of two independent variables, for example, E and N, which appear naturally from $dS = dE/T - \mu dN/T$. Since entropy is an extensive variable, $\lambda S = S(\lambda E, \lambda N)$. Differentiating this with respect to λ and evaluating the resulting expression at $\lambda = 1$ gives

$$S(E, N) = \left.\frac{\partial S}{\partial E}\right|_N E + \left.\frac{\partial S}{\partial N}\right|_E N = \frac{E}{T} - \frac{N\mu}{T},$$

leading to

$$\mu = \frac{E - TS}{N}.$$

(b) The heat capacity of crystalline quartz is approximately $C_X = \alpha T^3$, while that of glassy quartz is roughly $C_G = \beta T$, where α and β are constants.

Assuming that the third law of thermodynamics applies to both crystalline and glass phases, calculate the entropies of the two phases at temperatures $T \leq T_m$.

Finite-temperature entropies can be obtained by integrating đQ/T, starting from $S(T = 0) = 0$. Using the heat capacities to obtain the heat inputs, we find

$$\begin{cases} C_{\text{crystal}} = \alpha T^3 = \dfrac{T}{N}\dfrac{dS_{\text{crystal}}}{dT}, & \implies S_{\text{crystal}} = \dfrac{N\alpha T^3}{3}, \\[3mm] C_{\text{glass}} = \beta T = \dfrac{T}{N}\dfrac{dS_{\text{glass}}}{dT}, & \implies S_{\text{glass}} = \beta NT. \end{cases}$$

(c) At zero temperature the local bonding structure is similar in glass and crystalline quartz, so that they have approximately the same internal energy E_0. Calculate the internal energies of both phases at temperatures $T \leq T_m$.

Since $dE = TdS + \mu dN$, for $dN = 0$, we have

$$\begin{cases} dE = TdS = \alpha NT^3 \, dT & \text{(crystal)}, \\[2mm] dE = TdS = \beta NTdT & \text{(glass)}. \end{cases}$$

Integrating these expressions, starting with the same internal energy E_o at $T = 0$, yields

$$\begin{cases} E = E_o + \dfrac{\alpha N}{4}T^4 & \text{(crystal)}, \\[3mm] E = E_o + \dfrac{\beta N}{2}T^2 & \text{(glass)}. \end{cases}$$

(d) Use the condition of thermal equilibrium between two phases to compute the equilibrium melting temperature T_m in terms of α and β.

From the condition of chemical equilibrium between the two phases, $\mu_{\text{crystal}} = \mu_{\text{glass}}$, we obtain

$$\left(\frac{1}{3} - \frac{1}{4}\right) \cdot \alpha T^4 = \left(1 - \frac{1}{2}\right) \cdot \beta T^2, \Longrightarrow \frac{\alpha T^4}{12} = \frac{\beta T^2}{2},$$

resulting in a transition temperature

$$T_{\text{melt}} = \sqrt{\frac{6\beta}{\alpha}}.$$

(e) Compute the latent heat L in terms of α and β.

From the assumptions of the previous parts, we obtain the latent heats for the glass to crystal transition as

$$L = T_{\text{melt}} \left(S_{\text{glass}} - S_{\text{crystal}}\right) = NT_{\text{melt}} \left(\beta T_{\text{melt}} - \frac{\alpha T_{\text{melt}}^3}{3}\right)$$

$$= NT_{\text{melt}}^2 \left(\beta - \frac{\alpha T_{\text{melt}}^2}{3}\right) = NT_{\text{melt}}^2 (\beta - 2\beta) = -N\beta T_{\text{melt}}^2 < 0.$$

(f) Is the result in the previous part correct? If not, which of the steps leading to it is most likely to be incorrect?

The above result implies that the entropy of the crystal phase is larger than that of the glass phase. This is clearly unphysical, and one of the assumptions must be wrong. The questionable step is the assumption that the glass phase is subject to the third law of thermodynamics, and has zero entropy at $T = 0$. In fact, glass is a non-ergodic state of matter that does not have a unique ground state, and violates the third law.

****

Solutions to selected problems from chapter 2

1. *Characteristic functions*: calculate the characteristic function, the mean, and the variance of the following probability density functions:

   (a) *Uniform* $p(x) = \frac{1}{2a}$ for $-a < x < a$, and $p(x) = 0$ otherwise.

   A uniform probability distribution ,

   $$p(x) = \begin{cases} \frac{1}{2a} & \text{for } -a < x < a \\ 0 & \text{otherwise} \end{cases},$$

   for which there exist many examples, gives

   $$f(k) = \frac{1}{2a} \int_{-a}^{a} \exp(-ikx)\,\mathrm{d}x = \frac{1}{2a} \frac{1}{-ik} \exp(-ikx)\Big|_{-a}^{a}$$

   $$= \frac{1}{ak} \sin(ka) = \sum_{m=0}^{\infty} (-1)^m \frac{(ak)^{2m}}{(2m+1)!}.$$

   Therefore,

   $$m_1 = \langle x \rangle = 0, \quad \text{and} \quad m_2 = \langle x^2 \rangle = \frac{1}{3} a^2.$$

   (b) *Laplace* $p(x) = \frac{1}{2a} \exp\left(-\frac{|x|}{a}\right)$.

   The Laplace PDF,

   $$p(x) = \frac{1}{2a} \exp\left(-\frac{|x|}{a}\right),$$

   for example describing light absorption through a turbid medium, gives

   $$f(k) = \frac{1}{2a} \int_{-\infty}^{\infty} \mathrm{d}x \exp\left(-ikx - \frac{|x|}{a}\right)$$

   $$= \frac{1}{2a} \int_{0}^{\infty} \mathrm{d}x \exp(-ikx - x/a) + \frac{1}{2a} \int_{-\infty}^{0} \mathrm{d}x \exp(-ikx + x/a)$$

   $$= \frac{1}{2a} \left[\frac{1}{-ik + 1/a} - \frac{1}{-ik - 1/a} \right] = \frac{1}{1 + (ak)^2}$$

   $$= 1 - (ak)^2 + (ak)^4 - \cdots .$$

   Therefore,

   $$m_1 = \langle x \rangle = 0, \quad \text{and} \quad m_2 = \langle x^2 \rangle = 2a^2.$$

(c) *Cauchy* $p(x) = \frac{a}{\pi(x^2+a^2)}$.

The Cauchy, or Lorentz, PDF describes the spectrum of light scattered by diffusive modes, and is given by

$$p(x) = \frac{a}{\pi(x^2 + a^2)}.$$

For this distribution,

$$f(k) = \int_{-\infty}^{\infty} \exp(-ikx)\frac{a}{\pi(x^2 + a^2)}\, dx$$

$$= \frac{1}{2\pi i}\int_{-\infty}^{\infty} \exp(-ikx)\left[\frac{1}{x - ia} - \frac{1}{x + ia}\right] dx.$$

The easiest method for evaluating the above integrals is to close the integration contours in the complex plane, and evaluate the residue. The vanishing of the integrand at infinity determines whether the contour has to be closed in the upper or lower half of the complex plane, and leads to

$$f(k) = \left\{ \begin{array}{ll} -\dfrac{1}{2\pi i}\displaystyle\int_C \dfrac{\exp(-ikx)}{x + ia}\, dx = \exp(-ka) & \text{for } k \geq 0 \\[3mm] \dfrac{1}{2\pi i}\displaystyle\int_B \dfrac{\exp(-ikx)}{x - ia}\, dx = \exp(ka) & \text{for } k < 0 \end{array} \right\} = \exp(-|ka|).$$

Note that $f(k)$ is not an analytic function in this case, and hence does not have a Taylor expansion. The moments have to be determined by another method, for example, by direct evaluation, as

$$m_1 = \langle x \rangle = 0, \qquad \text{and} \qquad m_2 = \langle x^2 \rangle = \int dx \frac{\pi}{a} \cdot \frac{x^2}{x^2 + a^2} \to \infty.$$

The first moment vanishes by symmetry, while the second (and higher) moments diverge, explaining the non-analytic nature of $f(k)$.

The following two probability density functions are defined for $x \geq 0$. Compute only the mean and variance for each.

(d) *Rayleigh* $p(x) = \frac{x}{a^2}\exp(-\frac{x^2}{2a^2})$.

The Rayleigh distribution,

$$p(x) = \frac{x}{a^2}\exp\left(-\frac{x^2}{2a^2}\right), \qquad \text{for} \qquad x \geq 0,$$

can be used for the length of a random walk in two dimensions. Its characteristic function is

$$f(k) = \int_0^{\infty} \exp(-ikx)\frac{x}{a^2}\exp\left(-\frac{x^2}{2a^2}\right) dx$$

$$= \int_0^{\infty} [\cos(kx) - i\sin(kx)]\frac{x}{a^2}\exp\left(-\frac{x^2}{2a^2}\right) dx.$$

The integrals are not simple, but can be evaluated as

$$\int_0^{\infty} \cos(kx)\frac{x}{a^2}\exp\left(-\frac{x^2}{2a^2}\right) dx = \sum_{n=0}^{\infty} \frac{(-1)^n n!}{(2n)!}\left(2a^2 k^2\right)^n,$$

and

$$\int_0^\infty \sin(kx)\frac{x}{a^2}\exp\left(-\frac{x^2}{2a^2}\right)dx = \frac{1}{2}\int_{-\infty}^\infty \sin(kx)\frac{x}{a^2}\exp\left(-\frac{x^2}{2a^2}\right)dx$$

$$= \sqrt{\frac{\pi}{2}}\,ka\,\exp\left(-\frac{k^2a^2}{2}\right),$$

resulting in

$$f(k) = \sum_{n=0}^\infty \frac{(-1)^n n!}{(2n)!}\left(2a^2k^2\right)^n - i\sqrt{\frac{\pi}{2}}\,ka\,\exp\left(-\frac{k^2a^2}{2}\right).$$

The moments can also be calculated directly, from

$$m_1 = \langle x \rangle = \int_0^\infty \frac{x^2}{a^2}\exp\left(-\frac{x^2}{2a^2}\right)dx = \int_{-\infty}^\infty \frac{x^2}{2a^2}\exp\left(-\frac{x^2}{2a^2}\right)dx = \sqrt{\frac{\pi}{2}}\,a,$$

$$m_2 = \langle x^2 \rangle = \int_0^\infty \frac{x^3}{a^2}\exp\left(-\frac{x^2}{2a^2}\right)dx = 2a^2\int_0^\infty \frac{x^2}{2a^2}\exp\left(-\frac{x^2}{2a^2}\right)d\left(\frac{x^2}{2a^2}\right)$$

$$= 2a^2\int_0^\infty y\exp(-y)\,dy = 2a^2.$$

(e) *Maxwell* $p(x) = \sqrt{\frac{2}{\pi}}\frac{x^2}{a^3}\exp(-\frac{x^2}{2a^2})$.

It is difficult to calculate the characteristic function for the Maxwell distribution

$$p(x) = \sqrt{\frac{2}{\pi}}\frac{x^2}{a^3}\exp\left(-\frac{x^2}{2a^2}\right),$$

say describing the speed of a gas particle. However, we can directly evaluate the mean and variance, as

$$m_1 = \langle x \rangle = \sqrt{\frac{2}{\pi}}\int_0^\infty \frac{x^3}{a^3}\exp\left(-\frac{x^2}{2a^2}\right)dx$$

$$= 2\sqrt{\frac{2}{\pi}}\,a\int_0^\infty \frac{x^2}{2a^2}\exp\left(-\frac{x^2}{2a^2}\right)d\left(\frac{x^2}{2a^2}\right)$$

$$= 2\sqrt{\frac{2}{\pi}}\,a\int_0^\infty y\exp(-y)\,dy = 2\sqrt{\frac{2}{\pi}}\,a,$$

and

$$m_2 = \langle x^2 \rangle = \sqrt{\frac{2}{\pi}}\int_o^\infty \frac{x^4}{a^3}\exp\left(-\frac{x^2}{2a^2}\right)dx = 3a^2.$$

****

2. *Directed random walk*: the motion of a particle in three dimensions is a series of independent steps of length ℓ. Each step makes an angle θ with the z axis, with a probability density $p(\theta) = 2\cos^2(\theta/2)/\pi$, while the angle ϕ is uniformly distributed between 0 and 2π. (Note that the solid angle factor of $\sin\theta$ is already included in the definition of $p(\theta)$, which is correctly normalized to unity.) The particle (walker) starts at the origin and makes a large number of steps N.

(a) Calculate the expectation values $\langle z \rangle$, $\langle x \rangle$, $\langle y \rangle$, $\langle z^2 \rangle$, $\langle x^2 \rangle$, and $\langle y^2 \rangle$, and the covariances $\langle xy \rangle$, $\langle xz \rangle$, and $\langle yz \rangle$.

From symmetry arguments,

$$\langle x \rangle = \langle y \rangle = 0,$$

while along the z axis,

$$\langle z \rangle = \sum_i \langle z_i \rangle = N \langle z_i \rangle = Na \langle \cos\theta_i \rangle = \frac{Na}{2}.$$

The last equality follows from

$$\langle \cos\theta_i \rangle = \int p(\theta)\cos\theta \, d\theta = \int_0^\pi \frac{1}{\pi}\cos\theta \cdot (\cos\theta + 1)\, d\theta$$
$$= \int_0^\pi \frac{1}{2\pi}(\cos 2\theta + 1)\, d\theta = \frac{1}{2}.$$

The second moment of z is given by

$$\langle z^2 \rangle = \sum_{i,j}\langle z_i z_j \rangle = \sum_i \sum_{i\neq j}\langle z_i z_j \rangle + \sum_i \langle z_i^2 \rangle$$
$$= \sum_i \sum_{i\neq j}\langle z_i \rangle \langle z_j \rangle + \sum_i \langle z_i^2 \rangle$$
$$= N(N-1)\langle z_i \rangle^2 + N\langle z_i^2 \rangle.$$

Noting that

$$\frac{\langle z_i^2 \rangle}{a^2} = \int_0^\pi \frac{1}{\pi}\cos^2\theta(\cos\theta + 1)\, d\theta = \int_0^\pi \frac{1}{2\pi}(\cos 2\theta + 1)\, d\theta = \frac{1}{2},$$

we find

$$\langle z^2 \rangle = N(N-1)\left(\frac{a}{2}\right)^2 + N\frac{a^2}{2} = N(N+1)\frac{a^2}{4}.$$

The second moments in the x and y directions are equal, and given by

$$\langle x^2 \rangle = \sum_{i,j}\langle x_i x_j \rangle = \sum_i \sum_{i\neq j}\langle x_i x_j \rangle + \sum_i \langle x_i^2 \rangle = N\langle x_i^2 \rangle.$$

Using the result

$$\frac{\langle x_i^2 \rangle}{a^2} = \langle \sin^2\theta\cos^2\phi \rangle$$
$$= \frac{1}{2\pi^2}\int_0^{2\pi} d\phi\cos^2\phi \int_0^\pi d\theta\sin^2\theta(\cos\theta + 1) = \frac{1}{4},$$

we obtain

$$\langle x^2 \rangle = \langle y^2 \rangle = \frac{Na^2}{4}.$$

While the variables x, y, and z are not independent because of the constraint of unit length, simple symmetry considerations suffice to show that the three covariances are in fact zero, that is,

$$\langle xy \rangle = \langle xz \rangle = \langle yz \rangle = 0.$$

(b) Use the central limit theorem to estimate the probability density $p(x, y, z)$ for the particle to end up at the point (x, y, z).

From the central limit theorem, the probability density should be Gaussian. However, for correlated random variables we may expect cross terms that describe their covariance. However, we showed above that the covariances between x, y, and z are all zero. Hence we can treat them as three independent Gaussian variables, and write

$$p(x, y, z) \propto \exp\left[-\frac{(x - \langle x \rangle)^2}{2\sigma_x^2} - \frac{(y - \langle y \rangle)^2}{2\sigma_y^2} - \frac{(z - \langle z \rangle)^2}{2\sigma_z^2} \right].$$

(There will be correlations between x, y, and z appearing in higher cumulants, but all such cumulants become irrelevant in the $N \to \infty$ limit.) Using the moments

$$\langle x \rangle = \langle y \rangle = 0, \qquad \text{and} \qquad \langle z \rangle = N\frac{a}{2},$$

$$\sigma_x^2 = \langle x^2 \rangle - \langle x \rangle^2 = N\frac{a^2}{4} = \sigma_y^2,$$

$$\text{and} \qquad \sigma_z^2 = \langle z^2 \rangle - \langle z \rangle^2 = N(N+1)\frac{a^2}{4} - \left(\frac{Na}{2}\right)^2 = N\frac{a^2}{4},$$

we obtain

$$p(x, y, z) = \left(\frac{2}{\pi Na^2}\right)^{3/2} \exp\left[-\frac{x^2 + y^2 + (z - Na/2)^2}{Na^2/2} \right].$$

****

3. *Tchebycheff inequality*: consider any probability density $p(x)$ for $(-\infty < x < \infty)$, with mean λ, and variance σ^2. Show that the total probability of outcomes that are more than $n\sigma$ away from λ is less than $1/n^2$, that is,

$$\int_{|x-\lambda| \geq n\sigma} dx\, p(x) \leq \frac{1}{n^2}.$$

Hint. Start with the integral defining σ^2, and break it up into parts corresponding to $|x - \lambda| > n\sigma$, and $|x - \lambda| < n\sigma$.

By definition, for a system with a PDF $p(x)$, and average λ, the variance is

$$\sigma^2 = \int (x - \lambda)^2 p(x)\, dx.$$

Let us break the integral into two parts as

$$\sigma^2 = \int_{|x-\lambda| \geq n\sigma} (x-\lambda)^2 p(x)\,dx + \int_{|x-\lambda| < n\sigma} (x-\lambda)^2 p(x)\,dx,$$

resulting in

$$\sigma^2 - \int_{|x-\lambda| < n\sigma} (x-\lambda)^2 p(x)\,dx = \int_{|x-\lambda| \geq n\sigma} (x-\lambda)^2 p(x)\,dx.$$

Now since

$$\int_{|x-\lambda| \geq n\sigma} (x-\lambda)^2 p(x)\,dx \geq \int_{|x-\lambda| \geq n\sigma} (n\sigma)^2 p(x)\,dx,$$

we obtain

$$\int_{|x-\lambda| \geq n\sigma} (n\sigma)^2 p(x)\,dx \leq \sigma^2 - \int_{|x-\lambda| < n\sigma} (x-\lambda)^2 p(x)\,dx \leq \sigma^2,$$

and

$$\int_{|x-\lambda| \geq n\sigma} p(x)\,dx \leq \frac{1}{n^2}.$$

$$********$$

4. *Optimal selection*: in many specialized populations, there is little variability among the members. Is this a natural consequence of optimal selection?

(a) Let $\{r_\alpha\}$ be n random numbers, each independently chosen from a probability density $p(r)$, with $r \in [0, 1]$. Calculate the probability density $p_n(x)$ for the largest value of this set, that is, for $x = \max\{r_1, \cdots, r_n\}$.

The probability that the maximum of n random numbers falls between x and $x+dx$ is equal to the probability that one outcome is in this interval, while all the others are smaller than x, that is,

$$p_n(x) = p(r_1 = x, r_2 < x, r_3 < x, \cdots, r_n < x) \times \binom{n}{1},$$

where the second factor corresponds to the number of ways of choosing $r_\alpha = x$. As these events are independent,

$$p_n(x) = p(r_1 = x) \cdot p(r_2 < x) \cdot p(r_3 < x) \cdots p(r_n < x) \times \binom{n}{1}$$
$$= p(r = x)\,[p(r < x)]^{n-1} \times \binom{n}{1}.$$

The probability of $r < x$ is just a cumulative probability function, and

$$p_n(x) = n \cdot p(x) \cdot \left[\int_0^x p(r)\,dr \right]^{n-1}.$$

(b) If each r_α is uniformly distributed between 0 and 1, calculate the mean and variance of x as a function of n, and comment on their behavior at large n.

If each r_α is uniformly distributed between 0 and 1, $p(r) = 1$ ($\int_0^1 p(r)\,dr = \int_0^1 dr = 1$). With this PDF, we find

$$p_n(x) = n \cdot p(x) \cdot \left[\int_0^x p(r)\,dr\right]^{n-1} = n\left[\int_0^x dr\right]^{n-1} = nx^{n-1},$$

and the mean is now given by

$$\langle x \rangle = \int_0^1 x p_n(x)\,dx = n\int_0^1 x^n\,dx = \frac{n}{n+1}.$$

The second moment of the maximum is

$$\langle x^2 \rangle = n\int_0^1 x^{n+1}\,dx = \frac{n}{n+2},$$

resulting in a variance

$$\sigma^2 = \langle x^2 \rangle - \langle x \rangle^2 = \frac{n}{n+2} - \left(\frac{n}{n+1}\right)^2 = \frac{n}{(n+1)^2(n+2)}.$$

Note that for large n the mean approaches the limiting value of unity, while the variance vanishes as $1/n^2$. There is too little space at the top of the distribution for a wide variance.

$$**********$$

5. *Benford's law* describes the observed probabilities of the *first digit* in a great variety of data sets, such as stock prices. Rather counter-intuitively, the digits 1 through 9 occur with probabilities 0.301, 0.176, 0.125, 0.097, 0.079, 0.067, 0.058, 0.051, 0.046, respectively. The key observation is that this distribution is invariant under a change of scale, for example, if the stock prices were converted from dollars to Persian rials! Find a formula that fits the above probabilities on the basis of this observation.

Let us consider the observation that the probability distribution for first integers is unchanged under multiplication by any (i.e., a random) number. Presumably we can repeat such multiplications many times, and it is thus suggestive that we should consider the properties of the product of random numbers. (Why this should be a good model for stock prices is not entirely clear, but it seems to be as good an explanation as anything else!)

Consider the $x = \prod_{i=1}^N r_i$, where r_i are positive, random variables taken from some reasonably well-behaved probability distribution. The random variable $\ell \equiv \ln x = \sum_{i=1}^N \ln r_i$ is the sum of many random contributions, and according to the central limit theorem should have a Gaussian distribution in the limit of large N, that is,

$$\lim_{N \to \infty} p(\ell) = \exp\left[-\frac{(\ell - N\bar{\ell})^2}{2N\sigma^2}\right]\frac{1}{\sqrt{2\pi N\sigma^2}},$$

where $\bar{\ell}$ and σ^2 are the mean and variance of $\ln r$*, respectively. The product x is
distributed according to the log-normal distribution*

$$p(x) = p(\ell)\frac{d\ell}{dx} = \frac{1}{x}\exp\left[-\frac{\left(\ln(x) - N\bar{\ell}\right)^2}{2N\sigma^2}\right]\frac{1}{\sqrt{2\pi N\sigma^2}}.$$

*The probability that the first integer of x in a decimal representation is i is now
obtained approximately as follows:*

$$p_i = \sum_q \int_{10^q i}^{10^q (i+1)} dx\, p(x).$$

*The integral considers cases in which x is a number of magnitude 10^q (i.e., has $q+1$
digits before the decimal point). Since the number is quite widely distributed, we
then have to sum over possible magnitudes q. The range of the sum actually need
not be specified! The next stage is to change variables from x to $\ell = \ln x$, leading to*

$$p_i = \sum_q \int_{q+\ln i}^{q+\ln(i+1)} d\ell\, p(\ell) = \sum_q \int_{q+\ln i}^{q+\ln(i+1)} d\ell \exp\left[-\frac{\left(\ell - N\bar{\ell}\right)^2}{2N\sigma^2}\right]\frac{1}{\sqrt{2\pi N\sigma^2}}.$$

*We shall now make the approximation that over the range of integration ($q+\ln i$ to
$q+\ln(i+1)$), the integrand is approximately constant. (The approximation works
best for $q \approx N\bar{\ell}$ where the integral is largest.) This leads to*

$$p_i \approx \sum_q \exp\left[-\frac{\left(q - N\bar{\ell}\right)^2}{2N\sigma^2}\right]\frac{1}{\sqrt{2\pi N\sigma^2}}[\ln(i+1) - \ln i] \propto \ln\left(1 + \frac{1}{i}\right),$$

*where we have ignored the constants of proportionality that come from the sum
over q. We thus find that the distribution of the first digit is not uniform, and
the properly normalized proportions of $\ln(1 + 1/i)$ indeed reproduce the probabilities $p_1, \cdots, p_9$ of 0.301, 0.176, 0.125, 0.097, 0.079, 0.067, 0.058, 0.051, 0.046
according to Benford's law. (For further information check http://www.treasure-
troves.com/math/BenfordsLaw.html; and for an alternative view the article by
T. P. Hill, American Scientist **86**, 358 (1998).)*

$$\ast\ast\ast\ast\ast\ast\ast\ast$$

6. *Information*: consider the velocity of a gas particle in one dimension ($-\infty < v < \infty$).

   (a) Find the unbiased probability density $p_1(v)$, subject only to the constraint that the
   average *speed* is c, that is, $\langle|v|\rangle = c$.

   *For an unbiased probability estimation, we need to maximize entropy subject to
   the two constraints of normalization, and of given average speed ($\langle|v|\rangle = c$.). Using
   Lagrange multipliers α and β to impose these constraints, we need to maximize*

$$S = -\langle\ln p\rangle = -\int_{-\infty}^{\infty} p(v)\ln p(v)\, dv + \alpha\left(1 - \int_{-\infty}^{\infty} p(v)\, dv\right)$$

$$+ \beta\left(c - \int_{-\infty}^{\infty} p(v)|v|\, dv\right).$$

Extremizing the above expression yields

$$\frac{\partial S}{\partial p(v)} = -\ln p(v) - 1 - \alpha - \beta|v| = 0,$$

which is solved for

$$\ln p(v) = -1 - \alpha - \beta|v|,$$

or

$$p(v) = C e^{-\beta|v|}, \qquad \text{with} \quad C = e^{-1-\alpha}.$$

The constraints can now be used to fix the parameters C and β:

$$1 = \int_{-\infty}^{\infty} p(v)\, dv = \int_{-\infty}^{\infty} C e^{-\beta|v|}\, dv = 2C \int_{0}^{\infty} e^{-\beta v}\, dv = 2C \frac{1}{-\beta} e^{-\beta v} \Big|_{0}^{\infty} = \frac{2C}{\beta},$$

which implies

$$C = \frac{\beta}{2}.$$

From the second constraint, we have

$$c = \int_{-\infty}^{\infty} C e^{-\beta|v|} |v|\, dv = \beta \int_{0}^{\infty} e^{-\beta v} v\, dv,$$

which when integrated by parts yields

$$c = \beta \left[-\frac{1}{\beta} v e^{-\beta v} \Big|_{0}^{\infty} + \frac{1}{\beta} \int_{0}^{\infty} e^{-\beta v}\, dv \right] = \left[-\frac{1}{\beta} e^{-\beta v} \Big|_{0}^{\infty} \right] = \frac{1}{\beta},$$

or,

$$\beta = \frac{1}{c}.$$

The unbiased PDF is then given by

$$p(v) = C e^{-\beta|v|} = \frac{1}{2c} \exp\left(-\frac{|v|}{c} \right).$$

(b) Now find the probability density $p_2(v)$, given only the constraint of average kinetic energy, $\langle mv^2/2 \rangle = mc^2/2$.

When the second constraint is on the average kinetic energy, $\langle mv^2/2 \rangle = mc^2/2$, we have

$$S = -\int_{-\infty}^{\infty} p(v) \ln p(v)\, dv + \alpha \left(1 - \int_{-\infty}^{\infty} p(v)\, dv \right) + \beta \left(\frac{mc^2}{2} - \int_{-\infty}^{\infty} p(v) \frac{mv^2}{2}\, dv \right).$$

The corresponding extremization,

$$\frac{\partial S}{\partial p(v)} = -\ln p(v) - 1 - \alpha - \beta \frac{mv^2}{2} = 0,$$

results in

$$p(v) = C \exp\left(-\frac{\beta m v^2}{2}\right).$$

The normalization constraint implies

$$1 = \int_{-\infty}^{\infty} p(v)\, dv = C \int_{-\infty}^{\infty} e^{-\beta m v^2/2} = C\sqrt{2\pi/\beta m},$$

or

$$C = \sqrt{\beta m/2\pi}.$$

The second constraint,

$$\frac{mc^2}{2} = \int_{-\infty}^{\infty} p(v)\frac{mv^2}{2}\, dv = \frac{m}{2}\sqrt{\frac{\beta m}{2\pi}} \int_{-\infty}^{\infty} \exp\left(-\frac{\beta m v^2}{2}\right) v^2\, dv$$

$$= \frac{m}{2}\sqrt{\frac{\beta m}{2\pi}} \left[\frac{\sqrt{\pi}}{2}\left(\frac{2}{\beta m}\right)^{3/2}\right] = \frac{1}{2\beta},$$

gives

$$\beta = \frac{1}{mc^2},$$

for a full PDF of

$$p(v) = C \exp\left(-\frac{\beta m v^2}{2}\right) = \frac{1}{\sqrt{2\pi c^2}} \exp\left(-\frac{v^2}{2c^2}\right).$$

(c) Which of the above statements provides more information on the velocity? Quantify the difference in information in terms of $I_2 - I_1 \equiv (\langle \ln p_2 \rangle - \langle \ln p_1 \rangle)/\ln 2$.

The entropy of the first PDF is given by

$$S_1 = -\langle \ln p_1 \rangle = -\int_{-\infty}^{\infty} \frac{1}{2c} \exp\left(\frac{-|v|}{c}\right)\left[-\ln(2c) - \frac{|v|}{c}\right] dv$$

$$= \frac{\ln(2c)}{c}\int_0^{\infty} \exp\left(-\frac{v}{c}\right) dv + \frac{1}{c}\int_0^{\infty} \exp\left(-\frac{v}{c}\right)\frac{v}{c}\, dv$$

$$= -\ln(2c)\exp(-v/c)|_0^{\infty} + \frac{1}{c}\left[-c\exp(-v/c)|_0^{\infty}\right]$$

$$= \ln(2c) + 1 = 1 + \ln 2 + \ln c.$$

For the second distribution, we obtain

$$S_2 = -\langle \ln p_2 \rangle = -\frac{1}{\sqrt{2\pi c^2}}\int_{-\infty}^{\infty} \exp\left(-\frac{v^2}{2c^2}\right)\left[-\frac{1}{2}\ln\left(2\pi c^2\right) - \frac{v^2}{2c^2}\right]$$

$$= \frac{\ln\left(2\pi c^2\right)}{2\sqrt{2\pi c^2}}\int_{-\infty}^{\infty} \exp\left(-v^2/2c^2\right) dv + \frac{1}{\sqrt{2\pi c^2}}\int_{-\infty}^{\infty} \frac{v^2}{2c^2}\exp\left(-v^2/2c^2\right) dv$$

$$= \frac{1}{2}\ln\left(2\pi c^2\right) + \frac{1}{c^2\sqrt{2\pi c^2}}\left[\frac{\sqrt{2\pi c^2}c^2}{2}\right]$$

$$= \frac{1}{2}\ln\left(2\pi c^2\right) + \frac{1}{2} = \frac{1}{2} + \frac{1}{2}\ln(2\pi) + \ln c.$$

For a discrete probability, the information content is

$$I_\alpha = \ln_2 M - S_\alpha / \ln 2,$$

where M denotes the number of possible outcomes. While M, and also the proper measure of probability, are not well defined for a continuous PDF, the ambiguities disappear when we consider the difference

$$I_2 - I_1 = (-S_2 + S_1) / \ln 2$$
$$= -(S_2 - S_1) / \ln 2$$
$$= -\frac{(\ln \pi - \ln 2 - 1)}{2 \ln 2} \approx 0.3956.$$

Hence the constraint of constant energy provides 0.3956 more bits of information. (This is partly due to the larger variance of the distribution with constant speed.)

* * * * * * *

Solutions to selected problems from chapter 3

1. *One-dimensional gas*: a thermalized gas particle is suddenly confined to a one-dimensional trap. The corresponding mixed state is described by an initial density function $\rho(q, p, t = 0) = \delta(q)f(p)$, where $f(p) = \exp(-p^2/2mk_BT)/\sqrt{2\pi mk_BT}$.

   (a) Starting from Liouville's equation, derive $\rho(q, p, t)$ and sketch it in the (q, p) plane.

   Liouville's equation, describing the incompressible nature of phase space density, is

   $$\frac{\partial \rho}{\partial t} = -\dot{q}\frac{\partial \rho}{\partial q} - \dot{p}\frac{\partial \rho}{\partial p} = -\frac{\partial \mathcal{H}}{\partial p}\frac{\partial \rho}{\partial q} + \frac{\partial \mathcal{H}}{\partial q}\frac{\partial \rho}{\partial p} \equiv -\{\rho, \mathcal{H}\}.$$

   For the gas particle confined to a one-dimensional trap, the Hamiltonian can be written as

   $$\mathcal{H} = \frac{p^2}{2m} + V(q_x) = \frac{p^2}{2m},$$

   since $V_{q_x} = 0$, and there is no motion in the y and z directions. With this Hamiltonian, Liouville's equation becomes

   $$\frac{\partial \rho}{\partial t} = -\frac{p}{m}\frac{\partial \rho}{\partial q},$$

   whose solution, subject to the specified initial conditions, is

   $$\rho(q, p, t) = \rho\left(q - \frac{p}{m}t, p, 0\right) = \delta\left(q - \frac{p}{m}t\right)f(p).$$

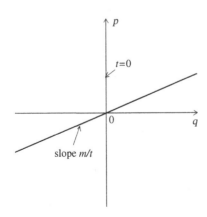

(b) Derive the expressions for the averages $\langle q^2 \rangle$ and $\langle p^2 \rangle$ at $t > 0$.

The expectation value for any observable $\mathcal{O}$ is

$$\langle \mathcal{O} \rangle = \int d\Gamma \mathcal{O} \rho(\Gamma, t),$$

and hence

$$\langle p^2 \rangle = \int p^2 f(p) \delta\left(q - \frac{p}{m}t\right) dp \, dq = \int p^2 f(p) dp$$

$$= \int_{-\infty}^{\infty} dp \, p^2 \frac{1}{\sqrt{2\pi m k_B T}} \exp\left(-\frac{p^2}{2m k_B T}\right) = m k_B T.$$

Likewise, we obtain

$$\langle q^2 \rangle = \int q^2 f(p) \delta\left(q - \frac{p}{m}t\right) dp \, dq = \int \left(\frac{p}{m}t\right)^2 f(p) dp$$

$$= \left(\frac{t}{m}\right)^2 \int p^2 f(p) dp = \frac{k_B T}{m} t^2.$$

(c) Suppose that hard walls are placed at $q = \pm Q$. Describe $\rho(q, p, t \gg \tau)$, where τ is an appropriately large relaxation time.

With hard walls placed at $q = \pm Q$, the appropriate relaxation time τ is the characteristic length between the containing walls divided by the characteristic velocity of the particle, that is,

$$\tau \sim \frac{2Q}{|\dot{q}|} = \frac{2Qm}{\sqrt{\langle p^2 \rangle}} = 2Q\sqrt{\frac{m}{k_B T}}.$$

Initially $\rho(q, p, t)$ resembles the distribution shown in part (a), but each time the particle hits the barrier, reflection changes p to $-p$. As time goes on, the slopes become less, and $\rho(q, p, t)$ becomes a set of closely spaced lines whose separation vanishes as $2mQ/t$.

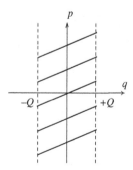

(d) A "coarse-grained" density $\tilde{\rho}$ is obtained by ignoring variations of ρ below some small resolution in the (q, p) plane; for example, by averaging ρ over cells of the resolution area. Find $\tilde{\rho}(q, p)$ for the situation in part (c), and show that it is stationary.

We can choose any resolution ε in (p, q) space, subdividing the plane into an array of pixels of this area. For any ε, after sufficiently long time many lines will pass through this area. Averaging over them leads to

$$\tilde{\rho}(q, p, t \gg \tau) = \frac{1}{2Q} f(p),$$

as (i) the density $f(p)$ at each p is always the same, and (ii) all points along $q \in [-Q, +Q]$ are equally likely. For the time variation of this coarse-grained density, we find

$$\frac{\partial \tilde{\rho}}{\partial t} = -\frac{p}{m} \frac{\partial \tilde{\rho}}{\partial q} = 0, \quad \text{that is,} \quad \tilde{\rho} \text{ is stationary.}$$

****

2. *Evolution of entropy*: the normalized ensemble density is a probability in the phase space Γ. This probability has an associated entropy $S(t) = -\int d\Gamma \rho(\Gamma, t) \ln \rho(\Gamma, t)$.

(a) Show that if $\rho(\Gamma, t)$ satisfies Liouville's equation for a Hamiltonian $\mathcal{H}$, $dS/dt = 0$.

The entropy associated with the phase space probability is

$$S(t) = -\int d\Gamma \rho(\Gamma, t) \ln \rho(\Gamma, t) = -\langle \ln \rho(\Gamma, t) \rangle.$$

Taking the derivative with respect to time gives

$$\frac{dS}{dt} = -\int d\Gamma \left(\frac{\partial \rho}{\partial t} \ln \rho + \rho \frac{1}{\rho} \frac{\partial \rho}{\partial t} \right) = -\int d\Gamma \frac{\partial \rho}{\partial t} (\ln \rho + 1).$$

Substituting the expression for $\partial \rho / \partial t$ obtained from Liouville's theorem gives

$$\frac{dS}{dt} = -\int d\Gamma \sum_{i=1}^{3N} \left(\frac{\partial \rho}{\partial p_i} \frac{\partial \mathcal{H}}{\partial q_i} - \frac{\partial \rho}{\partial q_i} \frac{\partial \mathcal{H}}{\partial p_i} \right) (\ln \rho + 1).$$

(Here the index i is used to label the three coordinates, as well as the N particles, and hence runs from 1 to $3N$.) Integrating the above expression by parts yields[1]

$$\frac{dS}{dt} = \int d\Gamma \sum_{i=1}^{3N} \left[\rho \frac{\partial}{\partial p_i} \left(\frac{\partial \mathcal{H}}{\partial q_i} (\ln \rho + 1) \right) - \rho \frac{\partial}{\partial q_i} \left(\frac{\partial \mathcal{H}}{\partial p_i} (\ln \rho + 1) \right) \right]$$

[1] *This is standard integration by parts, that is, $\int_b^a F dG = FG|_b^a - \int_b^a G dF$. Looking explicitly at one term in the expression to be integrated in this problem,*

$$\int \prod_{i=1}^{3N} dV_i \frac{\partial \rho}{\partial q_i} \frac{\partial \mathcal{H}}{\partial p_i} = \int dq_1 dp_1 \cdots dq_i dp_i \cdots dq_{3N} dp_{3N} \frac{\partial \rho}{\partial q_i} \frac{\partial \mathcal{H}}{\partial p_i},$$

we identify $dG = dq_i \frac{\partial \rho}{\partial q_i}$, and F with the remainder of the expression. Noting that $\rho(q_i) = 0$ at the boundaries of the box, we get

$$\int \prod_{i=1}^{3N} dV_i \frac{\partial \rho}{\partial q_i} \frac{\partial \mathcal{H}}{\partial p_i} = -\int \prod_{i=1}^{3N} dV_i \rho \frac{\partial}{\partial q_i} \frac{\partial \mathcal{H}}{\partial p_i}.$$

$$= \int d\Gamma \sum_{i=1}^{3N} \left[\rho \frac{\partial^2 \mathcal{H}}{\partial p_i \partial q_i}(\ln\rho+1) + \rho \frac{\partial \mathcal{H}}{\partial q_i}\frac{1}{\rho}\frac{\partial\rho}{\partial p_i} - \rho \frac{\partial^2 \mathcal{H}}{\partial q_i \partial p_i}(\ln\rho+1) - \rho \frac{\partial \mathcal{H}}{\partial p_i}\frac{1}{\rho}\frac{\partial\rho}{\partial q_i} \right]$$

$$= \int d\Gamma \sum_{i=1}^{3N} \left[\frac{\partial \mathcal{H}}{\partial q_i}\frac{\partial\rho}{\partial p_i} - \frac{\partial \mathcal{H}}{\partial p_i}\frac{\partial\rho}{\partial q_i} \right].$$

Integrating the final expression by parts gives

$$\frac{dS}{dt} = -\int d\Gamma \sum_{i=1}^{3N} \left[-\rho \frac{\partial^2 \mathcal{H}}{\partial p_i \partial q_i} + \rho \frac{\partial^2 \mathcal{H}}{\partial q_i \partial p_i} \right] = 0.$$

(b) Using the method of Lagrange multipliers, find the function $\rho_{max}(\Gamma)$ that maximizes the functional $S[\rho]$, subject to the constraint of fixed average energy, $\langle \mathcal{H} \rangle = \int d\Gamma \rho \mathcal{H} = E$.

There are two constraints, normalization and constant average energy, written respectively as

$$\int d\Gamma \rho(\Gamma) = 1, \qquad \text{and} \qquad \langle \mathcal{H} \rangle = \int d\Gamma \rho(\Gamma)\mathcal{H} = E.$$

Rewriting the expression for entropy,

$$S(t) = \int d\Gamma \rho(\Gamma)\left[-\ln\rho(\Gamma) - \alpha - \beta\mathcal{H} \right] + \alpha + \beta E,$$

where α and β are Lagrange multipliers used to enforce the two constraints. Extremizing the above expression with respect to the function $\rho(\Gamma)$ results in

$$\left.\frac{\partial S}{\partial \rho(\Gamma)}\right|_{\rho=\rho_{max}} = -\ln\rho_{max}(\Gamma) - \alpha - \beta\mathcal{H}(\Gamma) - 1 = 0.$$

The solution to this equation is

$$\ln\rho_{max} = -(\alpha+1) - \beta\mathcal{H},$$

which can be rewritten as

$$\rho_{max} = C\exp(-\beta\mathcal{H}), \quad \text{where} \quad C = e^{-(\alpha+1)}.$$

(c) Show that the solution to part (b) is stationary, that is, $\partial\rho_{max}/\partial t = 0$.

The density obtained in part (b) is stationary, as can easily be checked from

$$\frac{\partial\rho_{max}}{\partial t} = -\{\rho_{max}, \mathcal{H}\} = -\left\{ Ce^{-\beta\mathcal{H}}, \mathcal{H} \right\}$$

$$= \frac{\partial\mathcal{H}}{\partial p}C(-\beta)\frac{\partial\mathcal{H}}{\partial q}e^{-\beta\mathcal{H}} - \frac{\partial\mathcal{H}}{\partial q}C(-\beta)\frac{\partial\mathcal{H}}{\partial p}e^{-\beta\mathcal{H}} = 0.$$

(d) How can one reconcile the result in (a) with the observed increase in entropy as the system approaches the equilibrium density in (b)? (*Hint.* Think of the situation encountered in the previous problem.)

Liouville's equation preserves the information content of the PDF $\rho(\Gamma, t)$, and hence $S(t)$ does not increase in time. However, as illustrated in the example in Problem 1, the density becomes more finely dispersed in phase space. In the presence of any coarse graining of phase space, information disappears. The maximum entropy, corresponding to $\tilde{\rho}$, describes equilibrium in this sense.

****

3. *The Vlasov equation* is obtained in the limit of high particle density $n = N/V$, or large interparticle interaction range λ, such that $n\lambda^3 \gg 1$. In this limit, the collision terms are dropped from the left-hand side of the equations in the BBGKY hierarchy.

The BBGKY hierarchy

$$\left[\frac{\partial}{\partial t} + \sum_{n=1}^{s} \frac{\vec{p}_n}{m} \cdot \frac{\partial}{\partial \vec{q}_n} - \sum_{n=1}^{s} \left(\frac{\partial U}{\partial \vec{q}_n} + \sum_{l} \frac{\partial \mathcal{V}(\vec{q}_n - \vec{q}_l)}{\partial \vec{q}_n} \right) \cdot \frac{\partial}{\partial \vec{p}_n} \right] f_s$$

$$= \sum_{n=1}^{s} \int dV_{s+1} \frac{\partial \mathcal{V}(\vec{q}_n - \vec{q}_{s+1})}{\partial \vec{q}_n} \cdot \frac{\partial f_{s+1}}{\partial \vec{p}_n}$$

has the characteristic time scales

$$\begin{cases} \dfrac{1}{\tau_U} \sim \dfrac{\partial U}{\partial \vec{q}} \cdot \dfrac{\partial}{\partial \vec{p}} \sim \dfrac{v}{L}, \\[2mm] \dfrac{1}{\tau_c} \sim \dfrac{\partial \mathcal{V}}{\partial \vec{q}} \cdot \dfrac{\partial}{\partial \vec{p}} \sim \dfrac{v}{\lambda}, \\[2mm] \dfrac{1}{\tau_\times} \sim \displaystyle\int dx \dfrac{\partial \mathcal{V}}{\partial \vec{q}} \cdot \dfrac{\partial}{\partial \vec{p}} \dfrac{f_{s+1}}{f_s} \sim \dfrac{1}{\tau_c} \cdot n\lambda^3, \end{cases}$$

where $n\lambda^3$ is the number of particles within the interaction range λ, and v is a typical velocity. The Boltzmann equation is obtained in the dilute limit, $n\lambda^3 \ll 1$, by disregarding terms of order $1/\tau_\times \ll 1/\tau_c$. The Vlasov equation is obtained in the dense limit of $n\lambda^3 \gg 1$ by ignoring terms of order $1/\tau_c \ll 1/\tau_\times$.

(a) Assume that the N-body density is a product of one-particle densities, that is, $\rho = \prod_{i=1}^{N} \rho_1(\mathbf{x}_i, t)$, where $\mathbf{x}_i \equiv (\vec{p}_i, \vec{q}_i)$. Calculate the densities f_s, and their normalizations.

Let $\mathbf{x}_i$ denote the coordinates and momenta for particle i. Starting from the joint probability $\rho_N = \prod_{i=1}^{N} \rho_1(\mathbf{x}_i, t)$, for independent particles, we find

$$f_s = \frac{N!}{(N-s)!} \int \prod_{\alpha=s+1}^{N} dV_\alpha \rho_N = \frac{N!}{(N-s)!} \prod_{n=1}^{s} \rho_1(\mathbf{x}_n, t).$$

The normalizations follow from

$$\int d\Gamma \rho = 1, \quad \Longrightarrow \quad \int dV_1 \rho_1(\mathbf{x}, t) = 1,$$

and

$$\int \prod_{n=1}^{s} dV_n f_s = \frac{N!}{(N-s)!} \approx N^s \qquad \text{for} \qquad s \ll N.$$

(b) Show that once the collision terms are eliminated, all the equations in the BBGKY hierarchy are equivalent to the single equation

$$\left[\frac{\partial}{\partial t} + \frac{\vec{p}}{m} \cdot \frac{\partial}{\partial \vec{q}} - \frac{\partial U_{\text{eff}}}{\partial \vec{q}} \cdot \frac{\partial}{\partial \vec{p}} \right] f_1(\vec{p}, \vec{q}, t) = 0,$$

where

$$U_{\text{eff}}(\vec{q}, t) = U(\vec{q}) + \int d\mathbf{x}' \mathcal{V}(\vec{q} - \vec{q}') f_1(\mathbf{x}', t).$$

Noting that

$$\frac{f_{s+1}}{f_s} = \frac{(N-s)!}{(N-s-1)!}\rho_1(\mathbf{x}_{s+1}),$$

the reduced BBGKY hierarchy is

$$\left[\frac{\partial}{\partial t} + \sum_{n=1}^{s}\left(\frac{\vec{p}_n}{m}\cdot\frac{\partial}{\partial\vec{q}_n} - \frac{\partial U}{\partial\vec{q}_n}\cdot\frac{\partial}{\partial\vec{p}_n}\right)\right]f_s$$

$$\approx \sum_{n=1}^{s}\int dV_{s+1}\frac{\partial\mathcal{V}(\vec{q}_n-\vec{q}_{s+1})}{\partial\vec{q}_n}\cdot\frac{\partial}{\partial\vec{p}_n}\left[(N-s)f_s\rho_1(\mathbf{x}_{s+1})\right]$$

$$\approx \sum_{n=1}^{s}\frac{\partial}{\partial\vec{q}_n}\left[\int dV_{s+1}\rho_1(\mathbf{x}_{s+1})\mathcal{V}(\vec{q}_n-\vec{q}_{s+1})\cdot N\right]\frac{\partial}{\partial\vec{p}_n}f_s,$$

where we have used the approximation $(N-s)\approx N$ for $N \gg s$. Rewriting the above expression,

$$\left[\frac{\partial}{\partial t} + \sum_{n=1}^{s}\left(\frac{\vec{p}_n}{m}\cdot\frac{\partial}{\partial\vec{q}_n} - \frac{\partial U_{\text{eff}}}{\partial\vec{q}_n}\cdot\frac{\partial}{\partial\vec{p}_n}\right)\right]f_s = 0,$$

where

$$U_{\text{eff}} = U(\vec{q}) + N\int dV'\mathcal{V}(\vec{q}-\vec{q}')\rho_1(\mathbf{x}', t).$$

(c) Now consider N particles confined to a box of volume V, with no additional potential. Show that $f_1(\vec{q}, \vec{p}) = g(\vec{p})/V$ is a stationary solution to the Vlasov equation *for any* $g(\vec{p})$. Why is there no relaxation toward equilibrium for $g(\vec{p})$?

Starting from

$$\rho_1 = g(\vec{p})/V,$$

we obtain

$$\mathcal{H}_{\text{eff}} = \sum_{i=1}^{N}\left[\frac{\vec{p}_i^{\,2}}{2m} + U_{\text{eff}}(\vec{q}_i)\right],$$

with

$$U_{\text{eff}} = 0 + N\int dV'\mathcal{V}(\vec{q}-\vec{q}')\frac{1}{V}g(\vec{p}) = \frac{N}{V}\int d^3q\mathcal{V}(\vec{q}).$$

(We have taken advantage of the normalization $\int d^3 pg(\vec{p}) = 1$.) Substituting into the Vlasov equation yields

$$\left(\frac{\partial}{\partial t} + \frac{\vec{p}}{m}\cdot\frac{\partial}{\partial\vec{q}}\right)\rho_1 = 0.$$

There is no relaxation toward equilibrium because there are no collisions that allow $g(\vec{p})$ to relax. The momentum of each particle is conserved by $\mathcal{H}_{\text{eff}}$; that is, $\{\rho_1, \mathcal{H}_{\text{eff}}\} = 0$, preventing its change.

****

4. *Two-component plasma*: consider a *neutral* mixture of N ions of charge $+e$ and mass m_+, and N electrons of charge $-e$ and mass m_-, in a volume $V = N/n_0$.

(a) Show that the Vlasov equations for this two-component system are

$$
\begin{cases}
\left[\dfrac{\partial}{\partial t} + \dfrac{\vec{p}}{m_+} \cdot \dfrac{\partial}{\partial \vec{q}} + e \dfrac{\partial \Phi_{\mathrm{eff}}}{\partial \vec{q}} \cdot \dfrac{\partial}{\partial \vec{p}} \right] f_+(\vec{p}, \vec{q}, t) = 0 \\[2ex]
\left[\dfrac{\partial}{\partial t} + \dfrac{\vec{p}}{m_-} \cdot \dfrac{\partial}{\partial \vec{q}} - e \dfrac{\partial \Phi_{\mathrm{eff}}}{\partial \vec{q}} \cdot \dfrac{\partial}{\partial \vec{p}} \right] f_-(\vec{p}, \vec{q}, t) = 0
\end{cases},
$$

where the effective Coulomb potential is given by

$$
\Phi_{\mathrm{eff}}(\vec{q}, t) = \Phi_{\mathrm{ext}}(\vec{q}) + e \int d\mathbf{x}' \, C(\vec{q} - \vec{q}\,') \left[f_+(\mathbf{x}', t) - f_-(\mathbf{x}', t) \right].
$$

Here, Φ_{ext} is the potential set up by the external charges, and the Coulomb potential $C(\vec{q})$ satisfies the differential equation $\nabla^2 C = 4\pi \delta^3(\vec{q})$.

The Hamiltonian for the two-component mixture is

$$
\mathcal{H} = \sum_{i=1}^{N} \left[\frac{\vec{p}_i^{\,2}}{2m_+} + \frac{\vec{p}_i^{\,2}}{2m_-} \right] + \sum_{i,j=1}^{2N} e_i e_j \frac{1}{|\vec{q}_i - \vec{q}_j|} + \sum_{i=1}^{2N} e_i \Phi_{\mathrm{ext}}(\vec{q}_i),
$$

where $C(\vec{q}_i - \vec{q}_j) = 1/|\vec{q}_i - \vec{q}_j|$, resulting in

$$
\frac{\partial \mathcal{H}}{\partial \vec{q}_i} = e_i \frac{\partial \Phi_{\mathrm{ext}}}{\partial \vec{q}_i} + e_i \sum_{j \neq i} e_j \frac{\partial}{\partial \vec{q}_i} C(\vec{q}_i - \vec{q}_j).
$$

Substituting this into the Vlasov equation, we obtain

$$
\begin{cases}
\left[\dfrac{\partial}{\partial t} + \dfrac{\vec{p}}{m_+} \cdot \dfrac{\partial}{\partial \vec{q}} + e \dfrac{\partial \Phi_{\mathrm{eff}}}{\partial \vec{q}} \cdot \dfrac{\partial}{\partial \vec{p}} \right] f_+(\vec{p}, \vec{q}, t) = 0, \\[2ex]
\left[\dfrac{\partial}{\partial t} + \dfrac{\vec{p}}{m_-} \cdot \dfrac{\partial}{\partial \vec{q}} - e \dfrac{\partial \Phi_{\mathrm{eff}}}{\partial \vec{q}} \cdot \dfrac{\partial}{\partial \vec{p}} \right] f_-(\vec{p}, \vec{q}, t) = 0.
\end{cases}
$$

(b) Assume that the one-particle densities have the stationary forms $f_\pm = g_\pm(\vec{p}) n_\pm(\vec{q})$. Show that the effective potential satisfies the equation

$$
\nabla^2 \Phi_{\mathrm{eff}} = 4\pi \rho_{\mathrm{ext}} + 4\pi e \left(n_+(\vec{q}) - n_-(\vec{q}) \right),
$$

where ρ_{ext} is the external charge density.

Setting $f_\pm(\vec{p}, \vec{q}) = g_\pm(\vec{p}) n_\pm(\vec{q})$, and using $\int d^3 p \, g_\pm(\vec{p}) = 1$, the integrals in the effective potential simplify to

$$
\Phi_{\mathrm{eff}}(\vec{q}, t) = \Phi_{\mathrm{ext}}(\vec{q}) + e \int d^3 q' \, C(\vec{q} - \vec{q}\,') \left[n_+(\vec{q}\,') - n_-(\vec{q}\,') \right].
$$

Apply ∇^2 to the above equation, and use $\nabla^2 \Phi_{\mathrm{ext}} = 4\pi \rho_{\mathrm{ext}}$ and $\nabla^2 C(\vec{q} - \vec{q}\,') = 4\pi \delta^3 (\vec{q} - \vec{q}\,')$, to obtain

$$
\nabla^2 \phi_{\mathrm{eff}} = 4\pi \rho_{\mathrm{ext}} + 4\pi e \left[n_+(\vec{q}) - n_-(\vec{q}) \right].
$$

(c) Further assuming that the densities relax to the equilibrium Boltzmann weights
$n_\pm(\vec{q}) = n_0 \exp\left[\pm\beta e\Phi_{\text{eff}}(\vec{q})\right]$ leads to the self-consistency condition

$$\nabla^2\Phi_{\text{eff}} = 4\pi\left[\rho_{\text{ext}} + n_0 e\left(e^{\beta e\Phi_{\text{eff}}} - e^{-\beta e\Phi_{\text{eff}}}\right)\right],$$

known as the *Poisson–Boltzmann equation*. Due to its non-linear form, it is generally
not possible to solve the Poisson–Boltzmann equation. By linearizing the exponentials,
one obtains the simpler *Debye* equation

$$\nabla^2\Phi_{\text{eff}} = 4\pi\rho_{\text{ext}} + \Phi_{\text{eff}}/\lambda^2.$$

Give the expression for the *Debye screening length* λ.

Linearizing the Boltzmann weights gives

$$n_\pm = n_o \exp[\mp\beta e\Phi_{\text{eff}}(\vec{q})] \approx n_o\left[1 \mp \beta e\Phi_{\text{eff}}\right],$$

resulting in

$$\nabla^2\Phi_{\text{eff}} = 4\pi\rho_{\text{ext}} + \frac{1}{\lambda^2}\Phi_{\text{eff}},$$

with the screening length given by

$$\lambda^2 = \frac{k_B T}{8\pi n_o e^2}.$$

(d) Show that the Debye equation has the general solution

$$\Phi_{\text{eff}}(\vec{q}) = \int d^3\vec{q}' G(\vec{q} - \vec{q}')\rho_{\text{ext}}(\vec{q}'),$$

where $G(\vec{q}) = \exp(-|\vec{q}|/\lambda)/|\vec{q}|$ is the screened Coulomb potential.

We want to show that the Debye equation has the general solution

$$\Phi_{\text{eff}}(\vec{q}) = \int d^3\vec{q} G(\vec{q} - \vec{q}')\rho_{\text{ext}}(\vec{q}'),$$

where

$$G(\vec{q}) = \frac{\exp(-|q|/\lambda)}{|q|}.$$

*Effectively, we want to show that $\nabla^2 G = G/\lambda^2$ for $\vec{q} \neq 0$. In spherical coordinates,
$G = \exp(-r/\lambda)/r$. Evaluating ∇^2 in spherical coordinates gives*

$$\nabla^2 G = \frac{1}{r^2}\frac{\partial}{\partial r}\left(r^2\frac{\partial G}{\partial r}\right) = \frac{1}{r^2}\frac{\partial}{\partial r}r^2\left[-\frac{1}{\lambda}\frac{e^{-r/\lambda}}{r} - \frac{e^{-r/\lambda}}{r^2}\right]$$

$$= \frac{1}{r^2}\left[\frac{1}{\lambda^2}re^{-r/\lambda} - \frac{1}{\lambda}e^{-r/\lambda} + \frac{1}{\lambda}e^{-r/\lambda}\right] = \frac{1}{\lambda^2}\frac{e^{-r/\lambda}}{r} = \frac{G}{\lambda^2}.$$

(e) Give the condition for the self-consistency of the Vlasov approximation, and interpret
it in terms of the interparticle spacing.

The Vlasov equation assumes the limit $n_o \lambda^3 \gg 1$, which requires that

$$\frac{(k_B T)^{3/2}}{n_o^{1/2} e^3} \gg 1, \quad \implies \quad \frac{e^2}{k_B T} \ll n_o^{-1/3} \sim \ell,$$

where ℓ is the interparticle spacing. In terms of the interparticle spacing, the self-consistency condition is

$$\frac{e^2}{\ell} \ll k_B T,$$

that is, the interaction energy is much less than the kinetic (thermal) energy.

(f) Show that the characteristic relaxation time ($\tau \approx \lambda/v$) is temperature-independent. What property of the plasma is it related to?

A characteristic time is obtained from

$$\tau \sim \frac{\lambda}{v} \sim \sqrt{\frac{k_B T}{n_o e^2}} \cdot \sqrt{\frac{m}{k_B T}} \sim \sqrt{\frac{m}{n_o e^2}} \sim \frac{1}{\omega_p},$$

where ω_p is the plasma frequency.

$$* * * * * * * *$$

5. *Two-dimensional electron gas in a magnetic field*: when donor atoms (such as P or As) are added to a semiconductor (e.g., Si or Ge), their conduction electrons can be thermally excited to move freely in the host lattice. By growing layers of different materials, it is possible to generate a spatially varying potential (work-function) that traps electrons at the boundaries between layers. In the following, we shall treat the trapped electrons as a gas of classical particles *in two dimensions.*

If the layer of electrons is sufficiently separated from the donors, the main source of scattering is from electron–electron collisions.

(a) The Hamiltonian for non-interacting free electrons in a magnetic field has the form

$$\mathcal{H} = \sum_i \left[\frac{\left(\vec{p}_i - e\vec{A} \right)^2}{2m} \pm \mu_B |\vec{B}| \right].$$

(The two signs correspond to electron spins parallel or anti-parallel to the field.) The vector potential $\vec{A} = \vec{B} \times \vec{q}/2$ describes a uniform magnetic field $\vec{B}$. Obtain the classical equations of motion, and show that they describe rotation of electrons in cyclotron orbits in a plane orthogonal to $\vec{B}$.

The Hamiltonian for non-interacting free electrons in a magnetic field has the form

$$\mathcal{H} = \sum_i \left[\frac{\left(\vec{p}_i + e\vec{A} \right)^2}{2m} \pm \mu_B |\vec{B}| \right],$$

or in expanded form

$$\mathcal{H} = \frac{p^2}{2m} + \frac{e}{m} \vec{p} \cdot \vec{A} + \frac{e^2}{2m} \vec{A}^2 \pm \mu_B |\vec{B}|.$$

Substituting $\vec{A} = \vec{B} \times \vec{q}/2$, results in

$$\mathcal{H} = \frac{p^2}{2m} + \frac{e}{2m}\vec{p} \cdot \vec{B} \times \vec{q} + \frac{e^2}{8m}\left(\vec{B} \times \vec{q}\right)^2 \pm \mu_B |\vec{B}|$$

$$= \frac{p^2}{2m} + \frac{e}{2m}\vec{p} \times \vec{B} \cdot \vec{q} + \frac{e^2}{8m}\left(B^2 q^2 - (\vec{B} \cdot \vec{q})^2\right) \pm \mu_B |\vec{B}|.$$

Using the canonical equations, $\dot{\vec{q}} = \partial \mathcal{H}/\vec{p}$ and $\dot{\vec{p}} = -\partial \mathcal{H}/\vec{q}$, we find

$$\begin{cases} \dot{\vec{q}} = \dfrac{\partial \mathcal{H}}{\partial \vec{p}} = \dfrac{\vec{p}}{m} + \dfrac{e}{2m}\vec{B} \times \vec{q}, \quad \Longrightarrow \quad \vec{p} = m\dot{\vec{q}} - \dfrac{e}{2}\vec{B} \times \vec{q}, \\[2mm] \dot{\vec{p}} = -\dfrac{\partial \mathcal{H}}{\partial \vec{q}} = -\dfrac{e}{2m}\vec{p} \times \vec{B} - \dfrac{e^2}{4m}B^2\vec{q} + \dfrac{e^2}{4m}\left(\vec{B} \cdot \vec{q}\right)\vec{B}. \end{cases}$$

Differentiating the first expression obtained for $\vec{p}$, and setting it equal to the second expression for $\dot{\vec{p}}$ above, gives

$$m\ddot{\vec{q}} - \frac{e}{2}\vec{B} \times \dot{\vec{q}} = -\frac{e}{2m}\left(m\dot{\vec{q}} - \frac{e}{2}\vec{B} \times \vec{q}\right) \times \vec{B} - \frac{e^2}{4m}|\vec{B}|^2\vec{q} + \frac{e^2}{4m}\left(\vec{B} \cdot \vec{q}\right)\vec{B}.$$

Simplifying the above expression, using $\vec{B} \times \vec{q} \times \vec{B} = B^2\vec{q} - \left(\vec{B} \cdot \vec{q}\right)\vec{B}$, leads to

$$m\ddot{\vec{q}} = e\vec{B} \times \dot{\vec{q}}.$$

This describes the rotation of electrons in cyclotron orbits,

$$\ddot{\vec{q}} = \vec{\omega}_c \times \dot{\vec{q}},$$

where $\vec{\omega}_c = e\vec{B}/m$; that is, rotations are in the plane perpendicular to $\vec{B}$.

(b) Write down heuristically (i.e., not through a step-by-step derivation) the Boltzmann equations for the densities $f_\uparrow(\vec{p}, \vec{q}, t)$ and $f_\downarrow(\vec{p}, \vec{q}, t)$ of electrons with up and down spins, in terms of the two cross-sections $\sigma \equiv \sigma_{\uparrow\uparrow} = \sigma_{\downarrow\downarrow}$, and $\sigma_\times \equiv \sigma_{\uparrow\downarrow}$, of *spin-conserving* collisions.

Consider the classes of collisions described by cross-sections $\sigma \equiv \sigma_{\uparrow\uparrow} = \sigma_{\downarrow\downarrow}$, and $\sigma_\times \equiv \sigma_{\uparrow\downarrow}$. We can write the Boltzmann equations for the densities as

$$\frac{\partial f_\uparrow}{\partial t} - \{\mathcal{H}_\uparrow, f_\uparrow\} = \int d^2 p_2 d\Omega |v_1 - v_2| \left\{ \frac{d\sigma}{d\Omega}\left[f_\uparrow(\vec{p}_1')f_\uparrow(\vec{p}_2') - f_\uparrow(\vec{p}_1)f_\uparrow(\vec{p}_2)\right] \right.$$

$$\left. + \frac{d\sigma_\times}{d\Omega}\left[f_\uparrow(\vec{p}_1')f_\downarrow(\vec{p}_2') - f_\uparrow(\vec{p}_1)f_\downarrow(\vec{p}_2)\right] \right\},$$

and

$$\frac{\partial f_\downarrow}{\partial t} - \{\mathcal{H}_\downarrow, f_\downarrow\} = \int d^2 p_2 d\Omega |v_1 - v_2| \left\{ \frac{d\sigma}{d\Omega}\left[f_\downarrow(\vec{p}_1')f_\downarrow(\vec{p}_2') - f_\downarrow(\vec{p}_1)f_\downarrow(\vec{p}_2)\right] \right.$$

$$\left. + \frac{d\sigma_\times}{d\Omega}\left[f_\downarrow(\vec{p}_1')f_\uparrow(\vec{p}_2') - f_\downarrow(\vec{p}_1)f_\uparrow(\vec{p}_2)\right] \right\}.$$

(c) Show that $dH/dt \leq 0$, where $H = H_\uparrow + H_\downarrow$ is the sum of the corresponding H-functions.

The usual Boltzmann H-theorem states that $dH/dt \leq 0$, where

$$H = \int d^2q d^2p f(\vec{q}, \vec{p}, t) \ln f(\vec{q}, \vec{p}, t).$$

For the electron gas in a magnetic field, the H-function can be generalized to

$$H = \int d^2q d^2p \left[f_\uparrow \ln f_\uparrow + f_\downarrow \ln f_\downarrow \right],$$

where the condition $dH/dt \leq 0$ is proved as follows:

$$\frac{dH}{dt} = \int d^2q d^2p \left[\frac{\partial f_\uparrow}{\partial t} (\ln f_\uparrow + 1) + \frac{\partial f_\downarrow}{\partial t} (\ln f_\downarrow + 1) \right]$$

$$= \int d^2q d^2p \left[(\ln f_\uparrow + 1) \left(\{ f_\uparrow, \mathcal{H}_\uparrow \} + C_{\uparrow\uparrow} + C_{\uparrow\downarrow} \right) \right.$$

$$\left. + (\ln f_\downarrow + 1) \left(\{ f_\downarrow, \mathcal{H}_\downarrow \} + C_{\downarrow\downarrow} + C_{\downarrow\uparrow} \right) \right],$$

with $C_{\uparrow\uparrow}$, etc., defined via the right-hand side of the equations in part (b). Hence

$$\frac{dH}{dt} = \int d^2q d^2p \left(\ln f_\uparrow + 1 \right) \left(C_{\uparrow\uparrow} + C_{\uparrow\downarrow} \right) + \left(\ln f_\downarrow + 1 \right) \left(C_{\downarrow\downarrow} + C_{\downarrow\uparrow} \right)$$

$$= \int d^2q d^2p \left(\ln f_\uparrow + 1 \right) C_{\uparrow\uparrow} + \left(\ln f_\downarrow + 1 \right) C_{\downarrow\downarrow}$$

$$+ \left(\ln f_\uparrow + 1 \right) C_{\uparrow\downarrow} + \left(\ln f_\downarrow + 1 \right) C_{\downarrow\uparrow}$$

$$\equiv \frac{dH_{\uparrow\uparrow}}{dt} + \frac{dH_{\downarrow\downarrow}}{dt} + \frac{d}{dt} \left(H_{\uparrow\downarrow} + H_{\downarrow\uparrow} \right),$$

where the H's are in correspondence to the integrals for the collisions. We have also made use of the fact that $\int d^2p d^2q \{ f_\uparrow, \mathcal{H}_\uparrow \} = \int d^2p d^2q \{ f_\downarrow, \mathcal{H}_\downarrow \} = 0$. Dealing with each of the terms in the final equation individually,

$$\frac{dH_{\uparrow\uparrow}}{dt} = \int d^2q d^2p_1 d^2p_2 d\Omega |v_1 - v_2| \left(\ln f_\uparrow + 1 \right) \frac{d\sigma}{d\Omega} \left[f_\uparrow(\vec{p}_1{}') f_\uparrow(\vec{p}_2{}') \right.$$

$$\left. - f_\uparrow(\vec{p}_1) f_\uparrow(\vec{p}_2) \right]$$

After symmetrizing this equation, as done in the text,

$$\frac{dH_{\uparrow\uparrow}}{dt} = -\frac{1}{4} \int d^2q d^2p_1 d^2p_2 d\Omega |v_1 - v_2| \frac{d\sigma}{d\Omega} \left[\ln f_\uparrow(\vec{p}_1) f_\uparrow(\vec{p}_2) - \ln f_\uparrow(\vec{p}_1') f_\uparrow(\vec{p}_2') \right]$$

$$\cdot \left[f_\uparrow(\vec{p}_1) f_\uparrow(\vec{p}_2) - f_\uparrow(\vec{p}_1') f_\uparrow(\vec{p}_2') \right] \leq 0.$$

Similarly, $dH_{\downarrow\downarrow}/dt \leq 0$. Dealing with the two remaining terms,

$$\frac{dH_{\uparrow\downarrow}}{dt} = \int d^2q d^2p_1 d^2p_2 d\Omega |v_1 - v_2| \left[\ln f_\uparrow(\vec{p}_1) + 1 \right]$$

$$\frac{d\sigma_\times}{d\Omega} \left[f_\uparrow(\vec{p}_1') f_\downarrow(\vec{p}_2') - f_\uparrow(\vec{p}_1) f_\downarrow(\vec{p}_2) \right]$$

$$= \int d^2q d^2p_1 d^2p_2 d\Omega |v_1 - v_2| \left[\ln f_\uparrow(\vec{p}_1') + 1 \right]$$

$$\frac{d\sigma_\times}{d\Omega} \left[f_\uparrow(\vec{p}_1) f_\downarrow(\vec{p}_2) - f_\uparrow(\vec{p}_1') f_\downarrow(\vec{p}_2') \right],$$

where we have exchanged $(\vec{p}_1, \vec{p}_2 \leftrightarrow \vec{p}_1', \vec{p}_2')$. *Averaging these two expressions together,*

$$\frac{dH_{\uparrow\downarrow}}{dt} = -\frac{1}{2} \int d^2q\, d^2p_1\, d^2p_2\, d\Omega |v_1 - v_2| \frac{d\sigma_\times}{d\Omega} \left[\ln f_\uparrow(\vec{p}_1) - \ln f_\uparrow(\vec{p}_1') \right]$$
$$\cdot \left[f_\uparrow(\vec{p}_1) f_\downarrow(\vec{p}_2) - f_\uparrow(\vec{p}_1') f_\downarrow(\vec{p}_2') \right].$$

Likewise,

$$\frac{dH_{\downarrow\uparrow}}{dt} = -\frac{1}{2} \int d^2q\, d^2p_1\, d^2p_2\, d\Omega |v_1 - v_2| \frac{d\sigma_\times}{d\Omega} \left[\ln f_\downarrow(\vec{p}_2) - \ln f_\downarrow(\vec{p}_2') \right]$$
$$\cdot \left[f_\downarrow(\vec{p}_2) f_\uparrow(\vec{p}_1) - f_\downarrow(\vec{p}_2') f_\uparrow(\vec{p}_1') \right].$$

Combining these two expressions,

$$\frac{d}{dt} \left(H_{\uparrow\downarrow} + H_{\downarrow\uparrow} \right) = -\frac{1}{4} \int d^2q\, d^2p_1\, d^2p_2\, d\Omega |v_1 - v_2| \frac{d\sigma_\times}{d\Omega}$$
$$\left[\ln f_\uparrow(\vec{p}_1) f_\downarrow(\vec{p}_2) - \ln f_\uparrow(\vec{p}_1') f_\downarrow(\vec{p}_2') \right] \left[f_\uparrow(\vec{p}_1) f_\downarrow(\vec{p}_2) - f_\uparrow(\vec{p}_1') f_\downarrow(\vec{p}_2') \right] \le 0.$$

Since each contribution is separately negative, we have

$$\frac{dH}{dt} = \frac{dH_{\uparrow\uparrow}}{dt} + \frac{dH_{\downarrow\downarrow}}{dt} + \frac{d}{dt} \left(H_{\uparrow\downarrow} + H_{\downarrow\uparrow} \right) \le 0.$$

(d) Show that $dH/dt = 0$ for any $\ln f$ that is, *at each location*, a linear combination of quantities conserved in the collisions.

For $dH/dt = 0$ we need each of the three square brackets in the previous derivation to be zero. The first two contributions, from $dH_{\uparrow\downarrow}/dt$ and $dH_{\downarrow\downarrow}/dt$, are similar to those discussed in the notes for a single particle, and vanish for any $\ln f$ that is a linear combination of quantities conserved in collisions

$$\ln f_\alpha = \sum_i a_i^\alpha(\vec{q}) \chi_i(\vec{p}),$$

where $\alpha = (\uparrow$ or $\downarrow)$. Clearly at each location $\vec{q}$, for such f_α,

$$\ln f_\alpha(\vec{p}_1) + \ln f_\alpha(\vec{p}_2) = \ln f_\alpha(\vec{p}_1') + \ln f_\alpha(\vec{p}_2').$$

If we consider only the first two terms of $dH/dt = 0$, the coefficients $a_i^\alpha(\vec{q})$ can vary with both $\vec{q}$ and $\alpha = (\uparrow$ or $\downarrow)$. This changes when we consider the third term $d\left(H_{\uparrow\downarrow} + H_{\downarrow\uparrow}\right)/dt$. The conservations of momentum and kinetic energy constrain the corresponding four functions to be the same, that is, they require $a_i^\uparrow(\vec{q}) = a_i^\downarrow(\vec{q})$. There is, however, no similar constraint for the overall constant that comes from particle number conservation, as the numbers of spin-up and spin-down particles are separately conserved. This implies that the densities of up and down spins can be different in the final equilibrium, while the two systems must share the same velocity and temperature.

(e) Show that the streaming terms in the Boltzmann equation are zero for any function that depends only on the quantities conserved by the one-body Hamiltonians.

The Boltzmann equation is

$$\frac{\partial f_\alpha}{\partial t} = -\{f_\alpha, \mathcal{H}_\alpha\} + C_{\alpha\alpha} + C_{\alpha\beta},$$

where the right-hand side consists of streaming terms $\{f_\alpha, \mathcal{H}_\alpha\}$, *and collision terms C. Let* I_i *denote any quantity conserved by the one-body Hamiltonian, that is,* $\{I_i, \mathcal{H}_\alpha\} = 0$. *Consider* f_α, *which is a function only of the* I_i's

$$f_\alpha \equiv f_\alpha(I_1, I_2, \cdots).$$

Then

$$\{f_\alpha, \mathcal{H}_\alpha\} = \sum_j \frac{\partial f_\alpha}{\partial I_j} \{I_j, \mathcal{H}_\alpha\} = 0.$$

(f) Show that angular momentum $\vec{L} = \vec{q} \times \vec{p}$ is conserved during, and away from, collisions.

Conservation of momentum for a collision at $\vec{q}$

$$(\vec{p}_1 + \vec{p}_2) = (\vec{p}_1' + \vec{p}_2'),$$

implies

$$\vec{q} \times (\vec{p}_1 + \vec{p}_2) = \vec{q} \times (\vec{p}_1' + \vec{p}_2'),$$

or

$$\vec{L}_1 + \vec{L}_2 = \vec{L}_1' + \vec{L}_2',$$

where we have used $\vec{L}_i = \vec{q} \times \vec{p}_i$. *Hence angular momentum* $\vec{L}$ *is conserved during collisions. Note that only the z component* L_z *is present for electrons moving in two dimensions,* $\vec{q} \equiv (x_1, x_2)$, *as is the case for the electron gas studied in this problem. Consider the Hamiltonian discussed in (a)*

$$\mathcal{H} = \frac{p^2}{2m} + \frac{e}{2m} \vec{p} \times \vec{B} \cdot \vec{q} + \frac{e^2}{8m} \left(B^2 q^2 - (\vec{B} \cdot \vec{q})^2 \right) \pm \mu_B |\vec{B}|.$$

Let us evaluate the Poisson brackets of the individual terms with $L_z = \vec{q} \times \vec{p} \mid_z$. *The first term is*

$$-\{|\vec{p}|^2, \vec{q} \times \vec{p}\} = \varepsilon_{ijk} \{p_l p_l, x_j p_k\} = \varepsilon_{ijk} 2 p_l \frac{\partial}{\partial x_l}(x_j p_k) = 2\varepsilon_{ilk} p_l p_k = 0,$$

where we have used $\varepsilon_{ijk} p_j p_k = 0$ *since* $p_j p_k = p_k p_j$ *is symmetric. The second term is proportional to* L_z,

$$\left\{\vec{p} \times \vec{B} \cdot \vec{q}, L_z\right\} = \{B_z L_z, L_z\} = 0.$$

The final terms are proportional to q^2, *and* $\{q^2, \vec{q} \times \vec{p}\} = 0$ *for the same reason that* $\{p^2, \vec{q} \times \vec{p}\} = 0$, *leading to*

$$\{\mathcal{H}, \vec{q} \times \vec{p}\} = 0.$$

Hence angular momentum is conserved away from collisions as well.

(g) Write down the most general form for the equilibrium distribution functions for particles confined to a circularly symmetric potential.

The most general form of the equilibrium distribution functions must set both the collision terms, and the streaming terms, to zero. Based on the results of the previous parts, we thus obtain

$$f_\alpha = A_\alpha \exp\left[-\beta \mathcal{H}_\alpha - \gamma L_z\right].$$

The collision terms allow for the possibility of a term $-\vec{u} \cdot \vec{p}$ in the exponent, corresponding to an average velocity. Such a term will not commute with the potential set up by a stationary box, and is thus ruled out by the streaming terms. On the other hand, the angular momentum does commute with a circular potential $\{V(\vec{q}), L\} = 0$, and is allowed by the streaming terms. A non-zero γ describes the electron gas rotating in a circular box.

(h) How is the result in part (g) modified by including scattering from magnetic and non-magnetic impurities?

Scattering from any impurity removes the conservation of $\vec{p}$, and hence $\vec{L}$, in collisions. The γ term will no longer be needed. Scattering from magnetic impurities mixes populations of up and down spins, necessitating $A_\uparrow = A_\downarrow$; non-magnetic impurities do not have this effect.

(i) Do conservation of spin and angular momentum lead to new hydrodynamic equations?

Conservation of angular momentum is related to conservation of $\vec{p}$, as shown in (f), and hence does not lead to any new equation. In contrast, conservation of spin leads to an additional hydrodynamic equation involving the magnetization, which is proportional to $(n_\uparrow - n_\downarrow)$.

****

6. *The Lorentz gas* describes non-interacting particles colliding with a fixed set of scatterers. It is a good model for scattering of electrons from donor impurities. Consider a uniform two-dimensional density n_0 of fixed impurities, which are hard circles of radius a.

(a) Show that the differential cross-section of a hard circle scattering through an angle θ is

$$d\sigma = \frac{a}{2} \sin \frac{\theta}{2} d\theta,$$

and calculate the total cross-section.

Let b denote the impact parameter, which (see figure) is related to the angle θ between $\vec{p}'$ and $\vec{p}$ by

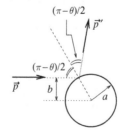

$$b(\theta) = a \sin \frac{\pi - \theta}{2} = a \cos \frac{\theta}{2}.$$

The differential cross-section is then given by

$$d\sigma = 2|db| = a \sin \frac{\theta}{2} d\theta.$$

Hence the total cross-section

$$\sigma_{\text{tot}} = \int_0^\pi d\theta a \sin \frac{\theta}{2} = 2a \left[-\cos \frac{\theta}{2} \right]_0^\pi = 2a.$$

(b) Write down the Boltzmann equation for the one-particle density $f(\vec{q}, \vec{p}, t)$ of the Lorentz gas (including only collisions with the fixed impurities). *(Ignore the electron spin.)*

   The corresponding Boltzmann equation is

$$\frac{\partial f}{\partial t} + \frac{\vec{p}}{m} \cdot \frac{\partial f}{\partial \vec{q}} + \vec{F} \cdot \frac{\partial f}{\partial \vec{p}}$$

$$= \int d\theta \frac{d\sigma}{d\theta} \frac{|\vec{p}|}{m} n_0 \left[-f(\vec{p}) + f(\vec{p}') \right]$$

$$= \frac{n_o |\vec{p}|}{m} \int d\theta \frac{d\sigma}{d\theta} \left[f(\vec{p}') - f(\vec{p}) \right] \equiv C \left[f(\vec{p}) \right].$$

(c) Find the eigenfunctions and eigenvalues of the collision operator.

(d) Using the definitions $\vec{F} \equiv -\partial U / \partial \vec{q}$, and

$$n(\vec{q}, t) = \int d^2 \vec{p} f(\vec{q}, \vec{p}, t), \qquad \text{and} \qquad \langle g(\vec{q}, t) \rangle = \frac{1}{n(\vec{q}, t)} \int d^2 \vec{p} f(\vec{q}, \vec{p}, t) g(\vec{q}, t),$$

   show that for any function $\chi(|\vec{p}|)$, we have

$$\frac{\partial}{\partial t} (n \langle \chi \rangle) + \frac{\partial}{\partial \vec{q}} \cdot \left(n \left\langle \frac{\vec{p}}{m} \chi \right\rangle \right) = \vec{F} \cdot \left(n \left\langle \frac{\partial \chi}{\partial \vec{p}} \right\rangle \right).$$

   Using the above definitions we can write

$$\frac{d}{dt} (n \langle \chi(|\vec{p}|) \rangle) = \int d^2 p \chi(|\vec{p}|) \left[-\frac{\vec{p}}{m} \cdot \frac{\partial f}{\partial \vec{q}} - \vec{F} \cdot \frac{\partial f}{\partial \vec{p}} \right.$$

$$\left. + \int d\theta \frac{d\sigma}{d\theta} \frac{|\vec{p}|}{m} n_o \left(f(\vec{p}) - f(\vec{p}') \right) \right]$$

$$= -\frac{\partial}{\partial \vec{q}} \cdot \left(n \left\langle \frac{\vec{p}}{m} \chi \right\rangle \right) + \vec{F} \cdot \left(n \left\langle \frac{\partial \chi}{\partial \vec{p}} \right\rangle \right).$$

   Rewriting this final expression gives the hydrodynamic equation

$$\frac{\partial}{\partial t} (n \langle \chi \rangle) + \frac{\partial}{\partial \vec{q}} \cdot \left(n \left\langle \frac{\vec{p}}{m} \chi \right\rangle \right) = \vec{F} \cdot \left(n \left\langle \frac{\partial \chi}{\partial \vec{p}} \right\rangle \right).$$

(e) Derive the conservation equation for local density $\rho \equiv mn(\vec{q}, t)$, in terms of the local velocity $\vec{u} \equiv \langle \vec{p}/m \rangle$.

   Using $\chi = 1$ in the above expression

$$\frac{\partial}{\partial t} n + \frac{\partial}{\partial \vec{q}} \cdot \left(n \left\langle \frac{\vec{p}}{m} \right\rangle \right) = 0.$$

In terms of the local density $\rho = mn$, and velocity $\vec{u} \equiv \langle \vec{p}/m \rangle$, we have

$$\frac{\partial}{\partial t}\rho + \frac{\partial}{\partial \vec{q}} \cdot (\rho \vec{u}) = 0.$$

(f) Since the magnitude of particle momentum is unchanged by impurity scattering, the Lorentz gas has an infinity of conserved quantities $|\vec{p}|^m$. This unrealistic feature is removed upon inclusion of particle–particle collisions. For the rest of this problem focus only on $p^2/2m$ as a conserved quantity. Derive the conservation equation for the energy density

$$\epsilon(\vec{q}, t) \equiv \frac{\rho}{2}\langle c^2 \rangle, \qquad \text{where} \qquad \vec{c} \equiv \frac{\vec{p}}{m} - \vec{u},$$

in terms of the energy flux $\vec{h} \equiv \rho\langle \vec{c}\, c^2 \rangle/2$, and the pressure tensor $P_{\alpha\beta} \equiv \rho\langle c_\alpha c_\beta \rangle$.

With the kinetic energy $\chi = p^2/2m$ as a conserved quantity, the general conservation equation reduces to

$$\frac{\partial}{\partial t}\left(\frac{n}{2m}\langle |\vec{p}|^2 \rangle\right) + \frac{\partial}{\partial \vec{q}} \cdot \left(\frac{n}{2}\left\langle \frac{\vec{p}}{m}\frac{p^2}{m} \right\rangle\right) = \vec{F} \cdot \left(n\frac{\langle \vec{p} \rangle}{m}\right).$$

Substituting $\vec{p}/m = \vec{u} + \vec{c}$, where $\langle \vec{c} \rangle = 0$, and using $\rho = nm$,

$$\frac{\partial}{\partial t}\left[\frac{\rho}{2}u^2 + \frac{\rho}{2}\langle c^2 \rangle\right] + \frac{\partial}{\partial \vec{q}} \cdot \left[\frac{\rho}{2}\langle (\vec{u}+\vec{c})(u^2 + c^2 + 2\vec{u}\cdot\vec{c}) \rangle\right] = \frac{\rho}{m}\vec{F}\cdot\vec{u}.$$

From the definition $\varepsilon = \rho\langle c^2 \rangle/2$, we have

$$\frac{\partial}{\partial t}\left[\frac{\rho}{2}u^2 + \varepsilon\right] + \frac{\partial}{\partial \vec{q}} \cdot \left[\frac{\rho}{2}\left(\vec{u}u^2 + \vec{u}\langle c^2 \rangle + \langle \vec{c}c^2 \rangle + 2\vec{u}\cdot\langle \vec{c}\,\vec{c} \rangle\right)\right] = \frac{\rho}{m}\vec{F}\cdot\vec{u}.$$

Finally, by substituting $\vec{h} \equiv \rho\langle \vec{c}\, c^2 \rangle/2$ and $P_{\alpha\beta} = \rho\langle c_\alpha c_\beta \rangle$, we get

$$\frac{\partial}{\partial t}\left[\frac{\rho}{2}u^2 + \varepsilon\right] + \frac{\partial}{\partial \vec{q}} \cdot \left[\vec{u}\left(\frac{\rho}{2}u^2 + \varepsilon\right) + \vec{h}\right] + \frac{\partial}{\partial q_\alpha}\left(u_\beta P_{\alpha\beta}\right) = \frac{\rho}{m}\vec{F}\cdot\vec{u}.$$

(g) Starting with a one-particle density

$$f^0(\vec{p}, \vec{q}, t) = n(\vec{q}, t)\exp\left[-\frac{p^2}{2mk_B T(\vec{q}, t)}\right]\frac{1}{2\pi m k_B T(\vec{q}, t)},$$

reflecting local equilibrium conditions, calculate $\vec{u}$, $\vec{h}$, and $P_{\alpha\beta}$. Hence obtain the zeroth-order hydrodynamic equations.

There are only two quantities, 1 and $p^2/2m$, conserved in collisions. Let us start with the one-particle density

$$f^0(\vec{p}, \vec{q}, t) = n(\vec{q}, t)\exp\left[-\frac{p^2}{2mk_B T(\vec{q}, t)}\right]\frac{1}{2\pi m k_B T(\vec{q}, t)}.$$

Then

$$\vec{u} = \left\langle \frac{\vec{p}}{m} \right\rangle_0 = 0, \qquad \text{and} \qquad \vec{h} = \left\langle \frac{\vec{p}}{m}\frac{p^2}{m} \right\rangle_0 \frac{\rho}{2} = 0,$$

since both are odd functions of $\vec{p}$, while f^o is an even function of $\vec{p}$, while

$$P_{\alpha\beta} = \rho\langle c_\alpha c_\beta\rangle = \frac{n}{m}\langle p_\alpha p_\beta\rangle = \frac{n}{m}\delta_{\alpha\beta}\cdot mk_BT = nk_BT\delta_{\alpha\beta}.$$

Substituting these expressions into the results for (c) and (d), we obtain the zeroth-order hydrodynamic equations

$$\begin{cases} \dfrac{\partial\rho}{\partial t} = 0, \\[2mm] \dfrac{\partial}{\partial t}\varepsilon = \dfrac{\partial}{\partial t}\dfrac{\rho}{2}\langle c^2\rangle = 0. \end{cases}$$

The above equations imply that ρ and ε are independent of time, that is,

$$\rho = n(\vec{q}), \qquad \text{and} \qquad \varepsilon = k_BT(\vec{q}),$$

or

$$f^0 = \frac{n(\vec{q})}{2\pi mk_BT(\vec{q})}\exp\left[-\frac{p^2}{2mk_BT(\vec{q})}\right].$$

(h) Show that in the single collision time approximation to the collision term in the Boltzmann equation, the first-order solution is

$$f^1(\vec{p},\vec{q},t) = f^0(\vec{p},\vec{q},t)\left[1 - \tau\frac{\vec{p}}{m}\cdot\left(\frac{\partial\ln\rho}{\partial\vec{q}} - \frac{\partial\ln T}{\partial\vec{q}} + \frac{p^2}{2mk_BT^2}\frac{\partial T}{\partial\vec{q}} - \frac{\vec{F}}{k_BT}\right)\right].$$

The single collision time approximation is

$$C[f] = \frac{f^0 - f}{\tau}.$$

The first-order solution to the Boltzmann equation

$$f = f^0(1+g)$$

is obtained from

$$\mathcal{L}[f^0] = -\frac{f^0 g}{\tau},$$

as

$$g = -\tau\frac{1}{f^0}\mathcal{L}[f^0] = -\tau\left[\frac{\partial}{\partial t}\ln f^0 + \frac{\vec{p}}{m}\cdot\frac{\partial}{\partial\vec{q}}\ln f^0 + \vec{F}\cdot\frac{\partial}{\partial\vec{p}}\ln f^0\right].$$

Noting that

$$\ln f^0 = -\frac{p^2}{2mk_BT} + \ln n - \ln T - \ln(2\pi mk_B),$$

where n and T are independent of t, we have $\partial\ln f^0/\partial t = 0$, and

$$g = -\tau\left\{\vec{F}\cdot\left(\frac{-\vec{p}}{mk_BT}\right) + \frac{\vec{p}}{m}\cdot\left[\frac{1}{n}\frac{\partial n}{\partial\vec{q}} - \frac{1}{T}\frac{\partial T}{\partial\vec{q}} + \frac{p^2}{2mk_BT^2}\frac{\partial T}{\partial\vec{q}}\right]\right\}$$

$$= -\tau\left\{\frac{\vec{p}}{m}\cdot\left(\frac{1}{\rho}\frac{\partial\rho}{\partial\vec{q}} - \frac{1}{T}\frac{\partial T}{\partial\vec{q}} + \frac{p^2}{2mk_BT^2}\frac{\partial T}{\partial\vec{q}} - \frac{\vec{F}}{k_BT}\right)\right\}.$$

(i) Show that using the first-order expression for f, we obtain

$$\rho\vec{u} = n\tau\left[\vec{F} - k_B T \nabla \ln(\rho T)\right].$$

Clearly $\int d^2q f^0(1+g) = \int d^2q f^0 = n$, and

$$u_\alpha = \left\langle\frac{p_\alpha}{m}\right\rangle = \frac{1}{n}\int d^2p \frac{p_\alpha}{m} f^0(1+g)$$

$$= \frac{1}{n}\int d^2p \frac{p_\alpha}{m}\left[-\tau\frac{p_\beta}{m}\left(\frac{\partial \ln\rho}{\partial q_\beta} - \frac{\partial \ln T}{\partial q_\beta} - \frac{F_\beta}{k_B T} + \frac{p^2}{2mk_B T^2}\frac{\partial T}{\partial q_\beta}\right)\right]f^0.$$

Wick's theorem can be used to check that

$$\langle p_\alpha p_\beta\rangle_0 = \delta_{\alpha\beta} m k_B T,$$

$$\langle p^2 p_\alpha p_\beta\rangle_0 = (mk_B T)^2\left[2\delta_{\alpha\beta} + 2\delta_{\alpha\beta}\right] = 4\delta_{\alpha\beta}(mk_B T)^2,$$

resulting in

$$u_\alpha = -\frac{n\tau}{\rho}\left[\delta_{\alpha\beta} k_B T\left(\frac{\partial}{\partial q_\beta}\ln\left(\frac{\rho}{T}\right) - \frac{1}{k_B T} F_\beta\right) + 2k_B\frac{\partial T}{\partial q_\beta}\delta_{\alpha\beta}\right].$$

Rearranging these terms yields

$$\rho u_\alpha = n\tau\left[F_\alpha - k_B T\frac{\partial}{\partial q_\alpha}\ln(\rho T)\right].$$

(j) From the above equation, calculate the velocity response function $\chi_{\alpha\beta} = \partial u_\alpha/\partial F_\beta$.

The velocity response function is now calculated easily as

$$\chi_{\alpha\beta} = \frac{\partial u_\alpha}{\partial F_\beta} = \frac{n\tau}{\rho}\delta_{\alpha\beta}.$$

(k) Calculate $P_{\alpha\beta}$, and $\vec{h}$, and hence write down the first-order hydrodynamic equations.

The first-order expressions for pressure tensor and heat flux are

$$P_{\alpha\beta} = \frac{\rho}{m^2}\langle p_\alpha p_\beta\rangle = \delta_{\alpha\beta} n k_B T, \qquad \text{and} \qquad \delta^1 P_{\alpha\beta} = 0,$$

$$h_\alpha = \frac{\rho}{2m^3}\langle p_\alpha p^2\rangle = -\frac{\tau\rho}{2m^3}\left\langle p_\alpha p^2\frac{p_i}{m}\left(a_i + b_i p^2\right)\right\rangle_0.$$

The latter is calculated from Wick's theorem results

$$\langle p_i p_\alpha p^2\rangle = 4\delta_{\alpha i}(mk_B T)^2, \qquad \text{and}$$

$$\langle p_i p_\alpha p^2 p^2\rangle = (mk_B T)^3\left[\delta_{\alpha i}(4+2) + 4\times 2\delta_{\alpha i} + 4\times 2\delta_{\alpha i}\right] = 22\delta_{\alpha i},$$

as

$$h_\alpha = -\frac{\rho\tau}{2m^3}\left[(mk_B T)^2\left(\frac{\partial}{\partial q_\alpha}\ln\frac{\rho}{T} - \vec{F}\right) + \frac{22(mk_B T)^3}{2mk_B T}\frac{\partial}{\partial q_\alpha}\ln T\right]$$

$$= -11 n k_B^2 T\tau\frac{\partial T}{\partial q_\alpha}.$$

Substitute these expressions for $P_{\alpha\beta}$ and h_α into the equation obtained in (e)

$$\frac{\partial}{\partial t}\left[\frac{\rho}{2}u^2 + \epsilon\right] + \frac{\partial}{\partial \vec{q}} \cdot \left[\vec{u}\left(\frac{\rho}{2}u^2 + \epsilon\right) - 11nk_B^2 T\tau\frac{\partial T}{\partial \vec{q}}\right] + \frac{\partial}{\partial \vec{q}}\left(\vec{u}\,nk_B T\right) = \frac{\rho}{m}\vec{F}\cdot\vec{u}.$$

****

7. *Thermal conductivity*: consider a classical gas between two plates separated by a distance w. One plate at $y = 0$ is maintained at a temperature T_1, while the other plate at $y = w$ is at a different temperature T_2. The gas velocity is zero, so that the initial zeroth-order approximation to the one-particle density is

$$f_1^0(\vec{p}, x, y, z) = \frac{n(y)}{[2\pi m k_B T(y)]^{3/2}} \exp\left[-\frac{\vec{p}\cdot\vec{p}}{2mk_B T(y)}\right].$$

(a) What is the necessary relation between $n(y)$ and $T(y)$ to ensure that the gas velocity $\vec{u}$ remains zero? (Use this relation between $n(y)$ and $T(y)$ in the remainder of this problem.)

Since there is no external force acting on the gas between plates, the gas can only flow locally if there are variations in pressure. Since the local pressure is $P(y) = n(y)k_B T(y)$, the condition for the fluid to be stationary is

$$n(y)T(y) = \text{constant}.$$

(b) Using Wick's theorem, or otherwise, show that

$$\langle p^2\rangle^0 \equiv \langle p_\alpha p_\alpha\rangle^0 = 3\,(mk_B T), \quad \text{and} \quad \langle p^4\rangle^0 \equiv \langle p_\alpha p_\alpha p_\beta p_\beta\rangle^0 = 15\,(mk_B T)^2,$$

where $\langle \mathcal{O}\rangle^0$ indicates local averages with the Gaussian weight f_1^0. Use the result $\langle p^6\rangle^0 = 105(mk_B T)^3$ in conjunction with symmetry arguments to conclude

$$\langle p_y^2 p^4\rangle^0 = 35\,(mk_B T)^3.$$

The Gaussian weight has a covariance $\langle p_\alpha p_\beta\rangle^0 = \delta_{\alpha\beta}(mk_B T)$. Using Wick's theorem gives

$$\langle p^2\rangle^0 = \langle p_\alpha p_\alpha\rangle^0 = (mk_B T)\,\delta_{\alpha\alpha} = 3\,(mk_B T).$$

Similarly

$$\langle p^4\rangle^0 = \langle p_\alpha p_\alpha p_\beta p_\beta\rangle^0 = (mk_B T)^2\left(\delta_{\alpha\alpha} + 2\delta_{\alpha\beta}\delta_{\alpha\beta}\right) = 15\,(mk_B T)^2.$$

The symmetry along the three directions implies

$$\langle p_x^2 p^4\rangle^0 = \langle p_y^2 p^4\rangle^0 = \langle p_z^2 p^4\rangle^0 = \frac{1}{3}\langle p^2 p^4\rangle^0 = \frac{1}{3}\times 105\,(mk_B T)^3 = 35\,(mk_B T)^3.$$

(c) The zeroth-order approximation does not lead to relaxation of temperature/density variations related as in part (a). Find a better (time-independent) approximation $f_1^1(\vec{p}, y)$, by linearizing the Boltzmann equation in the single collision time approximation to

$$\mathcal{L}\left[f_1^1\right] \approx \left[\frac{\partial}{\partial t} + \frac{p_y}{m}\frac{\partial}{\partial y}\right]f_1^0 \approx -\frac{f_1^1 - f_1^0}{\tau_K},$$

where τ_K is of the order of the mean time between collisions.

Since there are only variations in y, we have

$$\left[\frac{\partial}{\partial t} + \frac{p_y}{m}\frac{\partial}{\partial y}\right]f_1^0 = f_1^0\frac{p_y}{m}\partial_y \ln f_1^0 = f_1^0\frac{p_y}{m}\partial_y\left[\ln n - \frac{3}{2}\ln T - \frac{p^2}{2mk_BT} - \frac{3}{2}\ln(2\pi mk_B)\right]$$

$$= f_1^0\frac{p_y}{m}\left[\frac{\partial_y n}{n} - \frac{3}{2}\frac{\partial_y T}{T} + \frac{p^2}{2mk_BT}\frac{\partial T}{T}\right] = f_1^0\frac{p_y}{m}\left[-\frac{5}{2} + \frac{p^2}{2mk_BT}\right]\frac{\partial_y T}{T},$$

where in the last equality we have used $nT = $ constant to get $\partial_y n/n = -\partial_y T/T$. Hence the first-order result is

$$f_1^1(\vec{p}, y) = f_1^0(\vec{p}, y)\left[1 - \tau_K\frac{p_y}{m}\left(\frac{p^2}{2mk_BT} - \frac{5}{2}\right)\frac{\partial_y T}{T}\right].$$

(d) Use f_1^1, along with the averages obtained in part (b), to calculate h_y, the y component of the heat transfer vector, and hence find K, the coefficient of thermal conductivity.

Since the velocity $\vec{u}$ is zero, the heat transfer vector is

$$h_y = n\left\langle c_y\frac{mc^2}{2}\right\rangle^1 = \frac{n}{2m^2}\left\langle p_y p^2\right\rangle^1.$$

In the zeroth-order Gaussian weight all odd moments of p have zero average. The corrections in f_1^1, however, give a non-zero heat transfer

$$h_y = -\tau_K\frac{n}{2m^2}\frac{\partial_y T}{T}\left\langle\frac{p_y}{m}\left(\frac{p^2}{2mk_BT} - \frac{5}{2}\right)p_y p^2\right\rangle^0.$$

Note that we need the Gaussian averages of $\langle p_y^2 p^4\rangle^0$ and $\langle p_y^2 p^2\rangle^0$. From the results of part (b), these averages are equal to $35(mk_BT)^3$ and $5(mk_BT)^2$, respectively. Hence

$$h_y = -\tau_K\frac{n}{2m^3}\frac{\partial_y T}{T}(mk_BT)^2\left(\frac{35}{2} - \frac{5\times5}{2}\right) = -\frac{5}{2}\frac{n\tau_K k_B^2 T}{m}\partial_y T.$$

The coefficient of thermal conductivity relates the heat transferred to the temperature gradient by $\vec{h} = -K\nabla T$, and hence we can identify

$$K = \frac{5}{2}\frac{n\tau_K k_B^2 T}{m}.$$

(e) What is the temperature profile, $T(y)$, of the gas in steady state?

Since $\partial_t T$ is proportional to $-\partial_y h_y$, there will be no time variation if h_y is a constant. But $h_y = -K\partial_y T$, where K, which is proportional to the product nT, is *a constant in the situation under investigation. Hence $\partial_y T$ must be constant, and T(y) varies linearly between the two plates. Subject to the boundary conditions of $T(0) = T_1$, and $T(w) = T_2$, this gives*

$$T(y) = T_1 + \frac{T_2 - T_1}{w} y.$$

✳✳✳✳✳✳✳✳

Solutions to selected problems from chapter 4

1. *Classical harmonic oscillators*: consider N harmonic oscillators with coordinates and momenta $\{q_i, p_i\}$, and subject to a Hamiltonian

$$\mathcal{H}(\{q_i, p_i\}) = \sum_{i=1}^{N} \left[\frac{p_i^2}{2m} + \frac{m\omega^2 q_i^2}{2} \right].$$

(a) Calculate the entropy S, as a function of the total energy E.

(*Hint.* By appropriate change of scale, the surface of constant energy can be deformed into a sphere. You may then ignore the difference between the surface area and volume for $N \gg 1$. A more elegant method is to implement this deformation through a canonical transformation.)

The volume of accessible phase space for a given total energy is proportional to

$$\Omega = \frac{1}{h^N} \int_{\mathcal{H}=E} \mathrm{d}q_1 \, \mathrm{d}q_2 \cdots \mathrm{d}q_N \, \mathrm{d}p_1 \, \mathrm{d}p_2 \cdots \mathrm{d}p_N,$$

where the integration is carried out under the condition of constant energy,

$$E = \mathcal{H}(\{q_i, p_i\}) = \sum_{i=1}^{N} \left[\frac{p_i^2}{2m} + \frac{m\omega^2 q_i^2}{2} \right].$$

Note that Planck's constant h is included as a measure of phase space volume, so as to make the final result dimensionless.

The surface of constant energy is an ellipsoid in $2N$ dimensions, whose area is difficult to calculate. However, for $N \to \infty$ the difference between volume and area is subleading in N, and we shall instead calculate the volume of the ellipsoid, replacing the constraint $\mathcal{H} = E$ by $\mathcal{H} \le E$. The ellipsoid can be distorted into a sphere by a canonical transformation, changing coordinates to

$$q_i' \equiv \sqrt{m\omega} q_i, \qquad \text{and} \qquad p_i' \equiv \frac{p_i}{\sqrt{m\omega}}.$$

The Hamiltonian in this coordinate system is

$$E = \mathcal{H}(\{q_i', p_i'\}) = \frac{\omega}{2} \sum_{i=1}^{N} \left(p_i'^2 + q_i'^2 \right).$$

Since the canonical transformation preserves volume in phase space (the Jacobian is unity), we have

$$\Omega \approx \frac{1}{h^N} \int_{\mathcal{H} \le E} \mathrm{d}q_1' \cdots \mathrm{d}q_N' \, \mathrm{d}p_1' \cdots \mathrm{d}p_N',$$

where the integral is now over the 2N-dimensional (hyper-)sphere of radius $R = \sqrt{2E/\omega}$. As the volume of a d-dimensional sphere of radius R is $S_d R^d/d$, we obtain

$$\Omega \approx \frac{2\pi^N}{(N-1)!} \cdot \frac{1}{2N} \left(\frac{2E}{\hbar\omega}\right)^N = \left(\frac{2\pi E}{\hbar\omega}\right)^N \frac{1}{N!}.$$

The entropy is now given by

$$S \equiv k_B \ln \Omega \approx N k_B \ln \left(\frac{2\pi e E}{N\hbar\omega}\right).$$

(b) Calculate the energy E, and heat capacity C, as functions of temperature T, and N.

From the expression for temperature,

$$\frac{1}{T} \equiv \left.\frac{\partial S}{\partial E}\right|_N \approx \frac{N k_B}{E},$$

we obtain the energy

$$E = N k_B T,$$

and the heat capacity

$$C = N k_B.$$

(c) Find the joint probability density $P(p, q)$ for a single oscillator. Hence calculate the mean kinetic energy, and mean potential energy for each oscillator.

The single-particle distribution function is calculated by summing over the undesired coordinates and momenta of the other $N - 1$ particles. Keeping track of the units of h used to make phase space dimensionless gives

$$p(p_1, q_1)\, dp_1\, dq_1 = \frac{\int_{(\mathcal{H} \leq E_{N-1})} \frac{1}{\hbar^{N-1}} dq_2 \cdots dq_N\, dp_2 \cdots dp_N}{\int_{(\mathcal{H} \leq E)} \frac{1}{\hbar^N} dq_1\, dq_2 \cdots dq_N\, dp_1\, dp_2 \cdots dp_N} \times \frac{dp_1\, dq_1}{h},$$

where $E_{N-1} = E - p_1^2/2m - m\omega^2 q_1^2/2$. Using the results from part (a),

$$p(p_1, q_1) = \frac{\Omega(N-1, E_{N-1})}{h\Omega(N, E)}$$

$$= \frac{\left(\frac{2}{\hbar\omega}\right)^{N-1} \frac{\pi^{N-1}}{(N-1)!} \left(E - \frac{p_1^2}{2m} - \frac{m\omega^2}{2} q_1^2\right)^{N-1}}{h \left(\frac{2}{\hbar\omega}\right)^N \frac{\pi^N}{N!} E^N}$$

$$= \frac{\omega}{2\pi} \frac{N}{E} \left(1 - \frac{\frac{p_1^2}{2m} + \frac{m\omega^2}{2} q_1^2}{E}\right)^{N-1}.$$

Using the approximation $(N - 1) \sim N$, for $N \gg 1$, and setting $E = N k_B T$, we have

$$p(p_1, q_1) = \frac{\omega}{2\pi} \frac{N}{N k_B T} \left(1 - \frac{\frac{p_1^2}{2m} + \frac{m\omega^2}{2} q_1^2}{N k_B T}\right)^N$$

$$\approx \frac{\omega}{2\pi k_B T} \exp\left(-\frac{\frac{p_1^2}{2m} + \frac{m\omega^2}{2} q_1^2}{k_B T}\right).$$

Let us denote (p_1, q_1) by (p, q), then

$$p(p, q) = \frac{\omega}{2\pi k_B T} \exp\left(-\frac{p^2}{2mk_B T} - \frac{m\omega^2 q^2}{2k_B T}\right)$$

is a properly normalized product of two Gaussians. The mean kinetic energy is

$$\left\langle\frac{p^2}{2m}\right\rangle = \int p(p, q)\frac{p^2}{2m}\,dq\,dp = \frac{k_B T}{2},$$

while the mean potential energy is also

$$\left\langle\frac{m\omega^2 q^2}{2}\right\rangle = \int p(p, q)\frac{m\omega^2 q^2}{2}\,dq\,dp = \frac{k_B T}{2}.$$

****

2. *Quantum harmonic oscillators*: consider N independent quantum oscillators subject to a Hamiltonian

$$\mathcal{H}(\{n_i\}) = \sum_{i=1}^{N} \hbar\omega\left(n_i + \frac{1}{2}\right),$$

where $n_i = 0, 1, 2, \cdots$ is the quantum occupation number for the ith oscillator.

(a) Calculate the entropy S, as a function of the total energy E.

(*Hint.* $\Omega(E)$ can be regarded as the number of ways of rearranging $M = \sum_i n_i$ balls, and $N - 1$ partitions along a line.)

The total energy of the set of oscillators is

$$E = \hbar\omega\left(\sum_{i=1}^{N} n_i + \frac{N}{2}\right).$$

Let us set the sum over the individual quantum numbers to

$$M \equiv \sum_{i=1}^{N} n_i = \frac{E}{\hbar\omega} - \frac{N}{2}.$$

The number of configurations $\{n_i\}$ for a given energy (thus for a given value of M) is equal to the possible number of ways of distributing M energy units into N slots, or of partitioning M particles using $N - 1$ walls. This argument gives the number of states as

$$\Omega = \frac{(M + N - 1)!}{M!\,(N - 1)!},$$

and a corresponding entropy

$$S = k_B \ln \Omega \approx k_B\left[(M + N - 1)\ln\left(\frac{M + N - 1}{e}\right) - M\ln\frac{M}{e}\right.$$
$$\left. - (N - 1)\ln\left(\frac{N - 1}{e}\right)\right].$$

(b) Calculate the energy E, and heat capacity C, as functions of temperature T, and N.

The temperature is calculated by

$$\frac{1}{T} \equiv \left.\frac{\partial S}{\partial E}\right|_N \approx \frac{k_B}{\hbar\omega}\ln\left(\frac{M+N-1}{M}\right) = \frac{k_B}{\hbar\omega}\ln\left(\frac{\frac{E}{\hbar\omega}+\frac{N}{2}-1}{\frac{E}{\hbar\omega}-\frac{N}{2}}\right) \approx \frac{k_B}{\hbar\omega}\ln\left(\frac{E+\frac{N}{2}\hbar\omega}{E-\frac{N}{2}\hbar\omega}\right).$$

By inverting this equation, we get the energy

$$E = \frac{N}{2}\hbar\omega\frac{\exp(\hbar\omega/k_BT)+1}{\exp(\hbar\omega/k_BT)-1} = N\hbar\omega\left[\frac{1}{2}+\frac{1}{\exp(\hbar\omega/k_BT)-1}\right],$$

and a corresponding heat capacity

$$C \equiv \left.\frac{\partial E}{\partial T}\right|_N = Nk_B\left(\frac{\hbar\omega}{k_BT}\right)^2\frac{\exp(\hbar\omega/k_BT)}{[\exp(\hbar\omega/k_BT)-1]^2}.$$

(c) Find the probability $p(n)$ that a particular oscillator is in its nth quantum level.

The probability that a particular oscillator is in its nth quantum level is given by summing the joint probability over states for all the other oscillators, that is,

$$p(n) = \sum_{\{n_{i\neq1}\}} p(n_i) = \frac{\sum_{(E-(n+1/2)\hbar\omega)}1}{\sum_{(E)}1} = \frac{\Omega\left(N-1,E-(n+\frac{1}{2})\hbar\omega\right)}{\Omega(N,E)}$$

$$= \frac{[(M-n)+N-2]!}{(M-n)!\cdot(N-2)!}\cdot\frac{M!\cdot(N-1)!}{(M+N-1)!}$$

$$\approx \frac{M(M-1)\cdots(M-n+1)\cdot N}{(M+N-1)(M+N-2)\cdots(M+N-n-1)}$$

$$\approx N(M+N)^{-n-1}M^n,$$

where the approximations used are of the form $(I-1)\approx I$, for $I\gg1$. Hence,

$$p(n) = N\left(\frac{E}{\hbar\omega}-\frac{N}{2}+N\right)^{-n-1}\left(\frac{E}{\hbar\omega}-\frac{N}{2}\right)^n$$

$$- \frac{N}{\frac{E}{\hbar\omega}+\frac{N}{2}}\left(\frac{\frac{E}{\hbar\omega}-\frac{N}{2}}{\frac{E}{\hbar\omega}+\frac{N}{2}}\right)^n,$$

which using

$$\frac{1}{T} = \frac{k_B}{\hbar\omega}\ln\left(\frac{E+\frac{N\hbar\omega}{2}}{E-\frac{N\hbar\omega}{2}}\right), \implies \left(\frac{\frac{E}{\hbar\omega}+\frac{N}{2}}{\frac{E}{\hbar\omega}-\frac{N}{2}}\right)^n = \exp\left(-\frac{n\hbar\omega}{k_BT}\right),$$

leads to the probability

$$p(n) = \exp\left(-n\frac{\hbar\omega}{k_BT}\right)\left[1-\exp\left(-\frac{\hbar\omega}{k_BT}\right)\right].$$

(d) Comment on the difference between heat capacities for classical and quantum oscillators.

As found in part (b),

$$C_{\text{quantum}} = Nk_B\left(\frac{\hbar\omega}{k_BT}\right)^2\frac{\exp\left(\frac{\hbar\omega}{k_BT}\right)}{\left[\exp\left(\frac{\hbar\omega}{k_BT}\right)-1\right]^2}.$$

In the high-temperature limit, $\hbar\omega/(k_B T) \ll 1$, using the approximation $e^x \approx 1 + x$ for $x \ll 1$, gives

$$C_{\text{quantum}} = Nk_B = C_{\text{classical}}.$$

At low temperatures, the quantized nature of the energy levels of the quantum oscillators becomes noticeable. In the limit $T \to 0$, there is an energy gap between the ground state and the first excited state. This results in a heat capacity that goes to zero exponentially, as can be seen from the limit $\hbar\omega/(k_B T) \gg 1$,

$$C_{\text{quantum}} = Nk_B \left(\frac{\hbar\omega}{k_B T}\right)^2 \exp\left(-\frac{\hbar\omega}{k_B T}\right).$$

****

3. *Relativistic particles*: N *indistinguishable* relativistic particles move in *one dimension* subject to a Hamiltonian

$$\mathcal{H}(\{p_i, q_i\}) = \sum_{i=1}^{N} [c|p_i| + U(q_i)],$$

with $U(q_i) = 0$ for $0 \le q_i \le L$, and $U(q_i) = \infty$ otherwise. Consider a *microcanonical* ensemble of total energy E.

(a) Compute the contribution of the coordinates q_i to the available volume in phase space $\Omega(E, L, N)$.

Each of N coordinates explores a length L, for an overall contribution of $L^N/N!$. Division by $N!$ ensures no over-counting of phase space for indistinguishable particles.

(b) Compute the contribution of the momenta p_i to $\Omega(E, L, N)$.

(*Hint.* The volume of the hyper pyramid defined by $\sum_{i=1}^{d} x_i \le R$, and $x_i \ge 0$, in d dimensions is $R^d/d!$.)

The N momenta satisfy the constraint $\sum_{i=1}^{N} |p_i| = E/c$. For a particular choice of the signs of $\{p_i\}$, this constraint describes the surface of a hyper pyramid in N dimensions. If we ignore the difference between the surface area and volume in the large N limit, we can calculate the volume in momentum space from the expression given in the hint as

$$\Omega_{\mathbf{p}} = 2^N \cdot \frac{1}{N!} \cdot \left(\frac{E}{c}\right)^N.$$

The factor of 2^N takes into account the two possible signs for each p_i. The surface area of the pyramid is given by $\sqrt{d} R^{d-1}/(d-1)!$; the additional factor of $\sqrt{d}$ with respect to $d\text{volume}/dR$ is the ratio of the normal to the base to the side of the pyramid. Thus, the volume of a shell of energy uncertainty Δ_E, is

$$\Omega_{\mathbf{p}}' = 2^N \cdot \frac{\sqrt{N}}{(N-1)!} \cdot \left(\frac{E}{c}\right)^{N-1} \cdot \frac{\Delta_E}{c}.$$

We can use the two expressions interchangeably, as their difference is subleading in N.

(c) Compute the entropy $S(E, L, N)$.

Taking into account quantum modifications due to indistinguishability, and phase space measure, we have

$$\Omega(E, L, N) = \frac{1}{h^N} \cdot \frac{L^N}{N!} \cdot 2^N \cdot \frac{\sqrt{N}}{(N-1)!} \cdot \left(\frac{E}{c}\right)^{N-1} \cdot \frac{\Delta_E}{c}.$$

Ignoring subleading terms in the large N limit, the entropy is given by

$$S(E, L, N) = Nk_B \ln\left(\frac{2e^2}{hc} \cdot \frac{L}{N} \cdot \frac{E}{N}\right).$$

(d) Calculate the one-dimensional pressure P.

From $dE = TdS - PdV + \mu dN$, the pressure is given by

$$P = T \left.\frac{\partial S}{\partial L}\right|_{E,N} = \frac{Nk_B T}{L}.$$

(e) Obtain the heat capacities C_L and C_P.

Temperature and energy are related by

$$\frac{1}{T} = \left.\frac{\partial S}{\partial E}\right|_{L,N} = \frac{Nk_B}{E}, \implies E = Nk_B T, \implies C_L = \left.\frac{\partial E}{\partial T}\right|_{L,N} = Nk_B.$$

Including the work done against external pressure, and using the equation of state,

$$C_P = \left.\frac{\partial E}{\partial T}\right|_{P,N} + P \left.\frac{\partial L}{\partial T}\right|_{P,N} = 2Nk_B.$$

(f) What is the probability $p(p_1)$ of finding a particle with momentum p_1?

Having fixed p_1 for the first particle, the remaining $N-1$ particles are left to share an energy of $(E - c|p_1|)$. Since we are not interested in the coordinates, we can get the probability from the ratio of phase spaces for the momenta, that is,

$$p(p_1) = \frac{\Omega_\mathbf{p}(E - c|p_1|, N - 1)}{\Omega_\mathbf{p}(E, N)}$$

$$= \left[\frac{2^{N-1}}{(N-1)!} \cdot \left(\frac{E - c|p_1|}{c}\right)^{N-1}\right] \times \left[\frac{N!}{2^N} \cdot \left(\frac{c}{E}\right)^N\right]$$

$$\approx \frac{cN}{2E} \cdot \left(1 - \frac{c|p_1|}{E}\right)^N \approx \frac{cN}{2E} \cdot \exp\left(-\frac{cN|p_1|}{E}\right).$$

Substituting $E = Nk_B T$, we obtain the (properly normalized) Boltzmann weight

$$p(p_1) = \frac{c}{2k_B T} \cdot \exp\left(-\frac{c|p_1|}{k_B T}\right).$$

****

4. *Hard sphere gas*: consider a gas of N hard spheres in a box. A single sphere excludes a volume ω around it, while its center of mass can explore a volume V (if the box is otherwise empty). There are no other interactions between the spheres, except for the constraints of hard-core exclusion.

(a) Calculate the entropy S, as a function of the total energy E.

(*Hint.* $(V - a\omega)(V - (N - a)\omega) \approx (V - N\omega/2)^2$.)

The available phase space for N identical particles is given by

$$\Omega = \frac{1}{N! h^{3N}} \int_{H=E} d^3\vec{q}_1 \cdots d^3\vec{q}_N \, d^3\vec{p}_1 \cdots d^3\vec{p}_N,$$

where the integration is carried out under the condition

$$E = \mathcal{H}(\vec{q}_i, \vec{p}_i) = \sum_{i=1}^{N} \frac{p_i^2}{2m}, \qquad \text{or} \qquad \sum_{i=1}^{N} p_i^2 = 2mE.$$

The momentum integrals are now performed as in an ideal gas, yielding

$$\Omega = \frac{(2mE)^{3N/2-1}}{N! h^{3N}} \cdot \frac{2\pi^{3N}}{\left(\frac{3N}{2} - 1\right)!} \cdot \int d^3\vec{q}_1 \cdots d^3\vec{q}_N.$$

The joint integral over the spatial coordinates with excluded volume constraints is best performed by introducing particles one at a time. The first particle can explore a volume V, the second $V - \omega$, the third $V - 2\omega$, etc., resulting in

$$\int d^3\vec{q}_1 \cdots d^3\vec{q}_N = V(V - \omega)(V - 2\omega) \cdots (V - (N - 1)\omega).$$

Using the approximation $(V - a\omega)(V - (N - a)\omega) \approx (V - N\omega/2)^2$, we obtain

$$\int d^3\vec{q}_1 \cdots d^3\vec{q}_N \approx \left(V - \frac{N\omega}{2}\right)^N.$$

Thus the entropy of the system is

$$S = k_B \ln \Omega \approx N k_B \ln \left[\frac{e}{N} \left(V - \frac{N\omega}{2}\right) \left(\frac{4\pi m E e}{3N h^2}\right)^{3/2}\right].$$

(b) Calculate the equation of state of this gas.

We can obtain the equation of state by calculating the expression for the pressure of the gas,

$$\frac{P}{T} = \frac{\partial S}{\partial V}\bigg|_{E,N} \approx \frac{N k_B}{V - \frac{N\omega}{2}},$$

which is easily rearranged to

$$P\left(V - \frac{N\omega}{2}\right) = N k_B T.$$

Note that the joint effective excluded volume that appears in the above expressions is one half of the total volume excluded by N particles.

(c) Show that the isothermal compressibility, $\kappa_T = -V^{-1} \, \partial V/\partial P|_T$, is always positive.

The isothermal compressibility is calculated from

$$\kappa_T = -\frac{1}{V} \frac{\partial V}{\partial P}\bigg|_{T,N} = \frac{N k_B T}{P^2 V} > 0,$$

and is explicitly positive, as required by stability constraints.

****

5. *Non-harmonic gas*: let us re-examine the generalized ideal gas introduced in the previous section, using statistical mechanics rather than kinetic theory. Consider a gas of N non-interacting atoms in a d-dimensional box of "volume" V, with a kinetic energy

$$\mathcal{H} = \sum_{i=1}^{N} A \left| \vec{p}_i \right|^s,$$

where $\vec{p}_i$ is the momentum of the ith particle.

(a) Calculate the classical partition function $Z(N, T)$ at a temperature T. (*You don't have to keep track of numerical constants in the integration.*)

The partition function is given by

$$Z(N, T, V) = \frac{1}{N! h^{dN}} \int \cdots \int \mathrm{d}^d \vec{q}_1 \cdots \mathrm{d}^d \vec{q}_N \, \mathrm{d}^d \vec{p}_1 \cdots \mathrm{d}^d \vec{p}_N \exp\left[-\beta \sum_{i=1}^{N} A \left| \vec{p}_i \right|^s \right]$$

$$= \frac{1}{N! h^{dN}} \left[\int \int \mathrm{d}^d \vec{q} \, \mathrm{d}^d \vec{p} \exp\left(-\beta A \left| \vec{p} \right|^s \right) \right]^N.$$

Ignoring hard-core exclusion, each atom contributes a d-dimensional volume V to the integral over the spatial degrees of freedom, and

$$Z(N, T, V) = \frac{V^N}{N! h^{dN}} \left[\int \mathrm{d}^d \vec{p} \exp\left(-\beta A \left| \vec{p} \right|^s \right) \right]^N.$$

Observing that the integrand depends only on the magnitude $|\vec{p}| = p$, we can evaluate the integral in spherical coordinates using $\int \mathrm{d}^d \vec{p} = S_d \int \mathrm{d}p \, p^{d-1}$, where S_d denotes the surface area of a unit sphere in d dimensions, as

$$Z(N, T, V) = \frac{V^N}{N! h^{dN}} \left[S_d \int_0^\infty \mathrm{d}p \, p^{d-1} \exp\left(-\beta A p^s \right) \right]^N.$$

Introducing the variable $x \equiv (\beta A)^{1/s} p$, we have

$$Z(N, T, V) = \frac{V^N S_d^N}{N! h^{dN}} \left(\frac{A}{k_B T} \right)^{-dN/s} \left[\int_0^\infty \mathrm{d}x \, x^{d-1} \exp(-x^s) \right]^N$$

$$= C^N(d, s) \frac{1}{N!} \left(\frac{V S_d}{h^d} \right)^N \left(\frac{A}{k_B T} \right)^{-dN/s},$$

where C denotes the numerical value of the integral. (We assume that A and s are both real and positive. These conditions ensure that energy increases with increasing $|\vec{p}|$.) The integral is in fact equal to

$$C(d, s) = \int_0^\infty \mathrm{d}x \, x^{d-1} \exp(-x^s) \, \mathrm{d}x = \frac{1}{s} \Gamma\left(\frac{d}{s} \right),$$

and the partition function is

$$Z = \frac{1}{N!} \left(\frac{V S_d}{h^d s} \right)^N \left(\frac{A}{k_B T} \right)^{-dN/s} \left[\Gamma\left(\frac{d}{s} \right) \right]^N.$$

(b) Calculate the pressure and the internal energy of this gas. (Note how the usual equipartition theorem is modified for non-quadratic degrees of freedom.)

To calculate the pressure and internal energy, note that the Helmholtz free energy is

$$F = E - TS = -k_B T \ln Z,$$

and that

$$P = -\left.\frac{\partial F}{\partial V}\right|_T, \qquad \text{while} \qquad E = -\left.\frac{\partial \ln Z}{\partial \beta}\right|_V.$$

First calculating the pressure:

$$P = -\left.\frac{\partial F}{\partial V}\right|_T = k_B T \left.\frac{\partial \ln Z}{\partial V}\right|_T = \frac{N k_B T}{V}.$$

Now calculating the internal energy:

$$E = -\left.\frac{\partial \ln Z}{\partial \beta}\right|_V = -\frac{\partial}{\partial \beta}\left[-\frac{dN}{s}\ln\left(\frac{A}{k_B T}\right)\right] = \frac{d}{s} N k_B T.$$

Note that for each degree of freedom with energy $A|\vec{p}_i|^s$, we have the average value, $\langle A|\vec{p}_i|^s \rangle = \frac{d}{s} k_B T$. This evaluates to $\frac{3}{2} k_B T$ for the three-dimensional ideal gas.

(c) Now consider a diatomic gas of N molecules, each with energy

$$\mathcal{H}_i = A\left(\left|\vec{p}_i^{(1)}\right|^s + \left|\vec{p}_i^{(2)}\right|^s\right) + K\left|\vec{q}_i^{(1)} - \vec{q}_i^{(2)}\right|^t,$$

where the superscripts refer to the two particles in the molecule. (Note that this unrealistic potential allows the two atoms to occupy the same point.) Calculate the expectation value $\left\langle \left|\vec{q}_i^{(1)} - \vec{q}_i^{(2)}\right|^t \right\rangle$, at temperature T.

Now consider N diatomic molecules, with

$$\mathcal{H} = \sum_{i=1}^{N} \mathcal{H}_i, \qquad \text{where} \qquad \mathcal{H}_i = A\left(\left|\vec{p}_i^{(1)}\right|^s + \left|\vec{p}_i^{(2)}\right|^s\right) + K\left|\vec{q}_i^{(1)} - \vec{q}_i^{(2)}\right|^t.$$

The expectation value

$$\left\langle \left|\vec{q}_i^{(1)} - \vec{q}_i^{(2)}\right|^t \right\rangle = \frac{\frac{1}{N!}\int \prod_{i=1}^{N} d^d\vec{q}_i^{(1)}\, d^d\vec{q}_i^{(2)}\, d^d\vec{p}_i^{(1)}\, d^d\vec{p}_i^{(2)} \left|\vec{q}_i^{(1)} - \vec{q}_i^{(2)}\right|^t \exp\left[-\beta \sum_i \mathcal{H}_i\right]}{\frac{1}{N!}\int \prod_{i=1}^{N} d^d\vec{q}_i^{(1)}\, d^d\vec{q}_i^{(2)}\, d^d\vec{p}_i^{(1)}\, d^d\vec{p}_i^{(2)} \exp\left[-\beta \sum_i \mathcal{H}_i\right]}$$

is easily calculated by changing variables to

$$\vec{x} \equiv \vec{q}^{(1)} - \vec{q}^{(2)}, \qquad \text{and} \qquad \vec{y} \equiv \frac{\vec{q}^{(1)} + \vec{q}^{(2)}}{2},$$

as (note that the Jacobian of the transformation is unity)

$$\left\langle \left|\vec{q}^{(1)} - \vec{q}^{(2)}\right|^t \right\rangle = \frac{\int d^d\vec{x}\, d^d\vec{y} \cdot |\vec{x}|^t \cdot \exp\left[-\beta K |\vec{x}|^t\right]}{\int d^d\vec{x}\, d^d\vec{y} \cdot \exp\left[-\beta K |\vec{x}|^t\right]}$$

$$= \frac{\int d^d\vec{x} \cdot |\vec{x}|^t \cdot \exp\left[-\beta K |\vec{x}|^t\right]}{\int d^d\vec{x} \cdot \exp\left[-\beta K |\vec{x}|^t\right]}.$$

Further simplifying the algebra by introducing the variable $\vec{z} \equiv (\beta K)^{1/t}\vec{x}$, leads to

$$\left\langle \left| \vec{q}^{(1)} - \vec{q}^{(2)} \right|^t \right\rangle = \frac{(\beta K)^{-t/t} \int \mathrm{d}^d\vec{z} \cdot |\vec{z}|^t \cdot \exp\left[-|\vec{z}|^t\right]}{\int \mathrm{d}^d\vec{z} \cdot \exp\left[-|\vec{z}|^t\right]} = \frac{d}{t} \cdot \frac{k_B T}{K}.$$

Here we have assumed that the volume is large enough, so that the range of integration over the relative coordinate can be extended from 0 to ∞.

Alternatively, note that for the degree of freedom $\vec{x} = \vec{q}^{(1)} - \vec{q}^{(2)}$, the energy is $K|\vec{x}|^t$. Thus, from part (b) we know that

$$\left\langle K|\vec{x}|^t \right\rangle = \frac{d}{t} \cdot \frac{k_B T}{K},$$

that is,

$$\left\langle |\vec{x}|^t \right\rangle = \left\langle \left| \vec{q}^{(1)} - \vec{q}^{(2)} \right|^t \right\rangle = \frac{d}{t} \cdot \frac{k_B T}{K}.$$

And yet another way of calculating the expectation value is from

$$\sum_{i=1}^{N} \left\langle \left| \vec{q}_i^{(1)} - \vec{q}_i^{(2)} \right|^t \right\rangle = -\frac{1}{\beta} \frac{\partial \ln Z}{\partial K} = \frac{Nd}{t} \cdot \frac{k_B T}{K}$$

(note that the relevant part of Z is calculated in part (d) below).

(d) Calculate the heat capacity ratio $\gamma = C_P/C_V$, for the above diatomic gas.

For the ideal gas, the internal energy depends only on temperature T. The gas in part (c) is ideal in the sense that there are no molecule–molecule interactions. Therefore,

$$C_V = \left.\frac{\mathrm{d}Q}{\mathrm{d}T}\right|_V = \left.\frac{\mathrm{d}E + P\mathrm{d}V}{\mathrm{d}T}\right|_V = \frac{\mathrm{d}E(T)}{\mathrm{d}T},$$

and

$$C_P = \left.\frac{\mathrm{d}Q}{\mathrm{d}T}\right|_P = \left.\frac{\mathrm{d}E + P\mathrm{d}V}{\mathrm{d}T}\right|_P = \frac{\mathrm{d}E(T)}{\mathrm{d}T} + P\left.\frac{\partial V(T)}{\partial T}\right|_P.$$

Since $PV = Nk_B T$,

$$C_P = \frac{\mathrm{d}E(T)}{\mathrm{d}T} + Nk_B.$$

We now calculate the partition function

$$Z = \frac{1}{N!h^{dN}} \int \prod_{i=1}^{N} \mathrm{d}^d\vec{q}_i^{(1)} \, \mathrm{d}^d\vec{q}_i^{(2)} \, \mathrm{d}^d\vec{p}_i^{(1)} \, \mathrm{d}^d\vec{p}_i^{(2)} \exp\left[-\beta \sum_i \mathcal{H}_i\right]$$

$$= \frac{1}{N!h^{dN}} z_1^N,$$

where

$$z_1 = \int \mathrm{d}^d\vec{q}^{(1)} \, \mathrm{d}^d\vec{q}^{(2)} \, \mathrm{d}^d\vec{p}^{(1)} \, \mathrm{d}^d\vec{p}^{(2)} \exp\left[-\beta \cdot \left(A\left|\vec{p}^{(1)}\right|^s + A\left|\vec{p}^{(2)}\right|^s + K\left|\vec{q}^{(1)} - \vec{q}^{(2)}\right|^t \right)\right]$$

$$= \left[\int \mathrm{d}^d\vec{q}^{(1)} \, \mathrm{d}^d\vec{q}^{(2)} \exp\left(-\beta K\left|\vec{q}^{(1)} - \vec{q}^{(2)}\right|^t\right)\right] \cdot \left[\int \mathrm{d}^d\vec{p}^{(1)} \exp\left(-\beta A\left|\vec{p}^{(1)}\right|^s\right)\right]^2.$$

Introducing the variables, $\vec{x}$, $\vec{y}$, and $\vec{z}$, as in part (c),

$$Z \propto \frac{V^N}{N!} \left[(\beta K)^{-d/t} \int_0^\infty z^{d-1} \exp(-z^t) \, dz \right]^N \cdot \left[\int_0^\infty p^{d-1} \exp(-\beta A p^s) \, dp \right]^{2N}$$

$$= \frac{V^N}{N!} \left[(\beta K)^{-d/t} \frac{1}{t} \Gamma\left(\frac{d}{t}\right) \right]^N \cdot \left[\frac{1}{s}(\beta A)^{-d/s} \Gamma\left(\frac{d}{s}\right) \right]^{2N}$$

$$\propto \frac{V^N}{N!} (\beta K)^{-dN/t} (\beta A)^{-2Nd/s}.$$

Now we can calculate the internal energy as

$$\langle E \rangle = -\frac{\partial \ln Z}{\partial \beta} = \frac{d}{t} N k_B T + \frac{2d}{s} N k_B T = d N k_B T \left(\frac{1}{t} + \frac{2}{s} \right).$$

From this result, the heat capacities are obtained as

$$C_P = \left. \frac{\partial E}{\partial T} \right|_P + P \left. \frac{\partial V}{\partial T} \right|_P = N k_B \left(\frac{2d}{s} + \frac{d}{t} + 1 \right),$$

$$C_V = \left. \frac{\partial E}{\partial T} \right|_V = d N k_B \left(\frac{2}{s} + \frac{1}{t} \right),$$

resulting in the ratio

$$\gamma = \frac{C_P}{C_V} = \frac{2d/s + d/t + 1}{2d/s + d/t} = 1 + \frac{st}{d(2t+s)}.$$

$$* * * * * * * *$$

6. *Surfactant adsorption*: a dilute solution of surfactants can be regarded as an ideal three-dimensional gas. As surfactant molecules can reduce their energy by contact with air, a fraction of them migrate to the surface where they can be treated as a two-dimensional ideal gas. Surfactants are similarly adsorbed by other porous media such as polymers and gels with an affinity for them.

(a) Consider an ideal gas of classical particles of mass m in d dimensions, moving in a *uniform* attractive potential of strength ε_d. By calculating the partition function, or otherwise, show that the chemical potential at a temperature T, and particle density n_d, is given by

$$\mu_d = -\varepsilon_d + k_B T \ln \left[n_d \lambda(T)^d \right], \quad \text{where} \quad \lambda(T) = \frac{h}{\sqrt{2\pi m k_B T}}.$$

The partition function of a d-dimensional ideal gas is given by

$$Z_d = \frac{1}{N_d! h^{d N_d}} \int \cdots \int \prod_{i=1}^{N_d} d^d \vec{q}_i \, d^d \vec{p}_i \exp\left\{ -\beta \left[N_d \varepsilon_d + \sum_{i=1}^{N_d} \left(\frac{\vec{p}_i^2}{2m} \right) \right] \right\}$$

$$= \frac{1}{N_d!} \left(\frac{V_d}{\lambda^d} \right)^{N_d} e^{-\beta N_d \varepsilon_d},$$

where

$$\lambda \equiv \frac{h}{\sqrt{2\pi m k_B T}}.$$

The chemical potential is calculated from the Helmholtz free energy as

$$\mu_d = \frac{\partial F}{\partial N}\bigg|_{V,T} = -k_B T \frac{\partial \ln Z_d}{\partial N_d}\bigg|_{V,T}$$

$$= -\varepsilon_d + k_B T \ln\left(\frac{V_d}{N_d \lambda^d}\right).$$

(b) If a surfactant lowers its energy by ε_0 in moving from the solution to the surface, calculate the concentration of floating surfactants as a function of the solution concentration $n (= n_3)$, at a temperature T.

The density of particles can also be calculated from the grand canonical partition function, which for particles in a d-dimensional space is

$$\Xi(\mu, V_d, T) = \sum_{N_d=0}^{\infty} Z(N_d, V_d, T) e^{\beta N_d \mu}$$

$$= \sum_{N_d=0}^{\infty} \frac{1}{N_d!}\left(\frac{V_d}{\lambda^d}\right)^{N_d} e^{-\beta N_d \varepsilon_d} e^{\beta N_d \mu} = \exp\left[\left(\frac{V_d}{\lambda^d}\right) \cdot e^{\beta(\mu-\varepsilon_d)}\right].$$

The average number of particles absorbed in the space is

$$\langle N_d \rangle = \frac{1}{\beta}\frac{\partial}{\partial\mu}\ln\Xi = \frac{1}{\beta}\frac{\partial}{\partial\mu}\left[\left(\frac{V_d}{\lambda^d}\right) \cdot e^{\beta(\mu-\varepsilon_d)}\right] = \left(\frac{V_d}{\lambda^d}\right) \cdot e^{\beta(\mu-\varepsilon_d)}.$$

We are interested in the coexistence of surfactants between a $d = 3$ dimensional solution, and its $d = 2$ dimensional surface. Dividing the expressions for $\langle N_3 \rangle$ and $\langle N_2 \rangle$, and taking into account $\varepsilon_0 = \varepsilon_3 - \varepsilon_2$, gives

$$\frac{\langle N_2 \rangle}{\langle N_3 \rangle} = \frac{A\lambda}{V} e^{\beta\varepsilon_0},$$

which implies that

$$n_2 = \frac{\langle N_2 \rangle}{A} = n\lambda\, e^{\beta\varepsilon_0}.$$

(c) Gels are formed by cross-linking linear polymers. It has been suggested that the porous gel should be regarded as *fractal*, and the surfactants adsorbed on its surface treated as a gas in d_f-dimensional space, with a non-integer d_f. Can this assertion be tested by comparing the relative adsorption of surfactants to a gel, and to the individual polymers (presumably one-dimensional) before cross-linking, as a function of temperature?

Using the result found in part (b), but regarding the gel as a d_f-dimensional container, the adsorbed particle density is

$$\langle n_{\text{gel}} \rangle = n\lambda^{3-d_f} \exp\left[\beta(\varepsilon_3 - \varepsilon_{\text{gel}})\right].$$

Thus by studying the adsorption of particles as a function of temperature one can determine the fractal dimensionality, d_f, of the surface. The largest contribution comes from the difference in energies. If this leading part is accurately determined, there is a subleading dependence via λ^{3-d_f}, which depends on d_f.

$$* * * * * * * *$$

Solutions to selected problems from chapter 5

1. *Debye–Hückel theory and ring diagrams*: the virial expansion gives the gas pressure as an *analytic* expansion in the density $n = N/V$. Long-range interactions can result in *non-analytic* corrections to the ideal gas equation of state. A classic example is the Coulomb interaction in plasmas, whose treatment by Debye–Hückel theory is equivalent to summing all the *ring diagrams* in a cumulant expansion.

For simplicity, consider a gas of N electrons moving in a uniform background of positive charge density Ne/V to ensure overall charge neutrality. The Coulomb interaction takes the form

$$\mathcal{U}_Q = \sum_{i<j} \mathcal{V}(\vec{q}_i - \vec{q}_j), \quad \text{with} \quad \mathcal{V}(\vec{q}) = \frac{e^2}{4\pi|\vec{q}|} - c.$$

The constant c results from the background and ensures that the first-order correction vanishes, that is, $\int d^3\vec{q}\, \mathcal{V}(\vec{q}) = 0$.

(a) Show that the Fourier transform of $\mathcal{V}(\vec{q})$ takes the form

$$\tilde{\mathcal{V}}(\vec{\omega}) = \begin{cases} e^2/\omega^2 & \text{for } \vec{\omega} \neq 0 \\ 0 & \text{for } \vec{\omega} = 0 \end{cases}.$$

The Fourier transform of $\mathcal{V}(\vec{q})$ is singular at the origin, and can be defined explicitly as

$$\tilde{\mathcal{V}}(\vec{\omega}) = \lim_{\varepsilon \to 0} \int d^3\vec{q}\, \mathcal{V}(\vec{q})\, e^{i\vec{\omega}\cdot\vec{q} - \varepsilon q}.$$

The result at $\vec{\omega} = 0$ follows immediately from the definition of c. For $\vec{\omega} \neq 0$,

$$\tilde{\mathcal{V}}(\vec{\omega}) = \lim_{\varepsilon \to 0} \int d^3\vec{q} \left(\frac{e^2}{4\pi q} - c \right) e^{i\vec{\omega}\cdot\vec{q} - \varepsilon q} = \lim_{\varepsilon \to 0} \int d^3\vec{q} \left(\frac{e^2}{4\pi q} \right) e^{i\vec{\omega}\cdot\vec{q} - \varepsilon q}$$

$$= \lim_{\varepsilon \to 0} \left[2\pi \int_0^\pi \sin\theta\, d\theta \int_0^\infty q^2\, dq \left(\frac{e^2}{4\pi q} \right) e^{i\omega q \cos\theta - \varepsilon q} \right]$$

$$= -\frac{e^2}{2} \int_0^\infty \frac{e^{i\omega q - \varepsilon q} - e^{-i\omega q - \varepsilon q}}{i\omega}\, dq$$

$$= \lim_{\varepsilon \to 0} \frac{e^2}{2i\omega} \left(\frac{1}{i\omega - \varepsilon} + \frac{1}{i\omega + \varepsilon} \right) = \lim_{\varepsilon \to 0} \left(\frac{e^2}{\omega^2 + \varepsilon^2} \right) = \frac{e^2}{\omega^2}.$$

(b) In the cumulant expansion for $\langle \mathcal{U}_Q^\ell \rangle_c^0$, we shall retain only the diagrams forming a ring, which are proportional to

$$R_\ell = \int \frac{\mathrm{d}^3 \vec{q}_1}{V} \cdots \frac{\mathrm{d}^3 \vec{q}_\ell}{V} \, \mathcal{V}(\vec{q}_1 - \vec{q}_2)\mathcal{V}(\vec{q}_2 - \vec{q}_3) \cdots \mathcal{V}(\vec{q}_\ell - \vec{q}_1).$$

Use properties of Fourier transforms to show that

$$R_\ell = \frac{1}{V^{\ell-1}} \int \frac{\mathrm{d}^3 \vec{\omega}}{(2\pi)^3} \, \tilde{\mathcal{V}}(\vec{\omega})^\ell.$$

In the cumulant expansion for $\langle \mathcal{U}_Q^\ell \rangle_c^0$, we retain only the diagrams forming a ring. The contribution of these diagrams to the partition function is

$$R_\ell = \int \frac{\mathrm{d}^3 \vec{q}_1}{V} \frac{\mathrm{d}^3 \vec{q}_2}{V} \cdots \frac{\mathrm{d}^3 \vec{q}_\ell}{V} \mathcal{V}(\vec{q}_1 - \vec{q}_2)\mathcal{V}(\vec{q}_2 - \vec{q}_3) \cdots \mathcal{V}(\vec{q}_\ell - \vec{q}_1)$$

$$= \frac{1}{V^\ell} \int \cdots \int \mathrm{d}^3 \vec{x}_1 \, \mathrm{d}^3 \vec{x}_2 \cdots \mathrm{d}^3 \vec{x}_{\ell-1} \, \mathrm{d}^3 \vec{q}_\ell \mathcal{V}(\vec{x}_1)\mathcal{V}(\vec{x}_2) \cdots \mathcal{V}(\vec{x}_{\ell-1})\mathcal{V}\left(-\sum_{i=1}^{\ell-1} \vec{x}_i\right),$$

where we introduced the new set of variables $\{\vec{x}_i \equiv \vec{q}_i - \vec{q}_{i+1}\}$, for $i = 1, 2, \cdots, \ell - 1$. Note that since the integrand is independent of $\vec{q}_\ell$,

$$R_\ell = \frac{1}{V^{\ell-1}} \int \cdots \int \mathrm{d}^3 \vec{x}_1 \, \mathrm{d}^3 \vec{x}_2 \cdots \mathrm{d}^3 \vec{x}_{\ell-1} \mathcal{V}(\vec{x}_1)\mathcal{V}(\vec{x}_2) \cdots \mathcal{V}\left(-\sum_{i=1}^{\ell-1} \vec{x}_i\right).$$

Using the inverse Fourier transform

$$\mathcal{V}(\vec{q}) = \frac{1}{(2\pi)^3} \int \mathrm{d}^3 \vec{\omega} \cdot \tilde{\mathcal{V}}(\vec{\omega}) \mathrm{e}^{-\mathrm{i} \vec{q} \cdot \vec{\omega}},$$

the integral becomes

$$R_\ell = \frac{1}{(2\pi)^{3\ell} V^{\ell-1}} \int \cdots \int \mathrm{d}^3 \vec{x}_1 \cdots \mathrm{d}^3 \vec{x}_{\ell-1} \tilde{\mathcal{V}}(\vec{\omega}_1) \mathrm{e}^{-\mathrm{i} \vec{\omega}_1 \cdot \vec{x}_1} \tilde{\mathcal{V}}(\vec{\omega}_2) \mathrm{e}^{-\mathrm{i} \vec{\omega}_2 \cdot \vec{x}_2}$$

$$\cdots \tilde{\mathcal{V}}(\vec{\omega}_\ell) \exp\left(-\mathrm{i} \sum_{k=1}^{\ell-1} \vec{\omega}_\ell \cdot \vec{x}_k\right) \mathrm{d}^3 \vec{\omega}_1 \cdots \mathrm{d}^3 \vec{\omega}_\ell.$$

Since

$$\int \frac{\mathrm{d}^3 \vec{q}}{(2\pi)^3} \, \mathrm{e}^{-\mathrm{i} \vec{\omega} \cdot \vec{q}} = \delta^3(\vec{\omega}),$$

we have

$$R_\ell = \frac{1}{(2\pi)^3 V^{\ell-1}} \int \cdots \int \left(\prod_{k=1}^{\ell-1} \delta(\vec{\omega}_k - \vec{\omega}_\ell) \tilde{\mathcal{V}}(\vec{\omega}_k) \mathrm{d}^3 \vec{\omega}_k\right) \mathrm{d}^3 \vec{\omega}_\ell,$$

resulting finally in

$$R_\ell = \frac{1}{V^{\ell-1}} \int \frac{\mathrm{d}^3 \vec{\omega}}{(2\pi)^3} \tilde{\mathcal{V}}(\vec{\omega})^\ell.$$

(c) Show that the number of ring graphs generated in $\langle \mathcal{U}_Q^\ell \rangle_c^0$ is

$$S_\ell = \frac{N!}{(N-\ell)!} \times \frac{(\ell-1)!}{2} \approx \frac{(\ell-1)!}{2} N^\ell.$$

The number of ring graphs generated in $\langle \mathcal{U}_Q^\ell \rangle_c^0$ is given by the product of the number of ways to choose ℓ particles out of a total of N,

$$\frac{N!}{(N-\ell)!},$$

multiplied by the number of ways to arrange the ℓ particles in a ring,

$$\frac{\ell!}{2\ell}.$$

The numerator is the number of ways of distributing the ℓ labels on the ℓ points of the ring. This over-counts by the number of equivalent arrangements that appear in the denominator. The factor of $1/2$ comes from the equivalence of clockwise and counter-clockwise arrangements (reflection), and there are ℓ equivalent choices for the starting point of the ring (rotations). Hence

$$S_\ell = \frac{N!}{(N-\ell)!} \times \frac{\ell!}{2\ell} = \frac{N!}{(N-\ell)!} \times \frac{(\ell-1)!}{2}.$$

For $N \gg \ell$, we can approximate $N(N-1)\cdots(N-\ell+1) \approx N^\ell$, and

$$S_\ell \approx \frac{(\ell-1)!}{2} N^\ell.$$

Another way to justify this result is by induction: a ring of length $\ell+1$ can be created from a ring of ℓ links by inserting an additional point in between any of the existing ℓ nodes. Hence $S_{\ell+1} = S_\ell \times (N-\ell-1) \times \ell$, leading to the above result, when starting with $S_2 = N(N-1)/2$.

(d) Show that the contribution of the ring diagrams can be summed as

$$\ln Z_{\text{rings}} = \ln Z_0 + \sum_{\ell=2}^\infty \frac{(-\beta)^\ell}{\ell!} S_\ell R_\ell$$

$$\approx \ln Z_0 + \frac{V}{2} \int_0^\infty \frac{4\pi\omega^2 d\omega}{(2\pi)^3} \left[\left(\frac{\kappa}{\omega} \right)^2 - \ln\left(1 + \frac{\kappa^2}{\omega 2} \right) \right],$$

where $\kappa = \sqrt{\beta e^2 N/V}$ is the inverse Debye screening length.
(*Hint.* Use $\ln(1+x) = -\sum_{\ell=1}^\infty (-x)^\ell/\ell$.)

The contribution of the ring diagrams is summed as

$$\ln Z_{\text{rings}} = \ln Z_0 + \sum_{\ell=2}^\infty \frac{(-\beta)^\ell}{\ell!} S_\ell R_\ell$$

$$= \ln Z_0 + \sum_{\ell=2}^\infty \frac{(-\beta)^\ell}{\ell!} \frac{(\ell-1)!}{2} N^\ell \frac{1}{V^{\ell-1}} \int \frac{d^3\vec{\omega}}{(2\pi)^3} \tilde{\mathcal{V}}(\vec{\omega})^\ell$$

$$= \ln Z_0 + \frac{V}{2} \int_0^\infty \frac{4\pi\omega^2 \, d\omega}{(2\pi)^3} \sum_{\ell=2}^\infty \frac{1}{\ell} \left(-\frac{\beta N}{V} \tilde{\mathcal{V}}(\omega) \right)^\ell$$

$$= \ln Z_0 + \frac{V}{2} \int_0^\infty \frac{4\pi\omega^2 \, d\omega}{(2\pi)^3} \sum_{\ell=2}^\infty \frac{1}{\ell} \left(-\frac{\beta Ne^2}{V\omega^2} \right)^\ell$$

$$= \ln Z_0 + \frac{V}{2} \int_0^\infty \frac{4\pi\omega^2 \, d\omega}{(2\pi)^3} \left[\frac{\beta Ne^2}{V\omega^2} - \ln\left(1 + \frac{\beta Ne^2}{V\omega^2} \right) \right],$$

where we have used $\ln(1+x) = -\sum_{\ell=1}^\infty (-x)^\ell/\ell$. *Finally, substituting* $\kappa = \sqrt{\beta e^2 N/V}$
leads to

$$\ln Z_{\text{rings}} = \ln Z_0 + \frac{V}{2} \int_0^\infty \frac{4\pi\omega^2 \, d\omega}{(2\pi)^3} \left[\left(\frac{\kappa}{\omega} \right)^2 - \ln\left(1 + \frac{\kappa^2}{\omega^2} \right) \right].$$

(e) The integral in the previous part can be simplified by changing variables to $x = \kappa/\omega$,
and performing integration by parts. Show that the final result is

$$\ln Z_{\text{rings}} = \ln Z_0 + \frac{V}{12\pi} \kappa^3.$$

Changing variables to $x = \kappa/\omega$, *and integrating the integrand by parts, gives*

$$\int_0^\infty \omega^2 \, d\omega \left[\left(\frac{\kappa}{\omega} \right)^2 - \ln\left(1 + \frac{\kappa^2}{\omega^2} \right) \right] = \kappa^3 \int_0^\infty \frac{dx}{x^4} \left[x^2 - \ln(1+x^2) \right]$$

$$= \frac{\kappa^3}{3} \int_0^\infty \frac{dx}{x^3} \left[2x - \frac{2x}{1+x^2} \right] = \frac{2\kappa^3}{3} \int_0^\infty \frac{dx}{1+x^2} = \frac{\pi\kappa^3}{3},$$

resulting in

$$\ln Z_{\text{rings}} = \ln Z_0 + \frac{V}{4\pi^2} \cdot \frac{\pi\kappa^3}{3} = \ln Z_0 + \frac{V}{12\pi} \kappa^3.$$

(f) Calculate the correction to pressure from the above ring diagrams.

*The correction to the ideal gas pressure due to the Debye–Hückel approximation
is*

$$P = k_B T \left(\frac{\partial \ln Z_{\text{rings}}}{\partial V} \bigg|_{T,N} \right)$$

$$= P_0 + k_B T \frac{\partial}{\partial V} \left(\frac{V\kappa^3}{12\pi} \right) \bigg|_{T,N} = P_0 - \frac{k_B T}{24\pi} \kappa^3$$

$$= P_0 - \frac{k_B T}{24\pi} \left(\frac{e^2 N}{k_B TV} \right)^{3/2}.$$

*Note that the correction to the ideal gas behavior is non-analytic, and cannot be
expressed by a virial series. This is due to the long-range nature of the Coulomb
interaction.*

(g) We can introduce an effective potential $\overline{V}(\vec{q} - \vec{q}')$ between two particles by integrating
over the coordinates of all the other particles. This is equivalent to an expectation
value that can be calculated perturbatively in a cumulant expansion. If we include
only the loopless diagrams (the analog of the rings) between the particles, we have

$$\overline{V}(\vec{q} - \vec{q}') = V(\vec{q} - \vec{q}') + \sum_{\ell=1}^\infty (-\beta N)^\ell \int \frac{d^3\vec{q}_1}{V} \cdots \frac{d^3\vec{q}_\ell}{V} \, \mathcal{V}(\vec{q} - \vec{q}_1) \mathcal{V}(\vec{q}_1 - \vec{q}_2) \cdots$$

$$\mathcal{V}(\vec{q}_\ell - \vec{q}').$$

Show that this sum leads to the screened Coulomb interaction $\overline{V}(\vec{q}) = e^2 \exp(-\kappa|\vec{q}|)/(4\pi|\vec{q}|)$.

Introducing the effective potential $\overline{\mathcal{V}}(\vec{q} - \vec{q}')$, and summing over the loopless diagrams, gives

$$\overline{\mathcal{V}}(\vec{q} - \vec{q}') = \mathcal{V}(\vec{q} - \vec{q}') + \sum_{\ell=1}^{\infty} (-\beta N)^{\ell} \int \frac{\mathrm{d}^3 \vec{q}_1}{V} \cdots \frac{\mathrm{d}^3 \vec{q}_{\ell}}{V} \mathcal{V}(\vec{q} - \vec{q}_1) \mathcal{V}(\vec{q}_1 - \vec{q}_2) \cdots$$

$$\mathcal{V}(\vec{q}_{\ell} - \vec{q}')$$

$$= \mathcal{V}(\vec{q} - \vec{q}') - \beta N \int \frac{\mathrm{d}^3 \vec{q}_1}{V} \mathcal{V}(\vec{q} - \vec{q}_1) \mathcal{V}(\vec{q}_1 - \vec{q}')$$

$$+ (\beta N)^2 \int \frac{\mathrm{d}^3 \vec{q}_1}{V} \frac{\mathrm{d}^3 \vec{q}_2}{V} \mathcal{V}(\vec{q} - \vec{q}_1) \mathcal{V}(\vec{q}_1 - \vec{q}_2) \mathcal{V}(\vec{q}_2 - \vec{q}') - \cdots .$$

Using the changes of notation

$$\vec{x}_1 \equiv \vec{q}, \quad \vec{x}_2 \equiv \vec{q}', \quad \vec{x}_3 \equiv \vec{q}_1, \quad \vec{x}_4 \equiv \vec{q}_2, \quad \cdots \quad \vec{x}_{\ell} \equiv \vec{q}_{\ell},$$

$$\mathcal{V}_{12} \equiv \mathcal{V}(\vec{x}_1 - \vec{x}_2), \quad \text{and} \quad n \equiv N/V,$$

we can write

$$\overline{\mathcal{V}}_{12} = \mathcal{V}_{12} - \beta n \int \mathrm{d}^3 \vec{x}_3 \mathcal{V}_{13} \mathcal{V}_{32} + (\beta n)^2 \int \mathrm{d}^3 \vec{x}_3 \, \mathrm{d}^3 \vec{x}_4 \mathcal{V}_{13} \mathcal{V}_{34} \mathcal{V}_{42} - \cdots .$$

Using the inverse Fourier transform (as in part (a)), and the notation $\vec{x}_{ij} \equiv \vec{x}_i - \vec{x}_j$,

$$\overline{\mathcal{V}}_{12} = \mathcal{V}_{12} - \beta n \int \frac{\mathrm{d}^3 \vec{x}_3}{(2\pi)^6} \tilde{\mathcal{V}}(\vec{\omega}_{13}) \tilde{\mathcal{V}}(\vec{\omega}_{32}) \, \mathrm{e}^{-\mathrm{i}(\vec{x}_{13} \cdot \vec{\omega}_{13} + \vec{x}_{32} \cdot \vec{\omega}_{32})} \, \mathrm{d}^3 \vec{\omega}_{13} \, \mathrm{d}^3 \vec{\omega}_{32} + \cdots ,$$

and employing the delta function, as in part (a)

$$\overline{\mathcal{V}}_{12} = \mathcal{V}_{12} - \beta n \int \frac{\mathrm{d}^3 \vec{\omega}_{13} \, \mathrm{d}^3 \vec{\omega}_{32}}{(2\pi)^3} \delta^3 \left(\vec{\omega}_{13} - \vec{\omega}_{32} \right) \tilde{\mathcal{V}}(\vec{\omega}_{13}) \tilde{\mathcal{V}}(\vec{\omega}_{32}) \exp[\vec{x}_1 \cdot \vec{\omega}_{13} - \vec{x}_2 \cdot \vec{\omega}_{32}] + \cdots$$

$$= \mathcal{V}_{12} - \beta n \int \frac{\mathrm{d}^3 \vec{\omega}}{(2\pi)^3} \left[\tilde{\mathcal{V}}(\vec{\omega}) \right]^2 \exp[\vec{\omega} \cdot \vec{x}_{12}] + \cdots .$$

Generalizing this result and dropping the subscript such that $\vec{x} \equiv \vec{x}_{12}$,

$$\overline{\mathcal{V}}_{12} = \mathcal{V}_{12} + \sum_{\ell=1}^{\infty} \frac{(-\beta n)^{\ell}}{(2\pi)^3} \int \left[\tilde{\mathcal{V}}(\vec{\omega}) \right]^{\ell+1} \mathrm{e}^{\mathrm{i}\vec{x} \cdot \vec{\omega}} \, \mathrm{d}^3 \vec{\omega}.$$

Finally, including the Fourier transform of the direct potential (first term), gives

$$\overline{\mathcal{V}}_{12} = \sum_{\ell=0}^{\infty} \int \frac{\mathrm{d}^3 \vec{\omega}}{(2\pi)^3} (-\beta n)^{\ell} \frac{e^{2\ell+2}}{\omega^{2\ell+2}} \, \mathrm{e}^{\mathrm{i}\vec{x} \cdot \vec{\omega}} = \sum_{\ell=0}^{\infty} \int \frac{\mathrm{d}^3 \vec{\omega}}{(2\pi)^3} \frac{(-1)^{\ell} e^2 \kappa^{2\ell}}{\omega^{2\ell+2}} \, \mathrm{e}^{\mathrm{i}x\omega \cos \theta}$$

$$= \int_0^{\infty} \mathrm{d}\omega \sum_{\ell=0}^{\infty} \frac{(-1)^{\ell} e^2}{2\pi^2} \left(\frac{\kappa}{\omega} \right)^{2\ell} \int_{-1}^{1} \mathrm{e}^{\mathrm{i}x\omega \cos \theta} \, \mathrm{d} \cos \theta$$

$$= \int_0^{\infty} \mathrm{d}\omega \frac{e^2}{2\pi^2} \frac{2 \sin x\omega}{x\omega} \sum_{\ell=0}^{\infty} (-1)^{\ell} \left(\frac{\kappa}{\omega} \right)^{2\ell} .$$

Setting $y \equiv \omega/\kappa$ gives

$$\overline{\mathcal{V}}_{12} = \frac{1}{2} \int_{-\infty}^{\infty} \frac{e^2}{2\pi^2} \kappa \frac{e^{ix\kappa y} - e^{-ix\kappa y}}{2ix\kappa y} \frac{-1}{y^2 + 1} \, \mathrm{d}y.$$

Intergrating in the complex plane, via the residue theorem, gives

$$\overline{\mathcal{V}}_{12} = \frac{e^2}{4\pi^2} \left(\frac{e^{-\kappa x}}{2x} + \frac{e^{-\kappa x}}{2x} \right) \cdot \pi = \frac{e^2 e^{-\kappa x}}{4\pi x}.$$

Recalling our original notation, $x = |\vec{q} - \vec{q}'| \equiv |\vec{q}|$, we obtain the screened Coulomb potential

$$\overline{\mathcal{V}}(\vec{q}) = \frac{e^2}{4\pi} \frac{e^{-\kappa|\vec{q}|}}{|\vec{q}|}.$$

$$********$$

2. *Virial coefficients*: consider a gas of particles in d-dimensional space interacting through a pairwise central potential, $\mathcal{V}(r)$, where

$$\mathcal{V}(r) = \begin{cases} +\infty & \text{for } 0 < r < a, \\ -\varepsilon & \text{for } a < r < b, \\ 0 & \text{for } b < r < \infty. \end{cases}$$

(a) Calculate the second virial coefficient $B_2(T)$, and comment on its high- and low-temperature behaviors.

The second virial coefficient is obtained from

$$B_2 \equiv -\frac{1}{2} \int \mathrm{d}^d r_{12} \left\{ \exp[-\beta \mathcal{V}(r_{12})] - 1 \right\},$$

where $r_{12} \equiv |\vec{r}_1 - \vec{r}_2|$, as

$$B_2 = -\frac{1}{2} \left[\int_0^a \mathrm{d}^d r_{12}(-1) + \int_a^b \mathrm{d}^d r_{12} \left(e^{\beta \varepsilon} - 1 \right) \right]$$

$$= -\frac{1}{2} \left\{ V_d(a)(-1) + [V_d(b) - V_d(a)] \cdot [\exp(\beta \varepsilon) - 1] \right\},$$

where

$$V_d(r) = \frac{S_d}{d} r^d = \frac{2\pi^{d/2}}{d \, (d/2 - 1)!} r^d,$$

is the volume of a d-dimensional sphere of radius r. Thus,

$$B_2(T) = \frac{1}{2} V_d(b) - \frac{1}{2} \exp(\beta \varepsilon) \left[V_d(b) - V_d(a) \right].$$

For high temperatures $\exp(\beta \varepsilon) \approx 1 + \beta \varepsilon$, and

$$B_2(T) \approx \frac{1}{2} V_d(a) - \frac{\beta \varepsilon}{2} \left[V_d(b) - V_d(a) \right].$$

At the highest temperatures, $\beta\varepsilon \ll 1$, the hard-core part of the potential is dominant, and

$$B_2(T) \approx \frac{1}{2}V_d(a).$$

For low temperatures $\beta \gg 1$, the attractive component takes over, and

$$B_2 = -\frac{1}{2}\{V_d(a)(-1) + [V_d(b) - V_d(a)] \cdot [\exp(\beta\varepsilon) - 1]\}$$

$$\approx -\frac{1}{2}[V_d(b) - V_d(a)]\exp(\beta\varepsilon),$$

resulting in $B_2 < 0$.

(b) Calculate the first correction to isothermal compressibility

$$\kappa_T = -\frac{1}{V}\frac{\partial V}{\partial P}\bigg|_{T,N}.$$

The isothermal compressibility is defined by

$$\kappa_T \equiv -\frac{1}{V}\frac{\partial V}{\partial P}\bigg|_{T,N}.$$

From the expansion

$$\frac{P}{k_BT} = \frac{N}{V} + \frac{N^2}{V^2}B_2,$$

for constant temperature and particle number, we get

$$\frac{1}{k_BT}dP = -\frac{N}{V^2}dV - 2B_2\frac{N^2}{V^3}dV.$$

Thus

$$\frac{\partial V}{\partial P}\bigg|_{T,N} = -\frac{1}{k_BT}\frac{1}{N/V^2 + 2B_2N^2/V^3} = -\frac{V^2}{Nk_BT}\left(\frac{1}{1 + 2B_2N/V}\right),$$

and

$$\kappa_T = \frac{V}{Nk_BT}\left(\frac{1}{1 + 2B_2N/V}\right) \approx \frac{V}{Nk_BT}\left(1 - 2B_2\frac{N}{V}\right).$$

(c) In the high-temperature limit, reorganize the equation of state into the van der Waals form, and identify the van der Waals parameters.

Including the correction introduced by the second virial coefficient, the equation of state becomes

$$\frac{PV}{Nk_BT} = 1 + \frac{N}{V}B_2(T).$$

Using the expression for B_2 in the high-temperature limit,

$$\frac{PV}{Nk_BT} = 1 + \frac{N}{2V}\{V_d(a) - \beta\varepsilon[V_d(b) - V_d(a)]\},$$

and

$$P + \frac{N^2}{2V^2}\varepsilon[V_d(b) - V_d(a)] = k_BT\frac{N}{V}\left(1 + \frac{N}{2V}V_d(a)\right).$$

Using the variable $n = N/V$, and noting that for low concentrations

$$1 + \frac{n}{2}V_d(a) \approx \left(1 - \frac{n}{2}V_d(a)\right)^{-1} = V\left(V - \frac{N}{2}V_d(a)\right)^{-1},$$

the equation of state becomes

$$\left(P + \frac{n^2\varepsilon}{2}[V_d(b) - V_d(a)]\right) \cdot \left(V - \frac{N}{2}V_d(a)\right) = Nk_BT.$$

This can be recast in the usual van der Waals form

$$(P - an^2) \cdot (V - Nb) = Nk_BT,$$

with

$$a = \frac{\varepsilon}{2}[V_d(b) - V_d(a)], \quad \text{and} \quad b = \frac{1}{2}V_d(a).$$

(d) For $b = a$ (a hard sphere), and $d = 1$, calculate the third virial coefficient $B_3(T)$.

By definition, the third virial coefficient is

$$B_3 = -\frac{1}{3}\int d^d r \, d^d r' f(r)f(r')f(r - r'),$$

where, for a hard-core gas

$$f(r) \equiv \exp\left(-\frac{\mathcal{V}(r)}{k_BT}\right) - 1 = \begin{cases} -1 & \text{for } 0 < r < a, \\ 0 & \text{for } a < r < \infty \end{cases}.$$

In one dimension, the only contributions come from $0 < r$, and $r' < a$, where $f(r) = f(r') = -1$. Using the notations $|x| \equiv r$, $|y| \equiv r'$ (i.e., $-a < x$, and $y < a$),

$$B_3 = -\frac{1}{3}\int_{-a}^{a} dx \int_{-a}^{a} dy \cdot f(x - y) = \frac{1}{3}\int\int_{-a<x,y<a,-a<x-y<a} (-1) = \frac{1}{3}\frac{6}{8}(2a)^2 = a^2,$$

where the relevant integration area is plotted below.

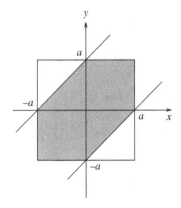

$$\ast\ast\ast\ast\ast\ast\ast\ast$$

3. *Dieterici's equation*: a gas obeys Dieterici's equation of state:

$$P(v - b) = k_B T \exp\left(-\frac{a}{k_B T v}\right),$$

where $v = V/N$.

(a) Find the ratio $Pv/k_B T$ at the critical point.

The critical point is the point of inflection, described by

$$\left.\frac{\partial P}{\partial v}\right|_{T_c, N} = 0, \quad \text{and} \quad \left.\frac{\partial^2 P}{\partial v^2}\right|_{T_c, N} = 0.$$

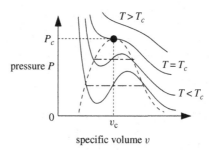

The first derivative of P is

$$\left.\frac{\partial P}{\partial v}\right|_{T_c, N} = \frac{\partial}{\partial v}\left[\frac{k_B T}{v - b}\exp\left(-\frac{a}{k_B T v}\right)\right] = \frac{k_B T}{v - b}\exp\left(-\frac{a}{k_B T v}\right)\left(\frac{a}{k_B T v^2} - \frac{1}{v - b}\right)$$

$$= P\left(\frac{a}{k_B T v^2} - \frac{1}{v - b}\right),$$

while a second derivative gives

$$\left.\frac{\partial^2 P}{\partial v^2}\right|_{T_c, N} = \frac{\partial}{\partial v}\left[P\left(\frac{a}{k_B T v^2} - \frac{1}{v - b}\right)\right]$$

$$= \frac{\partial P}{\partial v}\left(\frac{a}{k_B T v^2} - \frac{1}{v - b}\right) - P\left(\frac{2a}{k_B T v^3} - \frac{1}{(v - b)^2}\right).$$

Therefore v_c and T_c are determined by

$$\frac{a}{k_B T_c v_c^2} - \frac{1}{v_c - b} = 0, \quad \text{and} \quad \frac{2a}{k_B T_c v_c^3} - \frac{1}{(v_c - b)^2} = 0,$$

with the solutions

$$v_c = 2b, \quad \text{and} \quad k_B T_c = \frac{a}{4b}.$$

The critical pressure is

$$P_c = \frac{k_B T_c}{v_c - b}\exp\left(-\frac{a}{k_B T_c v_c}\right) = \frac{a}{4b^2}\,e^{-2},$$

resulting in the ratio

$$\frac{P_c v_c}{k_B T_c} = 2e^{-2} \approx 0.27.$$

Note that for the van der Waals gas

$$\frac{P_c v_c}{k_B T_c} = \frac{3}{8} = 0.375,$$

while for some actual gases

$$\left(\frac{P_c v_c}{k_B T_c}\right)_{\text{water}} = 0.230, \quad \text{and} \quad \left(\frac{P_c v_c}{k_B T_c}\right)_{\text{argon}} = 0.291.$$

(b) Calculate the isothermal compressibility κ_T for $v = v_c$ as a function of $T - T_c$.

   The isothermal compressibility is defined by

$$\kappa_T \equiv -\frac{1}{v} \left. \frac{\partial v}{\partial P} \right|_{T,N},$$

   and from part (a), given by

$$\left. \frac{\partial P}{\partial v} \right|_{T_c,N} = P\left(\frac{a}{k_B T v^2} - \frac{1}{v - b}\right).$$

   Expanding this expression, at $v = v_c$, in terms of $t \equiv k_B T - k_B T_c$ (for $T > T_c$), yields

$$\left. \frac{\partial P}{\partial v} \right|_{T_c,N} \approx P_c\left(\frac{a}{(a/4b + t)\, 4b^2} - \frac{1}{b}\right) \approx -\frac{P_c}{b} \frac{4bt}{a} = -\frac{2P_c}{v_c k_B T_c} t,$$

   and thus

$$\kappa_T = \frac{k_B T_c}{2P_c} \frac{1}{t} = \frac{be^2}{2k_B(T - T_c)}.$$

   Note that expanding any analytic equation of state will yield the same simple pole for the divergence of the compressibility.

(c) On the critical isotherm expand the pressure to the lowest non-zero order in $(v - v_c)$.

   Perform a Taylor series expansion along the critical isotherm $T = T_c$, as

$$P(v, T_c) = P_c + \left. \frac{\partial P}{\partial v} \right|_{T_c, v_c} (v - v_c) + \frac{1}{2!} \left. \frac{\partial^2 P}{\partial v^2} \right|_{T_c, v_c} (v - v_c)^2 + \frac{1}{3!} \left. \frac{\partial^3 P}{\partial v^3} \right|_{T_c, v_c} (v - v_c)^3 + \cdots.$$

   The first two terms are zero at the critical point, and

$$\begin{aligned}
\left. \frac{\partial^3 P}{\partial v^3} \right|_{T_c, v_c} &= -P_c \frac{\partial}{\partial v}\left(\frac{2a}{k_B T_c v^3} - \frac{1}{(v - b)^2}\right) \\
&= -P_c\left(\frac{6a}{k_B T_c v_c^4} - \frac{2}{(v_c - b)^3}\right) \\
&= -\frac{P_c}{2b^3}.
\end{aligned}$$

Substituting this into the Taylor expansion for $P(v, T_c)$ results in

$$P(v, T_c) = P_c \left(1 - \frac{(v - v_c)^3}{12b^3} \right),$$

which is equivalent to

$$\frac{P}{P_c} - 1 = \frac{2}{3} \left(\frac{v}{v_c} - 1 \right)^3.$$

****

4. *Two-dimensional Coulomb gas*: consider a classical mixture of N positive and N negative charged particles in a *two-dimensional* box of area $A = L \times L$. The Hamiltonian is

$$\mathcal{H} = \sum_{i=1}^{2N} \frac{\vec{p}_i^{\,2}}{2m} - \sum_{i<j}^{2N} c_i c_j \ln |\vec{q}_i - \vec{q}_j|,$$

where $c_i = +c_0$ for $i = 1, \cdots, N$, and $c_i = -c_0$ for $i = N+1, \cdots, 2N$, denote the charges of the particles; $\{\vec{q}_i\}$ and $\{\vec{p}_i\}$ their coordinates and momenta, respectively.

(a) Note that in the interaction term each pair appears only once, and that there is no self-interaction $i = j$. How many pairs have repulsive interactions, and how many have attractive interactions?

There are N positively charged particles, and N negatively charged particles. Hence there are $N \cdot N = N^2$ pairs of opposite charges, and $n_{\text{attractive}} = N^2$. For like charges, we can choose pairs from the N particles of positive charge, or from the N particles with negative charges. Hence the number of pairs of like pairs is

$$n_{\text{repulsive}} = 2 \times \binom{N}{2} = 2 \times \frac{N!}{2!(N-2)!} = N(N-1).$$

(b) Write down the expression for the partition function $Z(N, T, A)$ in terms of integrals over $\{\vec{q}_i\}$ and $\{\vec{p}_i\}$. Perform the integrals over the momenta, and rewrite the contribution of the coordinates as a product involving powers of $\{\vec{q}_i\}$, using the identity $e^{\ln x} = x$.

The partition function is

$$Z(N, T, A) = \frac{1}{(N!)^2 h^{4N}} \int \prod_{i=1}^{2N} d^2 \vec{q}_i \, d^2 \vec{p}_i \exp \left[-\beta \sum_{i=1}^{2N} \frac{p_i^2}{2m} + \beta \sum_{i<j} c_i c_j \ln |\vec{q}_i - \vec{q}_j| \right]$$

$$= \frac{1}{\lambda^{4N} (N!)^2} \int \prod_{i=1}^{2N} d^2 \vec{q}_i \exp \left[\beta \ln |\vec{q}_i - \vec{q}_j|^{c_i c_j} \right],$$

where $\lambda = h/\sqrt{2\pi m k_B T}$. Further simplifying the expression for the partition function

$$Z(N, T, A) = \frac{1}{\lambda^{4N} (N!)^2} \int \prod_{i=1}^{2N} d^2 \vec{q}_i \prod_{i<j}^{2N} |\vec{q}_i - \vec{q}_j|^{\beta c_i c_j},$$

where we have used the fact that $e^{\ln x} = x$.

(c) Although it is not possible to perform the integrals over $\{\vec{q}_i\}$ exactly, the dependence of Z on A can be obtained by the simple rescaling of coordinates, $\vec{q}_i{}' = \vec{q}_i/L$. Use the results in parts (a) and (b) to show that $Z \propto A^{2N - \beta c_0^2 N/2}$.

The only length scale appearing in the problem is set by the system size L. Rescaling the expression using $\vec{q}_i{}' = \vec{q}_i/L$ then yields

$$Z(N, T, A) = \frac{1}{\lambda^{4N}(N!)^2} \int \prod_{i=1}^{2N} \left(L^2\,d^2\vec{q}_i{}'\right) \prod_{i<j}^{2N} L^{\beta c_i c_j} |\vec{q}_i{}' - \vec{q}_j{}'|^{\beta c_i c_j}.$$

Note that there are N^2 terms for which the interaction is attractive $\left(\beta c_i c_j = -\beta c_0^2\right)$, and $N(N-1)$ terms for which the interaction is repulsive $\left(\beta c_i c_j = \beta c_0^2\right)$. Thus

$$Z(N, T, A) = L^{4N} \cdot L^{\beta c_0^2 N(N-1)} \cdot L^{-\beta c_0^2 N^2} \frac{1}{\lambda^{4N}(N!)^2} \int \prod_{i=1}^{2N} d^2\vec{q}_i{}' \prod_{i<j}^{2N} |\vec{q}_i{}' - \vec{q}_j{}'|^{\beta c_i c_j}$$

$$= L^{4N - \beta N c_0^2} Z_0(N, T, A' = L'^2 = 1) \propto A^{2N - \beta c_0^2 N/2},$$

since $A = L^2$.

(d) Calculate the two-dimensional pressure of this gas, and comment on its behavior at high and low temperatures.

The pressure is then calculated from

$$P = -\frac{1}{\beta} \left.\frac{\partial \ln Z}{\partial A}\right|_{N,T} = k_B T \frac{\partial}{\partial A} \ln\left(A^{2N - \beta c_0^2 N/2} Z_0\right)$$

$$= k_B T \left(2N - \beta c_0^2 N/2\right) \frac{\partial}{\partial A} \ln A = \frac{2N k_B T}{A} - \frac{N c_0^2}{2A}.$$

At high temperatures,

$$P = \frac{2N k_B T}{A},$$

which is the ideal gas behavior for $2N$ particles. The pressure becomes negative at temperatures below

$$T_c^0 = \frac{c_0^2}{4 k_B},$$

which is unphysical, indicating the collapse of the particles due to their attractions.

(e) The unphysical behavior at low temperatures is avoided by adding a hard core that prevents the coordinates of any two particles from coming closer than a distance a. The appearance of two length scales, a and L, makes the scaling analysis of part (c) questionable. By examining the partition function for $N = 1$, obtain an estimate for the temperature T_c at which the short distance scale a becomes important in calculating the partition function, invalidating the result of the previous part. What are the phases of this system at low and high temperatures?

A complete collapse of the system (to a single point) can be avoided by adding a hard-core repulsion that prevents any two particles from coming closer than a distance a. The partition function for two particles (i.e., $N = 1$) is now given by

$$Z(N = 1, T, A) = \frac{1}{\lambda^4} \int d^2\vec{q}_1\,d^2\vec{q}_2 \cdot |\vec{q}_1 - \vec{q}_2|^{-\beta c_0^2}.$$

To evaluate this integral, first change to center of mass and relative coordinates

$$\begin{cases} \vec{Q} = \frac{1}{2}(\vec{q}_1 + \vec{q}_2), \\ \vec{q} = \vec{q}_1 - \vec{q}_2. \end{cases}$$

Integrating over the center of mass gives

$$Z(N = 1, T, A) = \frac{A}{\lambda^4} \int d^2\vec{q}\, q^{-\beta c_0^2} \approx \frac{2\pi A}{\lambda^4} \int_a^L dq \cdot q^{1 - \beta c_0^2}$$

$$= \frac{2\pi A}{\lambda^4} \frac{q^{2-\beta c_0^2}}{2 - \beta c_0^2}\bigg|_a^L = \frac{2\pi A}{\lambda^4} \frac{L^{2-\beta c_0^2} - a^{2-\beta c_0^2}}{2 - \beta c_0^2}.$$

If $2 - \beta c_0^2 < 0$, as $L \to \infty$,

$$Z \approx \frac{2\pi A}{\lambda^4} \frac{a^{2-\beta c_0^2}}{2 - \beta c_0^2}$$

is controlled by the short distance cutoff a; while if $2 - \beta c_0^2 > 0$, the integral is controlled by the system size L, as assumed in part (c). Hence the critical temperature can be estimated by $\beta c_0^2 = 2$, giving

$$T_c = \frac{c_0^2}{2k_B},$$

which is larger than T_c^0 by a factor of 2. Thus the unphysical collapse at low temperatures is pre-empted at the higher temperature where the hard cores become important. The high-temperature phase $(T > T_c)$ is a dissociated plasma; while the low-temperature phase is a gas of paired dipoles.

$$* * * * * * * *$$

5. *Exact solutions for a one-dimensional gas*: in statistical mechanics, there are very few systems of interacting particles that can be solved *exactly*. Such exact solutions are very important as they provide a check for the reliability of various approximations. A one-dimensional gas with short-range interactions is one such solvable case.

(a) Show that for a potential with a hard core that screens the interactions from further neighbors, the Hamiltonian for N particles can be written as

$$\mathcal{H} = \sum_{i=1}^{N} \frac{p_i^2}{2m} + \sum_{i=2}^{N} \mathcal{V}(x_i - x_{i-1}).$$

The (indistinguishable) particles are labeled with coordinates $\{x_i\}$ such that

$$0 \le x_1 \le x_2 \le \cdots \le x_N \le L,$$

where L is the length of the box confining the particles.

   Each particle i interacts only with adjacent particles $i - 1$ and $i + 1$, as the hard cores from these nearest neighbors screen the interactions with any other particle. Thus we need only consider nearest-neighbor interactions, and, including the kinetic energies, the Hamiltonian is given by

$$\mathcal{H} = \sum_{i=1}^{N} \frac{p_i^2}{2m} + \sum_{i=2}^{N} \mathcal{V}(x_i - x_{i-1}), \quad \text{for} \quad 0 \le x_1 \le x_2 \le \cdots x_N \le L.$$

(b) Write the expression for the partition function $Z(T, N, L)$. Change variables to $\delta_1 = x_1$, $\delta_2 = x_2 - x_1$, $\cdots$, $\delta_N = x_N - x_{N-1}$, and carefully indicate the allowed ranges of integration and the constraints.

The partition function is

$$Z(T, N, L) = \frac{1}{h^N} \int_0^L dx_1 \int_{x_1}^L dx_2 \cdots \int_{x_{N-1}}^L dx_N \exp\left[-\beta \sum_{i=2}^N \mathcal{V}(x_i - x_{i-1})\right]$$

$$\cdot \int_{-\infty}^\infty dp_1 \cdots \int_{-\infty}^\infty dp_N \exp\left[-\beta \sum_{i=1}^N N \frac{p_i^2}{2m}\right]$$

$$= \frac{1}{\lambda^N} \int_0^L dx_1 \int_{x_1}^L dx_2 \cdots \int_{x_{N-1}}^L dx_N \exp\left[-\beta \sum_{i=2}^N \mathcal{V}(x_i - x_{i-1})\right],$$

where $\lambda = h/\sqrt{2\pi m k_B T}$. (Note that there is no $N!$ factor, as the ordering of the particles is specified.) Introducing a new set of variables

$$\delta_1 = x_1, \quad \delta_2 = x_2 - x_1, \quad \cdots \quad \delta_n = x_N - x_{N-1},$$

or equivalently

$$x_1 = \delta_1, \quad x_2 = \delta_1 + \delta_2, \quad \cdots \quad x_N = \sum_{i=1}^N \delta_i,$$

the integration becomes

$$Z(T, N, L) = \frac{1}{\lambda^N} \int_0^L d\delta_1 \int_0^{L-\delta_1} d\delta_2 \int_0^{L-(\delta_1+\delta_2)} d\delta_3 \cdots \int_0^{L-\sum_{i=1}^N \delta_i} d\delta_N \, e^{-\beta \sum_{i=2}^N \mathcal{V}(\delta_i)}.$$

This integration can also be expressed as

$$Z(T, N, L) = \frac{1}{\lambda^N} \left[\int d\delta_1 \, d\delta_2 \cdots d\delta_N\right]' \exp\left[-\beta \sum_{i=2}^N \mathcal{V}(\delta_i)\right],$$

with the constraint

$$0 \le \sum_{i=1}^N \delta_i \le L.$$

This constraint can be put into the equation explicitly with the use of the step function

$$\Theta(x) = \begin{cases} 0 & \text{for } x < 0, \\ 1 & \text{for } x \ge 0, \end{cases}$$

as

$$Z(T, N, L) = \frac{1}{\lambda^N} \int_0^\infty d\delta_1 \int_0^\infty d\delta_2 \cdots \int_0^\infty d\delta_N \exp\left[-\beta \sum_{i=2}^N \mathcal{V}(\delta_i)\right] \Theta\left(L - \sum_{i=1}^N \delta_i\right).$$

(c) Consider the Gibbs partition function obtained from the Laplace transformation

$$\mathcal{Z}(T, N, P) = \int_0^\infty dL \exp(-\beta P L) Z(T, N, L),$$

and by extremizing the integrand find the standard formula for P in the canonical ensemble.

The Gibbs partition function is

$$\mathcal{Z}(T, N, P) = \int_0^\infty dL \exp(-\beta PL) Z(T, N, L).$$

The saddle point is obtained by extremizing the integrand with respect to L,

$$\frac{\partial}{\partial L} \exp(-\beta PL) Z(T, N, L)\bigg|_{T,N} = 0,$$

which implies that

$$\beta P = \frac{\partial}{\partial L} \ln Z(T, N, L)\bigg|_{T,N}, \qquad \Longrightarrow \qquad P = k_B T \frac{\partial \ln Z}{\partial L}\bigg|_{T,N}.$$

From thermodynamics, for a one-dimensional gas we have

$$dF = -SdT - PdL, \Longrightarrow P = -\frac{\partial F}{\partial L}\bigg|_{T,N}.$$

Further noting that

$$F = -k_B T \ln Z$$

again results in

$$P_{\text{canonical}} = k_B T \frac{\partial \ln Z}{\partial L}\bigg|_{T,N}.$$

(d) Change variables from L to $\delta_{N+1} = L - \sum_{i=1}^{N} \delta_i$, and find the expression for $\mathcal{Z}(T, N, P)$ as a product over one-dimensional integrals over each δ_i.

The expression for the partition function given above is

$$Z(T, N, L) = \frac{1}{\lambda^N} \int_0^\infty d\delta_1 \int_0^\infty d\delta_2 \cdots \int_0^\infty d\delta_N \exp\left[-\beta \sum_{i=2}^{N} \mathcal{V}(\delta_i)\right] \Theta\left(L - \sum_{i=1}^{N} \delta_i\right).$$

The Laplace transform of this equation is

$$\mathcal{Z}(T, N, P) = \frac{1}{\lambda^N} \int_0^\infty dL \exp(-\beta PL) \int_0^\infty d\delta_1 \int_0^\infty d\delta_2 \cdots \int_0^\infty d\delta_N$$

$$\cdot \exp\left[-\beta \sum_{i=2}^{N} \mathcal{V}(\delta_i)\right] \Theta\left(L - \sum_{i=1}^{N} \delta_i\right)$$

$$= \frac{1}{\lambda^N \beta P} \int_0^\infty d\delta_1 \int_0^\infty d\delta_2 \cdots \int_0^\infty d\delta_N \exp\left[-\beta \sum_{i=2}^{N} \mathcal{V}(\delta_i)\right]$$

$$\exp\left[-\beta P \left(\sum_{i=1}^{N} \delta_1\right)\right]$$

$$= \frac{1}{\lambda^N (\beta P)^2} \int_0^\infty d\delta_2 \cdots \int_0^\infty d\delta_N \exp\left\{-\sum_{i=2}^{N} [\beta \mathcal{V}(\delta_i) + \beta P \delta_i]\right\}.$$

Since the integrals for different δ_i's are equivalent, we obtain

$$\mathcal{Z}(T, N, P) = \frac{1}{\lambda^N (\beta P)^2} \left\{ \int_0^\infty d\delta \exp\left[-\beta \left(\mathcal{V}(\delta) + P\delta\right)\right] \right\}^{N-1}.$$

This expression can also be obtained directly, without use of the step function as follows.

$$\mathcal{Z}(T, N, P) = \frac{1}{\lambda^N} \int_0^L d\delta_1 \int_0^{L-\delta_1} d\delta_2 \int_0^{L-(\delta_1+\delta_2)} d\delta_3 \cdots \int_0^{L-\sum_{i=1}^N \delta_i} d\delta_N$$

$$\cdot \int_0^\infty dL \exp\left[-\beta PL - \beta \left(\sum_{i=2}^N \mathcal{V}(\delta)\right)\right]$$

$$= \frac{1}{\lambda^N} \int_0^L d\delta_1 \int_0^{L-\delta_1} d\delta_2 \cdots \int_0^{L-\sum_{i=1}^N \delta_i} d\delta_N \int_{-\sum_{i=1}^N \delta_i}^\infty d\left(L - \sum_{i=1}^N \delta_i\right)$$

$$\cdot \exp\left\{-\beta P\left[\sum_{i=1}^N \delta_i + \left(L - \sum_{i=1}^N \delta_i\right)\right] - \beta\left(\sum_{i=2}^N \mathcal{V}(\delta_i)\right)\right\}.$$

Change variables to $\delta_{N+1} \equiv L - \sum_{i=1}^N \delta_i$, and note that each of the δ's indicates the distance between neighboring particles. The size of the gas L has been extended to any value, hence each δ can be varied independently from 0 to ∞. Thus the Gibbs partition function is

$$\mathcal{Z}(T, N, P) = \frac{1}{\lambda^N} \int_0^\infty d\delta_1 \int_0^\infty d\delta_2 \cdots \int_0^\infty d\delta_N \int_0^\infty d\delta_{N+1}$$

$$\cdot \exp\left[-\beta P\left(\sum_{i=1}^{N+1} \delta_i\right) - \beta\left(\sum_{i=2}^N \mathcal{V}(\delta_i)\right)\right]$$

$$= \frac{1}{\lambda^N} \left(\int_0^\infty d\delta \cdot \exp\left[-\beta \mathcal{V}(\delta) - \beta P\delta\right]\right)^{N-1} \int_0^\infty d\delta_1 \exp(-\beta P\delta_1)$$

$$\cdot \int_0^\infty d\delta_{N+1} \exp(-\beta P\delta_{N+1})$$

$$= \frac{1}{\lambda^N (\beta P)^2} \left\{\int_0^\infty d\delta \exp\left[-\beta \left(\mathcal{V}(\delta) + P\delta\right)\right]\right\}^{N-1}.$$

(e) At a fixed pressure P, find expressions for the mean length $L(T, N, P)$, and the density $n = N/L(T, N, P)$ (involving ratios of integrals that should be easy to interpret).

The mean length is

$$L(T, N, P) = -k_B T \left.\frac{\partial}{\partial(\beta P)} \ln \mathcal{Z}(T, N, P)\right|_{T,N}$$

$$= \frac{2}{\beta P} + (N-1)\frac{\int_0^\infty d\delta \cdot \delta \cdot \exp\left[-\beta \mathcal{V}(\delta) - \beta P\delta\right]}{\int_0^\infty d\delta \cdot \exp\left[-\beta \mathcal{V}(\delta) - \beta P\delta\right]},$$

and the density n is given by

$$n = \frac{N}{L(T, N, P)} = N \left\{\frac{2k_B T}{P} + (N-1)\frac{\int_0^\infty d\delta \cdot \delta \cdot \exp\left[-\beta \left(\mathcal{V}(\delta) - P\delta\right)\right]}{\int_0^\infty d\delta \cdot \exp\left[-\beta \left(\mathcal{V}(\delta) - P\delta\right)\right]}\right\}^{-1}.$$

Note that for an ideal gas $\mathcal{V}_{\text{i.g.}}(\delta) = 0$, and

$$L_{\text{i.g.}}(T, N, P) = \frac{(N+1)k_B T}{P},$$

leading to

$$n(p)_{\text{i.g.}} = \frac{N}{N+1} \frac{P}{k_B T}.$$

Since the expression for $n(T, P)$ in part (e) is continuous and non-singular for any choice of potential, there is in fact no condensation transition for the one-dimensional gas. By contrast, the approximate van der Waals equation (or the mean-field treatment) incorrectly predicts such a transition.

(f) For a hard-sphere gas, with minimum separation a between particles, calculate the equation of state $P(T, n)$. Compare the excluded volume factor with the approximate result obtained in earlier problems, and also obtain the general virial coefficient $B_\ell(T)$.

For a hard-sphere gas

$$\delta_i \geq a, \quad \text{for} \quad i = 2, 3, \cdots, N,$$

the Gibbs partition function is

$$\mathcal{Z}(T, N, P) = \frac{1}{\lambda^N (\beta P)^2} \left[\int_a^\infty d\delta \exp\left(-\beta \mathcal{V}(\delta) - \beta P \delta\right) \right]^{N-1}$$

$$= \frac{1}{\lambda^N (\beta P)^2} \left[\int_a^\infty d\delta \exp\left(-\beta P \delta\right) \right]^{N-1}$$

$$= \frac{1}{\lambda^N} \left(\frac{1}{\beta P} \right)^{N+1} \exp\left(-\beta P a\right)^{N-1}.$$

From the partition function, we can calculate the mean length

$$L = -k_B T \left. \frac{\partial \ln \mathcal{Z}}{\partial P} \right|_{T, N} = \frac{(N+1)k_B T}{P} + (N-1)a,$$

which after rearrangement yields

$$\beta P = \frac{(N+1)}{L - (N-1)a} = \frac{n + 1/L}{1 - (n-1/L)a} \approx (n + 1/l)(1 + (n - 1/L)a + (n - 1/L)^2 a^2 + \cdots).$$

For $N \gg 1$, $n \gg 1/L$, and

$$\beta P \approx n(1 + na + n^2 a^2 + \cdots) = n + an^2 + a^2 n^3 + \cdots,$$

which gives the virial coefficients

$$B_\ell(T) = a^{\ell-1}.$$

The value of $B_3 = a^2$ agrees with the result obtained in an earlier problem. Also note that the exact "excluded volume" is $(N-1)a$, as opposed to the estimate of $Na/2$ used in deriving the van der Waals equation.

****

Solutions to selected problems from chapter 6

1. *One-dimensional chain*: a chain of $N+1$ particles of mass m is connected by N massless springs of spring constant K and relaxed length a. The first and last particles are held fixed at the equilibrium separation of Na. Let us denote the longitudinal displacements of the particles from their equilibrium positions by $\{u_i\}$, with $u_0 = u_N = 0$ since the end particles are fixed. The Hamiltonian governing $\{u_i\}$, and the conjugate momenta $\{p_i\}$, is

$$\mathcal{H} = \sum_{i=1}^{N-1} \frac{p_i^2}{2m} + \frac{K}{2}\left[u_1^2 + \sum_{i=1}^{N-2} \left(u_{i+1} - u_i\right)^2 + u_{N-1}^2 \right].$$

(a) Using the appropriate (sine) Fourier transforms, find the normal modes $\{\tilde{u}_k\}$, and the corresponding frequencies $\{\omega_k\}$.

From the Hamiltonian

$$\mathcal{H} = \sum_{i=1}^{N-1} \frac{p_i^2}{2m} + \frac{K}{2}\left[u_1^2 + \sum_{i=2}^{N-1} \left(u_i - u_{i-1}\right)^2 + u_{N-1}^2 \right],$$

the classical equations of motion are obtained as

$$m\frac{\mathrm{d}^2 u_j}{\mathrm{d}t^2} = -K(u_j - u_{j-1}) - K(u_j - u_{j+1}) = K(u_{j-1} - 2u_j + u_{j+1}),$$

for $j = 1, 2, \cdots, N-1$, and with $u_0 = u_N = 0$. In a normal mode, the particles oscillate in phase. The usual procedure is to obtain the modes, and corresponding frequencies, by diagonalizing the matrix of coefficients coupling the displacements on the right-hand side of the equation of motion. For any linear system, we have $m\mathrm{d}^2 u_i/\mathrm{d}t^2 = \mathcal{K}_{ij}u_j$, and we must diagonalize $\mathcal{K}_{ij}$. In the above example, $\mathcal{K}_{ij}$ is only a function of the difference $i - j$. This is a consequence of translational symmetry, and allows us to diagonalize the matrix using Fourier modes. Due to the boundary conditions in this case, the sine transformation is appropriate, and the motion of the jth particle in a normal mode is given by

$$\tilde{u}_{k(n)}(j) = \sqrt{\frac{2}{N}} \, \mathrm{e}^{\pm i\omega_n t} \sin\left(k(n) \cdot j\right).$$

The origin of time is arbitrary, and to ensure that $u_N = 0$, we must set

$$k(n) \equiv \frac{n\pi}{N}, \quad \text{for} \quad n = 1, 2, \cdots, N-1.$$

Larger values of n give wavevectors that are simply shifted by a multiple of π, and hence coincide with one of the above normal modes. The number of normal modes thus equals the number of original displacement variables, as required. Furthermore, the amplitudes are chosen such that the normal modes are also orthonormal, that is,

$$\sum_{j=1}^{N-1} \tilde{u}_{k(n)}(j) \cdot \tilde{u}_{k(m)}(j) = \delta_{n,m}.$$

By substituting the normal modes into the equations of motion we obtain the dispersion relation

$$\omega_n^2 = 2\omega_0^2 \left[1 - \cos\left(\frac{n\pi}{N}\right)\right] = \omega_0^2 \sin^2\left(\frac{n\pi}{2N}\right),$$

where $\omega_0 \equiv \sqrt{K/m}$.

The potential energy for each normal mode is given by

$$U_n = \frac{K}{2}\sum_{i=1}^{N}|u_i - u_{i-1}|^2 = \frac{K}{N}\sum_{i=1}^{N}\left\{\sin\left(\frac{n\pi}{N}i\right) - \sin\left[\frac{n\pi}{N}(i-1)\right]\right\}^2$$

$$= \frac{4K}{N}\sin^2\left(\frac{n\pi}{2N}\right)\sum_{i=1}^{N}\cos^2\left[\frac{n\pi}{N}\left(i-\frac{1}{2}\right)\right].$$

Noting that

$$\sum_{i=1}^{N}\cos^2\left[\frac{n\pi}{N}\left(i-\frac{1}{2}\right)\right] = \frac{1}{2}\sum_{i=1}^{N}\left\{1+\cos\left[\frac{n\pi}{N}(2i-1)\right]\right\} = \frac{N}{2},$$

we have

$$U_{k(n)} = 2K\sin^2\left(\frac{n\pi}{2N}\right).$$

(b) Express the Hamiltonian in terms of the amplitudes of normal modes $\{\tilde{u}_k\}$, and evaluate the *classical* partition function. (You may integrate the $\{u_i\}$ from $-\infty$ to $+\infty$.)

Before evaluating the classical partition function, let's evaluate the potential energy by first expanding the displacement using the basis of normal modes, as

$$u_j = \sum_{n=1}^{N-1} a_n \cdot \tilde{u}_{k(n)}(j).$$

The expression for the total potential energy is

$$U = \frac{K}{2}\sum_{i=1}^{N}(u_i - u_{i-1})^2 = \frac{K}{2}\sum_{i=1}^{N}\left\{\sum_{n=1}^{N-1} a_n\left[\tilde{u}_{k(n)}(j) - \tilde{u}_{k(n)}(j-1)\right]\right\}^2.$$

Since

$$\sum_{j=1}^{N-1} \tilde{u}_{k(n)}(j) \cdot \tilde{u}_{k(m)}(j-1) = \frac{1}{N}\delta_{n,m}\sum_{j=1}^{N-1}\left\{-\cos\left[k(n)(2j-1)\right] + \cos k(n)\right\}$$

$$= \delta_{n,m}\cos k(n),$$

the total potential energy has the equivalent forms

$$U = \frac{K}{2} \sum_{i=1}^{N} (u_i - u_{i-1})^2 = K \sum_{n=1}^{N-1} a_n^2 \left(1 - \cos k(n)\right),$$

$$= \sum_{i=1}^{N-1} a_{k(n)}^2 U_{kn} = 2K \sum_{i=1}^{N-1} a_{k(n)}^2 \sin^2 \left(\frac{n\pi}{2N}\right).$$

The next step is to change the coordinates of phase space from u_j to a_n. The Jacobian associated with this change of variables is unity, and the classical partition function is now obtained from

$$Z = \frac{1}{\lambda^{N-1}} \int_{-\infty}^{\infty} da_1 \cdots \int_{-\infty}^{\infty} da_{N-1} \exp\left[-2\beta K \sum_{n=1}^{N-1} a_n^2 \sin^2 \left(\frac{n\pi}{2N}\right)\right],$$

where $\lambda = h/\sqrt{2\pi m k_B T}$ corresponds to the contribution to the partition function from each momentum coordinate. Performing the Gaussian integrals, we obtain

$$Z = \frac{1}{\lambda^{N-1}} \prod_{n=1}^{N-1} \left\{ \int_{-\infty}^{\infty} da_n \exp\left[-2\beta K a_n^2 \sin^2 \left(\frac{n\pi}{2N}\right)\right] \right\},$$

$$= \frac{1}{\lambda^{N-1}} \left(\frac{\pi k_B T}{2K}\right)^{\frac{N-1}{2}} \prod_{n=1}^{N-1} \left[\sin\left(\frac{n\pi}{2N}\right)\right]^{-1}.$$

(c) First evaluate $\left\langle |\tilde{u}_k|^2 \right\rangle$, and use the result to calculate $\langle u_i^2 \rangle$. Plot the resulting squared displacement of each particle as a function of its equilibrium position.

The average squared amplitude of each normal mode is

$$\langle a_n^2 \rangle = \frac{\int_{-\infty}^{\infty} da_n (a_n^2) \exp\left[-2\beta K a_n^2 \sin^2 \left(\frac{n\pi}{2N}\right)\right]}{\int_{-\infty}^{\infty} da_n \exp\left[-2\beta K a_n^2 \sin^2 \left(\frac{n\pi}{2N}\right)\right]}$$

$$= \left[4\beta K \sin^2 \left(\frac{n\pi}{2N}\right)\right]^{-1} = \frac{k_B T}{4K} \frac{1}{\sin^2 \left(\frac{n\pi}{2N}\right)}.$$

The variation of the displacement is then given by

$$\langle u_j^2 \rangle = \left\langle \left[\sum_{n=1}^{N-1} a_n \tilde{u}_n(j)\right]^2 \right\rangle = \sum_{n=1}^{N-1} \langle a_n^2 \rangle \tilde{u}_n^2(j)$$

$$= \frac{2}{N} \sum_{n=1}^{N-1} \langle a_n^2 \rangle \sin^2 \left(\frac{n\pi}{N} j\right) = \frac{k_B T}{2KN} \sum_{n=1}^{N-1} \frac{\sin^2 \left(\frac{n\pi}{N} j\right)}{\sin^2 \left(\frac{n\pi}{2N}\right)}.$$

The evaluation of the above sum is considerably simplified by considering the combination

$$\langle u_{j+1}^2 \rangle + \langle u_{j-1}^2 \rangle - 2 \langle u_j^2 \rangle = \frac{k_B T}{2KN} \sum_{n=1}^{N-1} \frac{2\cos\left[\frac{2n\pi}{N} j\right] - \cos\left[\frac{2n\pi}{N}(j+1)\right] - \cos\left[\frac{2n\pi}{N}(j-1)\right]}{1 - \cos\left(\frac{n\pi}{N}\right)}$$

$$= \frac{k_B T}{2KN} \sum_{n=1}^{N-1} \frac{2\cos\left(\frac{2n\pi}{N} j\right)\left[1 - \cos\left(\frac{n\pi}{N}\right)\right]}{1 - \cos\left(\frac{n\pi}{N}\right)} = -\frac{k_B T}{KN},$$

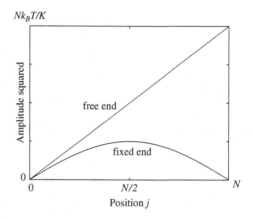

where we have used $\sum_{n=1}^{N-1}\cos(\pi n/N) = -1$. *It is easy to check that subject to the boundary conditions of* $\langle u_0^2\rangle = \langle u_N^2\rangle = 0$, *the solution to the above recursion relation is*

$$\langle u_j^2\rangle = \frac{k_B T}{K}\frac{j(N-j)}{N}.$$

(d) How are the results modified if only the first particle is fixed ($u_0 = 0$), while the other end is free ($u_N \neq 0$)? (Note that this is a much simpler problem as the partition function can be evaluated by changing variables to the $N-1$ spring extensions.)

When the last particle is free, the overall potential energy is the sum of the contributions of each spring, that is, $U = K\sum_{j=1}^{N-1}(u_j - u_{j-1})^2/2$. *Thus each extension can be treated independently, and we introduce a new set of independent variables* $\Delta u_j \equiv u_j - u_{j-1}$. *(In the previous case, where the two ends are fixed, these variables were not independent.) The partition function can be calculated separately for each spring as*

$$Z = \frac{1}{\lambda^{N-1}}\int_{-\infty}^{\infty}\mathrm{d}u_1\cdots\int_{-\infty}^{\infty}\mathrm{d}u_{N-1}\exp\left[-\frac{K}{2k_B T}\sum_{j=1}^{N-1}(u_j - u_{j-1})^2\right]$$

$$= \frac{1}{\lambda^{N-1}}\int_{-\infty}^{\infty}\mathrm{d}\Delta u_1\cdots\int_{-\infty}^{\infty}\mathrm{d}\Delta u_{N-1}\exp\left[-\frac{K}{2k_B T}\sum_{j=1}^{N-1}\Delta u_j^2\right] = \left(\frac{2\pi k_B T}{\lambda^2 K}\right)^{(N-1)/2}.$$

For each spring extension, we have

$$\langle\Delta u_j^2\rangle = \langle(u_j - u_{j-1})^2\rangle = \frac{k_B T}{K}.$$

The displacement

$$u_j = \sum_{i=1}^{j}\Delta u_i$$

is a sum of independent random variables, leading to the variance

$$\langle u_j^2\rangle = \left\langle\left(\sum_{i=1}^{j}\Delta u_i\right)^2\right\rangle = \sum_{i=1}^{j}(\Delta u_i)^2 = \frac{k_B T}{K}j.$$

The results for displacements of open and closed chains are compared in the above figure.

****

2. *Black hole thermodynamics*: according to Bekenstein and Hawking, the entropy of a black hole is proportional to its area A, and given by

$$S = \frac{k_B c^3}{4G\hbar} A.$$

(a) Calculate the escape velocity at a radius R from a mass M using classical mechanics. Find the relationship between the radius and mass of a black hole by setting this escape velocity to the speed of light c. (Relativistic calculations do not modify this result, which was originally obtained by Laplace.)

The classical escape velocity is obtained by equating the gravitational energy and the kinetic energy on the surface, as

$$G\frac{Mm}{R} = \frac{mv_E^2}{2},$$

leading to

$$v_E = \sqrt{\frac{2GM}{r}}.$$

Setting the escape velocity to the speed of light, we find

$$R = \frac{2G}{c^2} M.$$

For a mass larger than given by this ratio (i.e., $M > c^2 R/2G$), nothing will escape from distances closer than R.

(b) Does entropy increase or decrease when two black holes collapse into one? What is the entropy change for the Universe (in equivalent number of bits of information) when two solar mass black holes ($M_\odot \approx 2 \times 10^{30}$ kg) coalesce?

When two black holes of mass M collapse into one, the entropy change is

$$\Delta S = S_2 - 2S_1 = \frac{k_B c^3}{4G\hbar}(A_2 - 2A_1) = \frac{k_B c^3}{4G\hbar} 4\pi \left(R_2^2 - 2R_1^2\right)$$

$$= \frac{\pi k_B c^3}{G\hbar}\left[\left(\frac{2G}{c^2} 2M\right)^2 - 2\left(\frac{2G}{c^2}M\right)^2\right] = \frac{8\pi G k_B M^2}{c\hbar} > 0.$$

Thus the merging of black holes increases the entropy of the universe.

Consider the coalescence of two solar mass black holes. The entropy change is

$$\Delta S = \frac{8\pi G k_B M_\odot^2}{c\hbar}$$

$$\approx \frac{8\pi \cdot 6.7 \times 10^{-11} (\mathrm{N \cdot m^2\, kg^{-2}}) \cdot 1.38 \times 10^{-23} (\mathrm{J\,K^{-1}}) \cdot (2 \times 10^{30})^2\, \mathrm{kg^2}}{3 \times 10^8 (\mathrm{m\, s^{-1}}) \cdot 1.05 \times 10^{-34} (\mathrm{J \cdot s})}$$

$$\approx 3 \times 10^{54} (\mathrm{J/K}).$$

In units of bits, the information lost is

$$N_I = \frac{\Delta S \ln 2}{k_B} = 1.5 \times 10^{77}.$$

(c) The internal energy of the black hole is given by the Einstein relation, $E = Mc^2$. Find the temperature of the black hole in terms of its mass.

Using the thermodynamic definition of temperature $\frac{1}{T} = \frac{\partial S}{\partial E}$*, and the Einstein relation* $E = Mc^2$,

$$\frac{1}{T} = \frac{1}{c^2} \frac{\partial}{\partial M} \left[\frac{k_B c^3}{4G\hbar} 4\pi \left(\frac{2G}{c^2} M \right)^2 \right] = \frac{8\pi k_B G}{\hbar c^3} M, \qquad \Longrightarrow \qquad T = \frac{\hbar c^3}{8\pi k_B G} \frac{1}{M}.$$

(d) A "black hole" actually emits thermal radiation due to pair creation processes on its event horizon. Find the rate of energy loss due to such radiation.

The (quantum) vacuum undergoes fluctuations in which particle–anti-particle pairs are constantly created and destroyed. Near the boundary of a black hole, sometimes one member of a pair falls into the black hole while the other escapes. This is a hand-waving explanation for the emission of radiation from black holes. The decrease in energy E of a black body of area A at temperature T is given by the Stefan–Boltzmann law,

$$\frac{1}{A} \frac{\partial E}{\partial t} = -\sigma T^4, \quad \text{where} \quad \sigma = \frac{\pi^2 k_B^4}{60 \hbar^3 c^2}.$$

(e) Find the amount of time it takes an isolated black hole to evaporate. How long is this time for a black hole of solar mass?

Using the result in part (d) we can calculate the time it takes a black hole to evaporate. For a black hole

$$A = 4\pi R^2 = 4\pi \left(\frac{2G}{c^2} M \right)^2 = \frac{16\pi G^2}{c^4} M^2, \quad E = Mc^2, \quad \text{and} \quad T = \frac{\hbar c^3}{8\pi k_B G} \frac{1}{M}.$$

Hence

$$\frac{\mathrm{d}}{\mathrm{d}t} \left(Mc^2 \right) = -\frac{\pi^2 k_B^4}{60\hbar^3 c^2} \left(\frac{16\pi G^2}{c^4} M^2 \right) \left(\frac{\hbar c^3}{8\pi k_B G} \frac{1}{M} \right)^4,$$

which implies that

$$M^2 \frac{\mathrm{d}M}{\mathrm{d}t} = -\frac{\hbar c^4}{15360 G^2} \equiv -b.$$

This can be solved to give

$$M(t) = \left(M_0^3 - 3bt \right)^{1/3}.$$

The mass goes to zero, and the black hole evaporates after a time

$$\tau = \frac{M_0^3}{3b} = \frac{5120 G^2 M_\odot{}^3}{\hbar c^4} \approx 2.2 \times 10^{74} \, \text{s},$$

which is considerably longer than the current age of the Universe (approximately $\times 10^{18}$ s).

(f) What is the mass of a black hole that is in thermal equilibrium with the current cosmic background radiation at $T = 2.7\,\text{K}$?

The temperature and mass of a black hole are related by $M = \hbar c^3/(8\pi k_B G T)$. For a black hole in thermal equilibrium with the current cosmic background radiation at $T = 2.7\,\text{K}$,

$$M \approx \frac{1.05 \times 10^{-34}(\text{J}\cdot\text{s})(3 \times 10^8)^3(\text{m s}^{-1})^3}{8\pi \cdot 1.38 \times 10^{-23}(\text{J K}^{-1})\cdot 6.7 \times 10^{-11}(\text{N}\cdot\text{m}^2\,\text{kg}^{-2})\cdot 2.7\,\text{K}} \approx 4.5 \times 10^{22}\,\text{kg}.$$

(g) Consider a spherical volume of space of radius R. According to the recently formulated *Holographic Principle* there is a maximum to the amount of entropy that this volume of space can have, independent of its contents! What is this maximal entropy?

The mass inside the spherical volume of radius R must be less than the mass that would make a black hole that fills this volume. Bring in additional mass (from infinity) inside the volume, so as to make a volume-filling black hole. Clearly the entropy of the system will increase in the process, and the final entropy, which is the entropy of the black hole, is larger than the initial entropy in the volume, leading to the inequality

$$S \leq S_{BH} = \frac{k_B c^3}{4 G \hbar} A,$$

where $A = 4\pi R^2$ is the area enclosing the volume. The surprising observation is that the upper bound on the entropy is proportional to area, whereas for any system of particles we expect the entropy to be proportional to N. This should remain valid even at very high temperatures when interactions are unimportant. The "holographic principle" is an allusion to the observation that it appears as if the degrees of freedom are living on the surface of the system, rather than its volume. It was formulated in the context of string theory, which attempts to construct a consistent theory of quantum gravity, which replaces particles as degrees of freedom, with strings.

****

3. *Quantum harmonic oscillator:* consider a single harmonic oscillator with the Hamiltonian

$$\mathcal{H} = \frac{p^2}{2m} + \frac{m\omega^2 q^2}{2}, \quad \text{with} \quad p = \frac{\hbar}{i}\frac{d}{dq}.$$

(a) Find the partition function Z, at a temperature T, and calculate the energy $\langle \mathcal{H} \rangle$.

The partition function Z, at a temperature T, is given by

$$Z = \text{tr}\,\rho = \sum_n e^{-\beta E_n}.$$

As the energy levels for a harmonic oscillator are given by

$$\epsilon_n = \hbar\omega\left(n + \frac{1}{2}\right),$$

the partition function is

$$Z = \sum_n \exp\left[-\beta\hbar\omega\left(n+\frac{1}{2}\right)\right] = e^{-\beta\hbar\omega/2} + e^{-3\beta\hbar\omega/2} + \cdots$$

$$= \frac{1}{e^{\beta\hbar\omega/2} - e^{-\beta\hbar\omega/2}} = \frac{1}{2\sinh\left(\beta\hbar\omega/2\right)}.$$

The expectation value of the energy is

$$\langle \mathcal{H} \rangle = -\frac{\partial \ln Z}{\partial \beta} = \left(\frac{\hbar\omega}{2}\right)\frac{\cosh(\beta\hbar\omega/2)}{\sinh(\beta\hbar\omega/2)} = \left(\frac{\hbar\omega}{2}\right)\frac{1}{\tanh(\beta\hbar\omega/2)}.$$

(b) Write down the formal expression for the canonical density matrix ρ in terms of the eigenstates ($\{|n\rangle\}$), and energy levels ($\{\epsilon_n\}$) of $\mathcal{H}$.

Using the formal representation of the energy eigenstates, the density matrix ρ is

$$\rho = 2\sinh\left(\frac{\beta\hbar\omega}{2}\right)\left(\sum_n |n\rangle \exp\left[-\beta\hbar\omega\left(n+\frac{1}{2}\right)\right]\langle n|\right).$$

In the coordinate representation, the eigenfunctions are in fact given by

$$\langle n|q\rangle = \left(\frac{m\omega}{\pi\hbar}\right)^{1/4}\frac{H_n(\xi)}{\sqrt{2^n n!}}\exp\left(-\frac{\xi^2}{2}\right),$$

where

$$\xi \equiv \sqrt{\frac{m\omega}{\hbar}}\,q,$$

with

$$H_n(\xi) = (-1)^n \exp(\xi^2)\left(\frac{d}{d\xi}\right)^n \exp(-\xi^2)$$

$$= \frac{\exp(\xi^2)}{\pi}\int_{-\infty}^{\infty}(-2iu)^n \exp(-u^2+2i\xi u)du.$$

For example,

$$H_0(\xi) = 1, \quad \text{and} \quad H_1(\xi) = -\exp(\xi^2)\frac{d}{d\xi}\exp(-\xi^2) = 2\xi,$$

result in the eigenstates

$$\langle 0|q\rangle = \left(\frac{m\omega}{\pi\hbar}\right)^{1/4}\exp\left(-\frac{m\omega}{2\hbar}q^2\right),$$

and

$$\langle 1|q\rangle = \left(\frac{m\omega}{\pi\hbar}\right)^{1/4}\sqrt{\frac{2m\omega}{\hbar}}q\cdot\exp\left(-\frac{m\omega}{2\hbar}q^2\right).$$

Using the above expressions, the matrix elements are obtained as

$$\langle q'|\rho|q\rangle = \sum_{n,n'}\langle q'|n'\rangle\langle n'|\rho|n\rangle\langle n|q\rangle = \frac{\sum_n \exp\left[-\beta\hbar\omega\left(n+\frac{1}{2}\right)\right]\cdot\langle q'|n\rangle\langle n|q\rangle}{\sum_n \exp\left[-\beta\hbar\omega\left(n+\frac{1}{2}\right)\right]}$$

$$= 2\sinh\left(\frac{\beta\hbar\omega}{2}\right)\cdot\sum_n \exp\left[-\beta\hbar\omega\left(n+\frac{1}{2}\right)\right]\cdot\langle q'|n\rangle\langle n|q\rangle.$$

(c) Show that for a general operator $A(x)$,

$$\frac{\partial}{\partial x} \exp[A(x)] \neq \frac{\partial A}{\partial x} \exp[A(x)], \qquad \text{unless} \qquad \left[A, \frac{\partial A}{\partial x}\right] = 0,$$

while in all cases

$$\frac{\partial}{\partial x} \operatorname{tr}\{\exp[A(x)]\} = \operatorname{tr}\left\{\frac{\partial A}{\partial x} \exp[A(x)]\right\}.$$

By definition

$$e^A = \sum_{n=0}^{\infty} \frac{1}{n!} A^n,$$

and

$$\frac{\partial e^A}{\partial x} = \sum_{n=0}^{\infty} \frac{1}{n!} \frac{\partial A^n}{\partial x}.$$

But for a product of n operators,

$$\frac{\partial}{\partial x}(A \cdot A \cdots A) = \frac{\partial A}{\partial x} \cdot A \cdots A + A \cdot \frac{\partial A}{\partial x} \cdots A + \cdots + A \cdot A \cdots \frac{\partial A}{\partial x}.$$

The $\frac{\partial A}{\partial x}$ can be moved through the A's surrounding it only if $\left[A, \frac{\partial A}{\partial x}\right] = 0$, in which case

$$\frac{\partial A}{\partial x} = n \frac{\partial A}{\partial x} A^{n-1}, \qquad \text{and} \qquad \frac{\partial e^A}{\partial x} = \frac{\partial A}{\partial x} e^A.$$

However, as we can always reorder operators inside a trace, that is, $\operatorname{tr}(BC) = \operatorname{tr}(CB)$,

$$\operatorname{tr}\left(A \cdots A \cdots \frac{\partial A}{\partial x} \cdots A\right) = \operatorname{tr}\left(\frac{\partial A}{\partial x} \cdot A^{n-1}\right),$$

and the identity

$$\frac{\partial}{\partial x} \operatorname{tr}\left(e^A\right) = \operatorname{tr}\left(\frac{\partial A}{\partial x} \cdot e^A\right)$$

can always be satisfied, independent of any constraint on $\left[A, \frac{\partial A}{\partial x}\right]$.

(d) Note that the partition function calculated in part (a) does not depend on the mass m, that is, $\partial Z / \partial m = 0$. Use this information, along with the result in part (c), to show that

$$\left\langle \frac{p^2}{2m} \right\rangle = \left\langle \frac{m\omega^2 q^2}{2} \right\rangle.$$

The expectation values of the kinetic and potential energy are given by

$$\left\langle \frac{p^2}{2m} \right\rangle = \operatorname{tr}\left(\frac{p^2}{2m}\rho\right), \qquad \text{and} \qquad \left\langle \frac{m\omega^2 q^2}{2} \right\rangle = \operatorname{tr}\left(\frac{m\omega^2 q^2}{2}\rho\right).$$

Noting that the expression for the partition function derived in part (a) is independent of mass, we know that $\partial Z/\partial m = 0$. Starting with $Z = \mathrm{tr}\left(e^{-\beta\mathcal{H}}\right)$, and differentiating

$$\frac{\partial Z}{\partial m} = \frac{\partial}{\partial m}\,\mathrm{tr}\left(e^{-\beta\mathcal{H}}\right) = \mathrm{tr}\left[\frac{\partial}{\partial m}(-\beta\mathcal{H})e^{-\beta\mathcal{H}}\right] = 0,$$

where we have used the result in part (c). Differentiating the Hamiltonian, we find that

$$\mathrm{tr}\left[\beta\frac{p^2}{2m^2}e^{-\beta\mathcal{H}}\right] + \mathrm{tr}\left[-\beta\frac{\omega^2 q^2}{2}e^{-\beta\mathcal{H}}\right] = 0.$$

Equivalently,

$$\mathrm{tr}\left[\frac{p^2}{2m}e^{-\beta\mathcal{H}}\right] = \mathrm{tr}\left[\frac{m\omega^2 q^2}{2}e^{-\beta\mathcal{H}}\right],$$

which shows that the expectation values of kinetic and potential energies are equal.

(e) Using the results in parts (d) and (a), or otherwise, calculate $\langle q^2\rangle$. How are the results in Problem 1 modified at low temperatures by inclusion of quantum mechanical effects?

In part (a) it was found that $\langle\mathcal{H}\rangle = (\hbar\omega/2)\,(\tanh(\beta\hbar\omega/2))^{-1}$. Note that $\langle\mathcal{H}\rangle = \langle p^2/2m\rangle + \langle m\omega^2 q^2/2\rangle$, and that in part (d) it was determined that the contribution from the kinetic and potential energy terms are equal. Hence,

$$\langle m\omega^2 q^2/2\rangle = \frac{1}{2}\,(\hbar\omega/2)\,(\tanh(\beta\hbar\omega/2))^{-1}.$$

Solving for $\langle q^2\rangle$,

$$\langle q^2\rangle = \frac{\hbar}{2m\omega}\,(\tanh(\beta\hbar\omega/2))^{-1} = \frac{\hbar}{2m\omega}\,\coth(\beta\hbar\omega/2).$$

While the classical result $\langle q^2\rangle = k_B T/m\omega^2$ vanishes as $T\to 0$, the quantum result saturates at $T = 0$ to a constant value of $\langle q^2\rangle = \hbar/(2m\omega)$. The amplitudes of the displacement curves in Problem 1 are affected by exactly the same saturation factors.

(f) In a coordinate representation, calculate $\langle q'|\rho|q\rangle$ in the high-temperature limit. One approach is to use the result

$$\exp(\beta A)\exp(\beta B) = \exp\left[\beta(A+B) + \beta^2[A, B]/2 + \mathcal{O}(\beta^3)\right].$$

Using the general operator identity

$$\exp(\beta A)\exp(\beta B) = \exp\left[\beta(A+B) + \beta^2[A, B]/2 + \mathcal{O}(\beta^3)\right],$$

the Boltzmann operator can be decomposed in the high-temperature limit into those for kinetic and potential energy; to the lowest order as

$$\exp\left(-\beta\frac{p^2}{2m} - \beta\frac{m\omega^2 q^2}{2}\right) \approx \exp(-\beta p^2/2m)\cdot\exp(-\beta m\omega^2 q^2/2).$$

The first term is the Boltzmann operator for an ideal gas. The second term contains an operator diagonalized by $|q\rangle$. The density matrix element

$$\langle q'|\rho|q\rangle = \langle q'|\exp(-\beta p^2/2m)\exp(-\beta m\omega^2 q^2/2)|q\rangle$$

$$= \int dp' \langle q'|\exp(-\beta p^2/2m)|p'\rangle\langle p'|\exp(-\beta m\omega^2 q^2/2)|q\rangle$$

$$= \int dp' \langle q'|p'\rangle\langle p'|q\rangle \exp(-\beta p'^2/2m)\exp(-\beta q^2 m\omega^2/2).$$

Using the free particle basis $\langle q'|p'\rangle = \frac{1}{\sqrt{2\pi\hbar}}e^{-iq\cdot p/\hbar}$,

$$\langle q'|\rho|q\rangle = \frac{1}{2\pi\hbar}\int dp' e^{ip'(q-q')/\hbar}e^{-\beta p'^2/2m}e^{-\beta q^2 m\omega^2/2}$$

$$= e^{-\beta q^2 m\omega^2/2}\frac{1}{2\pi\hbar}\int dp' \exp\left[-\left(p'\sqrt{\frac{\beta}{2m}}+\frac{i}{2\hbar}\sqrt{\frac{2m}{\beta}}(q-q')\right)^2\right]\exp\left(-\frac{1}{4}\frac{2m}{\beta\hbar^2}(q-q')^2\right),$$

where we completed the square. Hence

$$\langle q'|\rho|q\rangle = \frac{1}{2\pi\hbar}e^{-\beta q^2 m\omega^2/2}\sqrt{2\pi m k_B T}\exp\left[-\frac{m k_B T}{2\hbar^2}(q-q')^2\right].$$

The proper normalization in the high-temperature limit is

$$Z = \int dq \langle q|e^{-\beta p^2/2m}\cdot e^{-\beta m\omega^2 q^2/2}|q\rangle$$

$$= \int dq \int dp' \langle q|e^{-\beta p^2/2m}|p'\rangle\langle p'|e^{-\beta m\omega^2 q^2/2}|q\rangle$$

$$= \int dq \int dp |\langle q|p\rangle|^2 e^{-\beta p'^2/2m}e^{-\beta m\omega^2 q^2/2} = \frac{k_B T}{\hbar\omega}.$$

Hence the properly normalized matrix element in the high-temperature limit is

$$\langle q'|\rho|q\rangle_{\lim T\to\infty} = \sqrt{\frac{m\omega^2}{2\pi k_B T}}\exp\left(-\frac{m\omega^2}{2k_B T}q^2\right)\exp\left[-\frac{m k_B T}{2\hbar^2}(q-q')^2\right].$$

(g) At low temperatures, ρ is dominated by low energy states. Use the ground state wave function to evaluate the limiting behavior of $\langle q'|\rho|q\rangle$ as $T\to 0$.

In the low-temperature limit, we retain only the first terms in the summation

$$\rho_{\lim T\to 0} \approx \frac{|0\rangle e^{-\beta\hbar\omega/2}\langle 0| + |1\rangle e^{-3\beta\hbar\omega/2}\langle 1| + \cdots}{e^{-\beta\hbar\omega/2} + e^{-3\beta\hbar\omega/2}}.$$

Retaining only the term for the ground state in the numerator, but evaluating the geometric series in the denominator,

$$\langle q'|\rho|q\rangle_{\lim T\to 0} \approx \langle q'|0\rangle\langle 0|q\rangle e^{-\beta\hbar\omega/2}\cdot\left(e^{\beta\hbar\omega/2} - e^{-\beta\hbar\omega/2}\right).$$

Using the expression for $\langle q|0\rangle$ *given in part (b),*

$$\langle q'|\rho|q\rangle_{\lim T\to 0} \approx \sqrt{\frac{m\omega}{\pi\hbar}}\exp\left[-\frac{m\omega}{2\hbar}\left(q^2 + q'^2\right)\right]\left(1 - e^{-\beta\hbar\omega}\right).$$

****

4. *Relativistic Coulomb gas*: consider a *quantum* system of N positive, and N negative charged relativistic particles in box of volume $V = L^3$. The Hamiltonian is

$$\mathcal{H} = \sum_{i=1}^{2N} c|\vec{p}_i| + \sum_{i<j}^{2N} \frac{e_i e_j}{|\vec{r}_i - \vec{r}_j|},$$

where $e_i = +e_0$ for $i = 1, \cdots, N$, and $e_i = -e_0$ for $i = N+1, \cdots, 2N$, denote the charges of the particles; $\{\vec{r}_i\}$ and $\{\vec{p}_i\}$ their coordinates and momenta, respectively. While this is too complicated a system to solve, we can nonetheless obtain some exact results.

(a) Write down the Schrödinger equation for the eigenvalues $\varepsilon_n(L)$, and (in coordinate space) eigenfunctions $\Psi_n(\{\vec{r}_i\})$. State the constraints imposed on $\Psi_n(\{\vec{r}_i\})$ if the particles are bosons or fermions.

In the coordinate representation $\vec{p}_i$ is replaced by $-i\hbar\nabla_i$, leading to the Schrödinger equation

$$\left[\sum_{i=1}^{2N} c|-i\hbar\nabla_i| + \sum_{i<j}^{2N} \frac{e_i e_j}{|\vec{r}_i - \vec{r}_j|} \right] \Psi_n(\{\vec{r}_i\}) = \varepsilon_n(L)\Psi_n(\{\vec{r}_i\}).$$

There are N identical particles of charge $+e_0$, and N identical particles of charge $-e_0$. We can examine the effect of permutation operators P_+ and P_- on these two sets. The symmetry constraints can be written compactly as

$$P_+ P_- \Psi_n(\{\vec{r}_i\}) = \eta_+^{P_+} \cdot \eta_-^{P_-} \Psi_n(\{\vec{r}_i\}),$$

where $\eta = +1$ for bosons, $\eta = -1$ for fermions, and $(-1)^P$ denotes the parity of the permutation. Note that there is no constraint associated with exchange of particles with opposite charge.

(b) By a change of scale $\vec{r}_i{}' = \vec{r}_i/L$, show that the eigenvalues satisfy a scaling relation $\varepsilon_n(L) = \varepsilon_n(1)/L$.

After the change of scale $\vec{r}_i{}' = \vec{r}_i/L$ (and corresponding change in the derivative $\nabla_i{}' = L\nabla_i$), the above Schrödinger equation becomes

$$\left[\sum_{i=1}^{2N} c\left|-i\hbar\frac{\nabla_i{}'}{L}\right| + \sum_{i<j}^{2N} \frac{e_i e_j}{L|\vec{r}_i{}' - \vec{r}_j{}'|} \right] \Psi_n\left(\left\{\frac{\vec{r}_i{}'}{L}\right\}\right) = \varepsilon_n(L)\Psi_n\left(\left\{\frac{\vec{r}_i{}'}{L}\right\}\right).$$

The coordinates in the above equation are confined to a box of unit size. We can regard it as the Schrödinger equation in such a box, with wave functions $\Psi_n'(\{\vec{r}_i\}) = \Psi_n(\{\vec{r}_i{}'/L\})$. The corresponding eigenvalues are $\varepsilon_n(1) = L\varepsilon_n(L)$ (obtained after multiplying both sides of the above equation by L). We thus obtain the scaling relation

$$\varepsilon_n(L) = \frac{\varepsilon_n(1)}{L}.$$

(c) Using the formal expression for the partition function $Z(N, V, T)$, in terms of the eigenvalues $\{\varepsilon_n(L)\}$, show that Z does not depend on T and V separately, but only on a specific scaling combination of them.

The formal expression for the partition function is

$$Z(N, V, T) = \text{tr}\left(e^{-\beta \mathcal{H}}\right) = \sum_n \exp\left(-\frac{\varepsilon_n(L)}{k_B T}\right)$$

$$= \sum_n \exp\left(-\frac{\varepsilon_n(1)}{k_B TL}\right),$$

where we have used the scaling form of the energy levels. Clearly, in the above sum T and L always occur in the combination TL. Since $V = L^3$, the appropriate scaling variable is VT^3, and

$$Z(N, V, T) = \mathcal{Z}(N, VT^3).$$

(d) Relate the energy E, and pressure P, of the gas to variations of the partition function. Prove the exact result $E = 3PV$.

The average energy in the canonical ensemble is given by

$$E = -\frac{\partial \ln Z}{\partial \beta} = k_B T^2 \frac{\partial \ln Z}{\partial T} = k_B T^2 (3VT^2) \frac{\partial \ln \mathcal{Z}}{\partial (VT^3)} = 3k_B VT^4 \frac{\partial \ln \mathcal{Z}}{\partial (VT^3)}.$$

The free energy is $F = -k_B T \ln Z$, and its variations are $dF = -SdT - PdV + \mu dN$. Hence the gas pressure is given by

$$P = -\frac{\partial F}{\partial V} = k_B T \frac{\partial \ln Z}{\partial V} = k_B T^4 \frac{\partial \ln \mathcal{Z}}{\partial (VT^3)}.$$

The ratio of the above expressions gives the exact identity $E = 3PV$.

(e) The Coulomb interaction between charges in d-dimensional space falls off with separation as $e_i e_j / \left|\vec{r}_i - \vec{r}_j\right|^{d-2}$. (In $d = 2$ there is a logarithmic interaction.) In what dimension d can you construct an exact relation between E and P for *non-relativistic* particles (kinetic energy $\sum_i \vec{p}_i^2 / 2m$)? What is the corresponding exact relation between energy and pressure?

The above exact result is a consequence of the simple scaling law relating the energy eigenvalues $\varepsilon_n(L)$ to the system size. We could obtain the scaling form in part (b) since the kinetic and potential energies scaled in the same way. The kinetic energy for non-relativistic particles $\sum_i \vec{p}_i^2 / 2m = -\sum_i \hbar^2 \nabla_i^2 / 2m$ scales as $1/L^2$ under the change of scale $\vec{r}_i' = \vec{r}_i / L$, while the interaction energy $\sum_{i<j}^{2N} e_i e_j / \left|\vec{r}_i - \vec{r}_j\right|^{d-2}$ in d space dimensions scales as $1/L^{d-2}$. The two forms will scale the same way in $d = 4$ dimensions, leading to

$$\varepsilon_n(L) = \frac{\varepsilon_n(1)}{L^2}.$$

The partition function now has the scaling form

$$Z(N, V = L^4, T) = \mathcal{Z}\left(N, (TL^2)^2\right) = \mathcal{Z}\left(N, VT^2\right).$$

Following steps in the previous part, we obtain the exact relationship $E = 2PV$.

(f) Why are the above "exact" scaling laws not expected to hold in dense (liquid or solid) Coulomb mixtures?

The scaling results were obtained based on the assumption of the existence of a single scaling length L, relevant to the statistical mechanics of the problem. This is a good approximation in a gas phase. In a dense (liquid or solid) phase, the short-range repulsion between particles is expected to be important, and the particle size a is another relevant length scale that will enter in the solution to the Schrödinger equation, and invalidate the scaling results.

* * * * * * * *

5. *The virial theorem* is a consequence of the invariance of the phase space for a system of N (classical or quantum) particles under canonical transformations, such as a change of scale. In the following, consider N particles with coordinates $\{\vec{q}_i\}$, and conjugate momenta $\{\vec{p}_i\}$ (with $i = 1, \cdots, N$), and subject to a Hamiltonian $\mathcal{H}(\{\vec{p}_i\}, \{\vec{q}_i\})$.

(a) *Classical version*: write down the expression for the classical partition function, $Z \equiv Z[\mathcal{H}]$. Show that it is invariant under the rescaling $\vec{q}_1 \rightarrow \lambda \vec{q}_1$, $\vec{p}_1 \rightarrow \vec{p}_1/\lambda$ of a pair of conjugate variables, that is, $Z[\mathcal{H}_\lambda]$ is independent of λ, where $\mathcal{H}_\lambda$ is the Hamiltonian obtained after the above rescaling.

The classical partition function is obtained by appropriate integrations over phase space as

$$Z = \frac{1}{N! h^{3N}} \int \left(\prod_i \mathrm{d}^3 p_i \mathrm{d}^3 q_i \right) e^{-\beta \mathcal{H}}.$$

The rescaled Hamiltonian $\mathcal{H}_\lambda = \mathcal{H}(\vec{p}_1/\lambda, \{\vec{p}_{i \neq 1}\}, \lambda \vec{q}_1, \{\vec{q}_{i \neq 1}\})$ leads to a rescaled partition function

$$Z[\mathcal{H}_\lambda] = \frac{1}{N! h^{3N}} \int \left(\prod_i \mathrm{d}^3 p_i \mathrm{d}^3 q_i \right) e^{-\beta \mathcal{H}_\lambda},$$

which reduces to

$$Z[\mathcal{H}_\lambda] = \frac{1}{N! h^{3N}} \int \left(\lambda^3 \mathrm{d}^3 p_1' \right) \left(\lambda^{-3} \mathrm{d}^3 q_1' \right) \left(\prod_i \mathrm{d}^3 p_i \mathrm{d}^3 q_i \right) e^{-\beta \mathcal{H}} = Z,$$

under the change of variables $\vec{q}_1' = \lambda \vec{q}_1$, $\vec{p}_1' = \vec{p}_1/\lambda$.

(b) *Quantum mechanical version*: write down the expression for the quantum partition function. Show that it is also invariant under the rescalings $\vec{q}_1 \rightarrow \lambda \vec{q}_1$, $\vec{p}_1 \rightarrow \vec{p}_1/\lambda$, where $\vec{p}_i$ and $\vec{q}_i$ are now quantum mechanical operators. (*Hint.* Start with the time-independent Schrödinger equation.)

Using the energy basis

$$Z = \mathrm{tr}\left(e^{-\beta \mathcal{H}} \right) = \sum_n e^{-\beta E_n},$$

where E_n are the energy eigenstates of the system, obtained from the Schrödinger equation

$$\mathcal{H}(\{\vec{p}_i\}, \{\vec{q}_i\}) |\psi_n\rangle = E_n |\psi_n\rangle,$$

where $|\psi_n\rangle$ are the eigenstates. After the rescaling transformation, the corresponding equation is

$$\mathcal{H}\left(\vec{p}_1/\lambda, \left\{\vec{p}_{i\neq 1}\right\}, \lambda\vec{q}_1, \left\{\vec{q}_{i\neq 1}\right\}\right)\left|\psi_n^{(\lambda)}\right\rangle = E_n^{(\lambda)}\left|\psi_n^{(\lambda)}\right\rangle.$$

In the coordinate representation, the momentum operator is $\vec{p}_i = -i\hbar\partial/\partial\vec{q}_i$, and therefore $\psi_\lambda\left(\left\{\vec{q}_i\right\}\right) = \psi\left(\left\{\lambda\vec{q}_i\right\}\right)$ is a solution of the rescaled equation with eigenvalue $E_n^{(\lambda)} = E_n$. Since the eigenenergies are invariant under the transformation, so is the partition function, which is simply the sum of corresponding exponentials.

(c) Now assume a Hamiltonian of the form

$$\mathcal{H} = \sum_i \frac{\vec{p}_i^{\;2}}{2m} + V\left(\left\{\vec{q}_i\right\}\right).$$

Use the result that $Z\left[\mathcal{H}_\lambda\right]$ is independent of λ to prove the *virial* relation

$$\left\langle\frac{\vec{p}_1^{\;2}}{m}\right\rangle = \left\langle\frac{\partial V}{\partial\vec{q}_1}\cdot\vec{q}_1\right\rangle,$$

where the brackets denote thermal averages.

Differentiating the free energy with respect to λ at $\lambda = 1$, we obtain

$$0 = \left.\frac{\partial\ln Z_\lambda}{\partial\lambda}\right|_{\lambda=1} = -\beta\left\langle\left.\frac{\partial H_\lambda}{\partial\lambda}\right|_{\lambda=1}\right\rangle = -\beta\left\langle-\frac{\vec{p}_1^{\;2}}{m} + \frac{\partial V}{\partial\vec{q}_1}\cdot\vec{q}_1\right\rangle,$$

that is

$$\left\langle\frac{\vec{p}_1^{\;2}}{m}\right\rangle = \left\langle\frac{\partial V}{\partial\vec{q}_1}\cdot\vec{q}_1\right\rangle.$$

(d) The above relation is sometimes used to estimate the mass of distant galaxies. The stars on the outer boundary of the G-8.333 galaxy have been measured to move with velocity $v \approx 200\,\text{km s}^{-1}$. Give a numerical estimate of the ratio of the G-8.333's mass to its size.

The virial relation applied to a gravitational system gives

$$\langle mv^2\rangle = \left\langle\frac{GMm}{R}\right\rangle.$$

Assuming that the kinetic and potential energies of the stars in the galaxy have reached some form of equilibrium gives

$$\frac{M}{R} \approx \frac{v^2}{G} \approx 6\times 10^{20}\,\text{kg m}^{-1}.$$

****

Solutions to selected problems from chapter 7

1. *Identical particle pair*: let $Z_1(m)$ denote the partition function for a single quantum particle of mass m in a volume V.

   (a) Calculate the partition function of two such particles, if they are bosons, and also if they are (spinless) fermions.

   A two-particle wave function is constructed from one-particle states k_1, and k_2. Depending on the statistics of the two particles, we have

   Bosons:

   $$|k_1, k_2\rangle_B = \begin{cases} (|k_1\rangle |k_2\rangle + |k_2\rangle |k_1\rangle)/\sqrt{2} & \text{for } k_1 \neq k_2 \\ |k_1\rangle |k_2\rangle & \text{for } k_1 = k_2 \end{cases},$$

   Fermions:

   $$|k_1, k_2\rangle_F = \begin{cases} (|k_1\rangle |k_2\rangle - |k_2\rangle |k_1\rangle)/\sqrt{2} & \text{for } k_1 \neq k_2 \\ \text{no state} & \text{for } k_1 = k_2 \end{cases}.$$

   For the bose system, the partition function is given by

   $$\begin{aligned} Z_2^B = \text{tr}\left(e^{-\beta H}\right) &= \sum_{k_1, k_2} \langle k_1, k_2 | e^{-\beta H} | k_1, k_2 \rangle_B \\ &= \sum_{k_1 > k_2} \frac{\langle k_1 | \langle k_2 | + \langle k_2 | \langle k_1 |}{\sqrt{2}} e^{-\beta H} \frac{|k_1\rangle |k_2\rangle + |k_2\rangle |k_1\rangle}{\sqrt{2}} + \sum_k \langle k | \langle k | e^{-\beta H} |k\rangle |k\rangle \\ &= \sum_{k_1 > k_2} \exp\left[-\frac{\beta \hbar^2}{2m}(k_1^2 + k_2^2)\right] + \sum_k \exp\left(-\frac{2\beta \hbar^2 k^2}{2m}\right) \\ &= \frac{1}{2} \sum_{k_1, k_2} \exp\left[-\frac{\beta \hbar^2}{2m}(k_1^2 + k_2^2)\right] + \frac{1}{2} \sum_k \exp\left(-\frac{\beta \hbar^2 k^2}{m}\right), \end{aligned}$$

   and thus

   $$Z_2^B = \frac{1}{2}\left[Z_1^2(m) + Z_1\left(\frac{m}{2}\right)\right].$$

   For the fermi system,

   $$Z_2^F = \text{tr}\left(e^{-\beta H}\right) = \sum_{k_1, k_2} \langle k_1, k_2 | e^{-\beta H} | k_1, k_2 \rangle_F$$

$$= \sum_{k_1 > k_2} \frac{\langle k_1| \langle k_2| - \langle k_2| \langle k_1|}{\sqrt{2}} e^{-\beta H} \frac{|k_1\rangle |k_2\rangle - |k_2\rangle |k_1\rangle}{\sqrt{2}}$$

$$= \frac{1}{2} \sum_{k_1,k_2} \exp\left[-\frac{\beta\hbar^2}{2m}(k_1^2 + k_2^2)\right] - \frac{1}{2}\sum_k \exp\left(-\frac{\beta\hbar^2 k^2}{m}\right),$$

and thus

$$Z_2^F = \frac{1}{2}\left[Z_1^2(m) - Z_1\left(\frac{m}{2}\right)\right].$$

Note that classical Boltzmann particles have a partition function

$$Z_2^{\text{classical}} = \frac{1}{2}Z_1(m)^2.$$

(b) Use the classical approximation $Z_1(m) = V/\lambda^3$ with $\lambda = h/\sqrt{2\pi m k_B T}$. Calculate the corrections to the energy E, and the heat capacity C, due to bose or fermi statistics.

If the system is non-degenerate, the correction term is much smaller than the classical term, and

$$\ln Z_2^\pm = \ln\left\{\left[Z_1(m)^2 \pm Z_1(m/2)\right]/2\right\}$$

$$= 2\ln Z_1(m) + \ln\left\{1 \pm \frac{Z_1(m/2)}{Z_1(m)^2}\right\} - \ln 2$$

$$\approx 2\ln Z_1(m) \pm \frac{Z_1(m/2)}{Z_1(m)^2} - \ln 2.$$

Using

$$Z_1(m) = \frac{V}{\lambda(m)^3}, \quad \text{where} \quad \lambda(m) \equiv \frac{h}{\sqrt{2\pi m k_B T}},$$

we can write

$$\ln Z_2^\pm \approx \ln Z_2^{\text{classical}} \pm \frac{\lambda(m)^3}{V}\left(\frac{\lambda(m)}{\lambda(m/2)}\right)^3 = \ln Z_2^{\text{classical}} \pm 2^{-3/2}\frac{\lambda(m)^3}{V},$$

$$\implies \quad \Delta\ln Z_2^\pm = \pm 2^{-3/2}\frac{h^3}{V(2\pi m k_B T)^{3/2}} = \pm 2^{-3/2}\frac{h^3\beta^{3/2}}{V(2\pi m)^{3/2}}.$$

Thus the energy differences are

$$\Delta E^\pm = -\frac{\partial}{\partial\beta}\Delta\ln Z_2 = \mp\frac{3}{2^{5/2}}\frac{h^3\beta^{3/2}}{V(2\pi m)^{3/2}} = \mp\frac{3}{2^{5/2}}k_B T\left(\frac{\lambda^3}{V}\right),$$

resulting in heat capacity differences of

$$\Delta C_V^\pm = \frac{\partial\Delta E}{\partial T}\bigg|_V = \pm\frac{3}{2^{7/2}}\frac{h^3 k_B}{V(2\pi m k_B T)^{3/2}} = \pm\frac{3}{2^{7/2}}k_B\left(\frac{\lambda^3}{V}\right).$$

This approximation holds only when the thermal volume is much smaller than the volume of the gas where the ratio constitutes a small correction.

(c) At what temperature does the approximation used above break down?

The approximation is invalid if the correction terms are comparable to the first term. This occurs when the thermal wavelength becomes comparable to the size of the box, that is, for

$$\lambda = \frac{h}{\sqrt{2\pi m k_B T}} \geq L \sim V^{1/3}, \quad \text{or} \quad T \leq \frac{h^2}{2\pi m k_B L^2}.$$

$$********$$

2. *Generalized ideal gas*: consider a gas of non-interacting identical (spinless) quantum particles with an energy spectrum $\varepsilon = |\vec{p}/\hbar|^s$, contained in a box of "volume" V in d dimensions.

(a) Calculate the grand potential $\mathcal{G}_\eta = -k_B T \ln \mathcal{Q}_\eta$, and the density $n = N/V$, at a chemical potential μ. Express your answers in terms of s, d, and $f_m^\eta(z)$, where $z = e^{\beta\mu}$, and

$$f_m^\eta(z) = \frac{1}{\Gamma(m)} \int_0^\infty \frac{dx\, x^{m-1}}{z^{-1}e^x - \eta}.$$

(*Hint.* Use integration by parts on the expression for $\ln \mathcal{Q}_\eta$.)

The grand partition function is given by

Fermions:

$$Q_F = \sum_{N=0}^\infty e^{\beta\mu N} \sum_{\{n_\nu\}} e^{-\beta \sum_\nu n_\nu \varepsilon_\nu} = \prod_\nu \sum_{n_\nu=0}^1 e^{\beta(\mu-\varepsilon_\nu)n_\nu} = \prod_\nu \left[1 + e^{\beta(\mu-\varepsilon_\nu)}\right].$$

Bosons:

$$Q_B = \sum_{N=0}^\infty e^{\beta\mu N} \sum_{\{n_\nu\}} e^{-\beta \sum_\nu n_\nu \varepsilon_\nu} = \prod_\nu \sum_{\{n_\nu\}} e^{\beta(\mu-\varepsilon_\nu)n_\nu} = \prod_\nu \left[1 - e^{\beta(\mu-\varepsilon_\nu)}\right]^{-1}.$$

Hence we obtain

$$\ln Q_\eta = -\eta \sum_\nu \ln\left[1 - \eta e^{\beta(\mu-\varepsilon_\nu)}\right],$$

where $\eta = +1$ for bosons, and $\eta = -1$ for fermions.

Changing the summation into an integration in d dimensions yields

$$\sum_\nu \quad \to \quad \int d^d n \quad \to \quad \frac{V}{(2\pi)^d} \int d^d k = \frac{VS_d}{(2\pi)^d} \int k^{d-1}\, dk,$$

where

$$S_d = \frac{2\pi^{d/2}}{(d/2 - 1)!}.$$

Then

$$\ln Q_\eta = -\eta \frac{VS_d}{(2\pi)^d} \int k^{d-1} dk \, \ln\left[1 - \eta e^{\beta(\mu-\varepsilon_\nu)}\right]$$

$$= -\eta \frac{VS_d}{(2\pi)^d} \int k^{d-1} dk \, \ln\left[1 - \eta z e^{-\beta k^s}\right],$$

where

$$z \equiv e^{\beta\mu}.$$

Introducing a new variable

$$x \equiv \beta k^s, \quad \rightarrow \quad k = \left(\frac{x}{\beta}\right)^{1/s}, \quad \text{and} \quad dk = \frac{1}{s}\left(\frac{x}{\beta}\right)^{1/s-1}\frac{dx}{\beta}$$

results in

$$\ln Q_\eta = -\eta \frac{VS_d}{(2\pi)^d}\frac{\beta^{-d/s}}{s}\int dx\, x^{d/s-1}\ln\left(1 - \eta z e^{-x}\right).$$

Integrating by parts,

$$\ln Q_\eta = \frac{VS_d}{(2\pi)^d}\frac{\beta^{-d/s}}{d}\int dx\, x^{d/s}\frac{z e^{-x}}{1 - \eta z e^{-x}} = \frac{VS_d}{(2\pi)^d}\frac{\beta^{-d/s}}{d}\int dx\, \frac{x^{d/s}}{z^{-1}e^x - \eta}.$$

Finally,

$$\ln Q_\eta = \frac{VS_d}{(2\pi)^d}\beta^{-d/s}\Gamma\left(\frac{d}{s}+1\right)\cdot f_{d/s+1}^\eta(z),$$

where

$$f_n^\eta(z) \equiv \frac{1}{\Gamma(n)}\int dx\, \frac{x^{n-1}}{z^{-1}e^x - \eta}.$$

Using similar notation, the total particle number is obtained as

$$N = \frac{\partial}{\partial(\beta\mu)}\ln Q_\eta\bigg|_\beta = -\eta\frac{VS_d}{(2\pi)^d}\frac{\beta^{d/s}}{s}\frac{\partial}{\partial(\beta\mu)}\int dx\, x^{d/s-1}\ln\left(1 - \eta z e^{-x}\right)$$

$$= \frac{VS_d}{(2\pi)^d}\frac{\beta^{d/s}}{s}\int dx\, \frac{x^{d/s-1}z e^{-x}}{1 - \eta z e^{-x}} = \frac{VS_d}{(2\pi)^d}\frac{\beta^{d/s}}{s}z\int dx\, \frac{x^{d/s-1}}{z^{-1}e^{-x} - \eta},$$

and thus

$$n = \frac{N}{V} = \frac{S_d}{(2\pi)^d s}(k_B T)^{d/s}\,\Gamma\left(\frac{d}{s}\right)\cdot f_{d/s}^\eta(z).$$

(b) Find the ratio PV/E, and compare it to the classical result obtained previously.

The gas pressure is given by

$$PV = -k_B T\ln Q_\eta = \frac{VS_d}{(2\pi)^d}\frac{(k_B T)^{d/s+1}}{d}\Gamma\left(\frac{d}{s}+1\right)\cdot f_{d/s+1}(z).$$

Note that since $\ln Q_\eta \propto \beta^{-d/s}$,

$$E = -\frac{\partial}{\partial\beta}\ln Q_\eta\bigg|_z = -\frac{d}{s}\frac{\ln Q_\eta}{\beta} = -\frac{d}{s}k_B T\ln Q_\eta, \quad \Longrightarrow \quad \frac{PV}{E} = \frac{s}{d},$$

which is the same result obtained classically.

(c) *For fermions,* calculate the dependence of E/N, and P, on the density $n = N/V$, at zero temperature. (*Hint.* $f_m(z) \to (\ln z)^m/m!$ as $z \to \infty$.)

For fermions at low T,

$$\frac{1}{e^{\beta(\varepsilon - \varepsilon_F)} + 1} \approx \theta(\varepsilon - \varepsilon_F), \quad \Longrightarrow \quad f_m(z) \approx \frac{(\beta\varepsilon_F)^m}{n\Gamma(n)},$$

resulting in

$$E \approx \frac{VS_d}{(2\pi)^d} \frac{\varepsilon_F^{d/s+1}}{d}, \quad PV \approx \frac{VS_d}{(2\pi)^d} \frac{\varepsilon_F^{d/s+1}}{s}, \quad n \approx \frac{S_d}{(2\pi)^d} \frac{\varepsilon_F^{d/s}}{s}.$$

Hence, we obtain

$$\frac{E}{N} \approx \frac{1}{n} \cdot \frac{S_d}{(2\pi)^d} \frac{\varepsilon_F^{d/s+1}}{d} \approx \frac{s}{d}\varepsilon_F \propto n^{s/d}, \quad \text{and}$$

$$P \propto \varepsilon_F^{d/s+1} \propto n^{s/d+1}.$$

(d) *For bosons,* find the dimension $d_\ell(s)$, below which there is no Bose condensation. Is there condensation for $s = 2$ at $d = 2$?

To see if Bose–Einstein condensation occurs, check whether a density n can be found such that $z = z_{\max} = 1$. Since $f^+_{d/s}(z)$ is a monotonic function of z for $0 \le z \le 1$, the maximum value for the right-hand side of the equation for $n = N/V$, given in part (a) is

$$\frac{S_d}{(2\pi)^d s}(k_B T)^{d/s} \Gamma\left(\frac{d}{s}\right) f^+_{d/s}(1).$$

If this value is larger than n, we can always find $z < 1$ that satisfies the above equation. If this value is smaller than n, then the remaining portion of the particles should condense into the ground state. Thus the question is whether $f^+_{d/s}(1)$ diverges at $z = 1$, where

$$f^+_{d/s}(z_{\max} = 1) = \frac{1}{\Gamma(d/s)} \int dx \frac{x^{d/s-1}}{e^x - 1}.$$

The integrand may diverge near $x = 0$; the contribution for $x \sim 0$ is

$$\int_0^\varepsilon dx \frac{x^{d/s-1}}{e^x - 1} \simeq \int_0^\varepsilon dx\, x^{d/s-2},$$

which converges for $d/s - 2 > -1$, or $d > s$. Therefore, Bose–Einstein condensation occurs for $d > s$. For a two-dimensional gas, $d = s = 2$, the integral diverges logarithmically, and hence Bose–Einstein condensation does not occur.

****

3. *Pauli paramagnetism:* calculate the contribution of electron spin to its magnetic susceptibility as follows. Consider non-interacting electrons, each subject to a Hamiltonian

$$\mathcal{H}_1 = \frac{\vec{p}^{\,2}}{2m} - \mu_0 \vec{\sigma} \cdot \vec{B},$$

where $\mu_0 = e\hbar/2mc$, and the eigenvalues of $\vec{\sigma} \cdot \vec{B}$ are $\pm B$.
(The orbital effect, $\vec{p} \to \vec{p} - e\vec{A}$, has been ignored.)

(a) Calculate the grand potential $\mathcal{G}_- = -k_B T \ln \mathcal{Q}_-$, at a chemical potential μ.

The energy of the electron gas is given by

$$E \equiv \sum_p E_p(n_p^+, n_p^-),$$

where $n_p^{\pm}(= 0 \text{ or } 1)$ denotes the number of particles having $\pm$ spins and momentum p, and

$$E_p(n_p^+, n_p^-) \equiv \left(\frac{p^2}{2m} - \mu_0 B\right) n_p^+ + \left(\frac{p^2}{2m} + \mu_0 B\right) n_p^-$$

$$= (n_p^+ + n_p^-)\frac{p^2}{2m} - (n_p^+ - n_p^-)\mu_0 B.$$

The grand partition function of the system is

$$Q = \sum_{N=0}^{\infty} \exp(-\beta\mu N) \overset{N=\sum(n_p^+ + n_p^-)}{\underset{\{n_p^+, n_p^-\}}{\sum}} \exp\left[-\beta E_p(n_p^+, n_p^-)\right]$$

$$= \sum_{\{n_p^+, n_p^-\}} \exp\left[\beta\mu\left(n_p^+ + n_p^-\right) - \beta E_p\left(n_p^+, n_p^-\right)\right]$$

$$= \prod_p \sum_{\{n_p^+, n_p^-\}} \exp\left\{\beta\left[\left(\mu - \mu_0 B - \frac{p^2}{2m}\right) n_p^+ + \left(\mu + \mu_0 B - \frac{p^2}{2m}\right) n_p^-\right]\right\}$$

$$= \prod_p \left\{1 + \exp\left[\beta\left(\mu - \mu_0 B - \frac{p^2}{2m}\right)\right]\right\} \cdot \left\{1 + \exp\left[\beta\left(\mu + \mu_0 B - \frac{p^2}{2m}\right)\right]\right\}$$

$$= Q_0\left(\mu + \mu_0 B\right) \cdot Q_0\left(\mu - \mu_0 B\right),$$

where

$$Q_0(\mu) \equiv \prod_p \left\{1 + \exp\left[\beta\left(\mu - \frac{p^2}{2m}\right)\right]\right\}.$$

Thus

$$\ln Q = \ln\left\{Q_0(\mu + \mu_0 B)\right\} + \ln\left\{Q_0(\mu - \mu_0 B)\right\}.$$

Each contribution is given by

$$\ln Q_0(\mu) = \sum_p \ln\left(1 + \exp\left[\beta(\mu - \frac{p^2}{2m})\right]\right) = \frac{V}{(2\pi\hbar)^3} \int d^3p \ln\left(1 + ze^{-\beta\frac{p^2}{2m}}\right)$$

$$= \frac{V}{h^3} \frac{4\pi m}{\beta} \sqrt{\frac{2m}{\beta}} \int dx \sqrt{x} \ln(1 + ze^{-x}), \quad \text{where} \quad z \equiv e^{\beta\mu},$$

and integrating by parts yields

$$\ln Q_0(\mu) = V\left(\frac{2\pi m k_B T}{h^2}\right)^{3/2} \frac{2}{\sqrt{\pi}} \frac{2}{3} \int dx \frac{x^{3/2}}{z^{-1}e^x + 1} = \frac{V}{\lambda^3} f_{5/2}^-(z).$$

The total grand free energy is obtained from

$$\ln Q(\mu) = \frac{V}{\lambda^3} \left[f_{5/2}^- \left(z e^{\beta \mu_0 B} \right) + f_{5/2}^- \left(z e^{-\beta \mu_0 B} \right) \right],$$

as

$$G = -k_B T \ln Q(\mu) = -k_B T \frac{V}{\lambda^3} \left[f_{5/2}^- \left(z e^{\beta \mu_0 B} \right) + f_{5/2}^- \left(z e^{-\beta \mu_0 B} \right) \right].$$

(b) Calculate the densities $n_+ = N_+/V$, and $n_- = N_-/V$, of electrons pointing parallel and anti-parallel to the field.

   The number densities of electrons with up or down spins are given by

$$\frac{N_\pm}{V} = z \frac{\partial}{\partial z} \ln Q_\pm = \frac{V}{\lambda^3} f_{3/2}^- \left(z e^{\pm \beta \mu_0 B} \right),$$

where we used

$$z \frac{\partial}{\partial z} f_n^- (z) = f_{n-1}^- (z).$$

The total number of electrons is the sum of these, that is,

$$N = N_+ + N_- = \frac{V}{\lambda^3} \left[f_{3/2}^- \left(z e^{\beta \mu_0 B} \right) + f_{3/2}^- \left(z e^{-\beta \mu_0 B} \right) \right].$$

(c) Obtain the expression for the magnetization $M = \mu_0(N_+ - N_-)$, and expand the result for small B.

   The magnetization is related to the difference between numbers of spin-up and spin-down electrons as

$$M = \mu_0(N_+ - N_-) = \mu_0 \frac{V}{\lambda^3} \left[f_{3/2}^- \left(z e^{\beta \mu_0 B} \right) - f_{3/2}^- \left(z e^{-\beta \mu_0 B} \right) \right].$$

Expanding the results for small B gives

$$f_{3/2}^- \left(z e^{\pm \beta \mu_0 B} \right) \approx f_{3/2}^- \left[z (1 \pm \beta \mu_0 B) \right] \approx f_{3/2}^-(z) \pm z \cdot \beta \mu_0 B \frac{\partial}{\partial z} f_{3/2}^-(z),$$

which results in

$$M = \mu_0 \frac{V}{\lambda^3} (2 \beta \mu_0 B) \cdot f_{1/2}^-(z) = \frac{2 \mu_0^2}{k_B T} \frac{V}{\lambda^3} \cdot B \cdot f_{1/2}^-(z).$$

(d) Sketch the zero-field susceptibility $\chi(T) = \partial M / \partial B|_{B=0}$, and indicate its behavior at low and high temperatures.

   The magnetic susceptibility is

$$\chi \equiv \left. \frac{\partial M}{\partial B} \right|_{B=0} = \frac{2 \mu_0^2}{k_B T} \frac{V}{\lambda^3} \cdot f_{1/2}^-(z),$$

with z given by

$$N = 2 \frac{V}{\lambda^3} \cdot f_{3/2}^-(z).$$

In the low-temperature limit $(\ln z = \beta\mu \to \infty)$,

$$f_n^-(z) = \frac{1}{\Gamma(n)} \int_0^\infty \mathrm{d}x\, \frac{x^{n-1}}{1+e^{x-\ln(z)}} \underset{T \to 0}{\approx} \frac{1}{\Gamma(n)} \int_0^{\ln(z)} \mathrm{d}x\, x^{n-1} = \frac{[\ln(z)]^n}{n\Gamma(n)},$$

$$N = 2\frac{V}{\lambda^3} \cdot \frac{4(\ln z)^{3/2}}{3\sqrt{\pi}}, \quad \Longrightarrow \quad \ln z = \left(\frac{3N\sqrt{\pi}}{8V}\lambda^3\right)^{2/3},$$

$$\chi = \frac{4\mu_0^2 V}{\sqrt{\pi}k_B T \lambda^3} \cdot \left(\frac{3N\sqrt{\pi}}{8V}\lambda^3\right)^{1/3} = \frac{2\mu_o^2 V}{k_B T \lambda^2} \cdot \left(\frac{3N}{\pi V}\right)^{1/3} = \frac{4\pi m \mu_0^2 V}{h^2} \cdot \left(\frac{3N}{\pi V}\right)^{1/3}.$$

The ratio of the last two expressions gives

$$\left.\frac{\chi}{N}\right|_{T\to 0} = \frac{\mu_0^2}{k_B T}\frac{f_{1/2}^-}{f_{3/2}^-} = \frac{3\mu_0^2}{2k_B T}\frac{1}{\ln(z)} = \frac{3\mu_0^2}{2k_B T}\frac{1}{\beta\varepsilon_F} = \frac{3\mu_0^2}{2k_B T_F}.$$

In the high-temperature limit $(z \to 0)$,

$$f_n(z) \underset{z\to 0}{\to} \frac{z}{\Gamma(n)} \int_0^\infty \mathrm{d}x\, x^{n-1}e^{-x} = z,$$

and thus

$$N \underset{\beta\to 0}{\to} \frac{2V}{\lambda^3} \cdot z, \quad \Longrightarrow \quad z \approx \frac{N}{2V} \cdot \lambda^3 = \frac{N}{2V} \cdot \left(\frac{h^2}{2\pi m k_B T}\right)^{3/2} \to 0,$$

which is consistent with $\beta \to 0$. *Using this result,*

$$\chi \approx \frac{2\mu_0^2 V}{k_B T \lambda^3} \cdot z = \frac{N\mu_0^2}{k_B T}.$$

The result

$$\left(\frac{\chi}{N}\right)_{T\to\infty} = \frac{\mu_0^2}{k_B T}$$

is known as the Curie susceptibility.

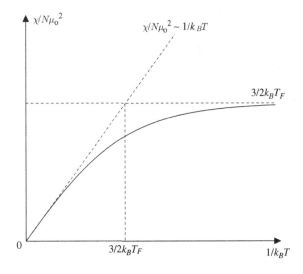

(e) Estimate the magnitude of χ/N for a typical metal at room temperature.

Since $T_{\text{room}} \ll T_F \approx 10^4 \text{K}$, we can take the low T limit for χ (see (d)), and

$$\frac{\chi}{N} = \frac{3\mu_0^2}{2k_B T_F} \approx \frac{3 \times (9.3 \times 10^{-24})^2}{2 \times 1.38 \times 10^{-23}} \approx 9.4 \times 10^{-24} \,\text{J}\,\text{T}^{-2},$$

where we used

$$\mu_0 = \frac{eh}{2mc} \simeq 9.3 \times 10^{-24} \,\text{J}\,\text{T}^{-1}.$$

****

4. *Freezing of He^3*: at low temperatures He^3 can be converted from liquid to solid by application of pressure. A peculiar feature of its phase boundary is that $(dP/dT)_{\text{melting}}$ is negative at temperatures below 0.3 K $[(dP/dT)_m \approx -30 \,\text{atm} \,\text{K}^{-1}$ at $T \approx 0.1 \,\text{K}]$. We will use a simple model of liquid and solid phases of He^3 to account for this feature.

(a) In the solid phase, the He^3 atoms form a crystal lattice. Each atom has nuclear spin of 1/2. Ignoring the interaction between spins, what is the entropy per particle s_s, due to the spin degrees of freedom?

Entropy of solid He^3 comes from the nuclear spin degeneracies, and is given by

$$s_s = \frac{S_s}{N} = \frac{k_B \ln(2^N)}{N} = k_B \ln 2.$$

(b) Liquid He^3 is modeled as an ideal fermi gas, with a volume of $46 \,\text{Å}^3$ per atom. What is its fermi temperature T_F, in degrees Kelvin?

The fermi temperature for liquid He^3 may be obtained from its density as

$$T_F = \frac{\varepsilon_F}{k_B} = \frac{h^2}{2mk_B}\left(\frac{3N}{8\pi V}\right)^{2/3}$$

$$\approx \frac{(6.7 \times 10^{-34})^2}{2 \cdot (6.8 \times 10^{-27})(1.38 \times 10^{-23})}\left(\frac{3}{8\pi \times 46 \times 10^{-30}}\right)^{2/3} \approx 9.2 \,\text{K}.$$

(c) How does the heat capacity of liquid He^3 behave at low temperatures? Write down an expression for C_V in terms of N, T, k_B, T_F, up to a numerical constant, that is valid for $T \ll T_F$.

The heat capacity comes from the excited states at the fermi surface, and is given by

$$C_V = k_B \frac{\pi^2}{6} k_B T \, D(\varepsilon_F) = \frac{\pi^2}{6}k_B^2 T \frac{3N}{2k_B T_F} = \frac{\pi^2}{4}Nk_B \frac{T}{T_F}.$$

(d) Using the result in (c), calculate the entropy per particle s_ℓ, in the liquid at low temperatures. For $T \ll T_F$, which phase (solid or liquid) has the higher entropy?

The entropy can be obtained from the heat capacity as

$$C_V = \frac{T dS}{dT}, \quad \Rightarrow \quad s_\ell = \frac{1}{N}\int_0^T \frac{C_V dT}{T} = \frac{\pi^2}{4}k_B \frac{T}{T_F}.$$

As $T \to 0$, $s_\ell \to 0$, while s_s *remains finite. This is an unusual situation in which the solid has more entropy than the liquid! (The finite entropy is due to treating the nuclear spins as independent. There is actually a weak coupling between spins, which causes magnetic ordering at a much lower temperature, removing the finite entropy.)*

(e) By equating chemical potentials, or by any other technique, prove the Clausius–Clapeyron equation $(\mathrm{d}P/\mathrm{d}T)_{\text{melting}} = (s_\ell - s_s)/(v_\ell - v_s)$, where v_ℓ and v_s are the volumes per particle in the liquid and solid phases, respectively.

The Clausius–Clapeyron equation can be obtained by equating the chemical potentials at the phase boundary,

$$\mu_\ell(T, P) = \mu_s(T, P), \quad \text{and} \quad \mu_\ell(T + \Delta T, P + \Delta P) = \mu_s(T + \Delta T, P + \Delta P).$$

Expanding the second equation, and using the thermodynamic identities

$$\left(\frac{\partial \mu}{\partial T}\right)_P = S, \text{ and } \left(\frac{\partial \mu}{\partial P}\right)_T = -V,$$

results in

$$\left(\frac{\partial P}{\partial T}\right)_{\text{melting}} = \frac{s_\ell - s_s}{v_\ell - v_s}.$$

(f) It is found experimentally that $v_\ell - v_s = 3 \, \text{Å}^3$ per atom. Using this information, plus the results obtained in previous parts, estimate $(\mathrm{d}P/\mathrm{d}T)_{\text{melting}}$ at $T \ll T_F$.

The negative slope of the phase boundary results from the solid having more entropy than the liquid, and can be calculated from the Clausius–Clapeyron relation

$$\left(\frac{\partial P}{\partial T}\right)_{\text{melting}} = \frac{s_\ell - s_s}{v_\ell - v_s} \approx k_B \frac{\frac{\pi^2}{4}\left(\frac{T}{T_F}\right) - \ln 2}{v_\ell - v_s}.$$

Using the values, $T = 0.1 \, \text{K}$, $T_F = 9.2 \, \text{K}$, and $v_\ell - v_s = 3 \, \text{Å}^3$, we estimate

$$\left(\frac{\partial P}{\partial T}\right)_{\text{melting}} \approx -2.7 \times 10^6 \, \text{Pa K}^{-1},$$

in reasonable agreement with the observations.

$$********$$

5. *Non-interacting fermions*: consider a grand canonical ensemble of non-interacting *fermions* with chemical potential μ. The one-particle states are labeled by a wave vector $\vec{k}$, and have energies $\mathcal{E}(\vec{k})$.

(a) What is the joint probability $P(\{n_{\vec{k}}\})$ of finding a set of occupation numbers $\{n_{\vec{k}}\}$ of the one-particle states?

In the grand canonical ensemble with chemical potential μ, the joint probability of finding a set of occupation numbers $\{n_{\vec{k}}\}$ for one-particle states of energies $\mathcal{E}(\vec{k})$ is given by the fermi distribution

$$P(\{n_{\vec{k}}\}) = \prod_{\vec{k}} \frac{\exp\left[\beta(\mu - \mathcal{E}(\vec{k}))n_{\vec{k}}\right]}{1 + \exp\left[\beta(\mu - \mathcal{E}(\vec{k}))\right]}, \quad \text{where} \quad n_{\vec{k}} = 0 \text{ or } 1, \quad \text{for each } \vec{k}.$$

(b) Express your answer to part (a) in terms of the average occupation numbers $\{\langle n_{\vec{k}}\rangle_-\}$.

The average occupation numbers are given by

$$\langle n_{\vec{k}}\rangle_- = \frac{\exp\left[\beta(\mu - \mathcal{E}(\vec{k}))\right]}{1 + \exp\left[\beta(\mu - \mathcal{E}(\vec{k}))\right]},$$

from which we obtain

$$\exp\left[\beta(\mu - \mathcal{E}(\vec{k}))\right] = \frac{\langle n_{\vec{k}}\rangle_-}{1 - \langle n_{\vec{k}}\rangle_-}.$$

This enables us to write the joint probability as

$$P(\{n_{\vec{k}}\}) = \prod_{\vec{k}}\left[(\langle n_{\vec{k}}\rangle_-)^{n_{\vec{k}}}\left(1 - \langle n_{\vec{k}}\rangle_-\right)^{1-n_{\vec{k}}}\right].$$

(c)   A random variable has a set of ℓ discrete outcomes with probabilities p_n, where $n = 1, 2, \cdots, \ell$. What is the entropy of this probability distribution? What is the maximum possible entropy?

A random variable has a set of ℓ discrete outcomes with probabilities p_n. The entropy of this probability distribution is calculated from

$$S = -k_B \sum_{n=1}^{\ell} p_n \ln p_n.$$

The maximum entropy is obtained if all probabilities are equal, $p_n = 1/\ell$, and given by $S_{\max} = k_B \ln \ell$.

(d) Calculate the entropy of the probability distribution for fermion occupation numbers in part (b), and comment on its zero-temperature limit.

Since the occupation numbers of different one-particle states are independent, the corresponding entropies are additive, and given by

$$S = -k_B \sum_{\vec{k}}\left[\langle n_{\vec{k}}\rangle_- \ln \langle n_{\vec{k}}\rangle_- + \left(1 - \langle n_{\vec{k}}\rangle_-\right) \ln\left(1 - \langle n_{\vec{k}}\rangle_-\right)\right].$$

In the zero-temperature limit all occupation numbers are either 0 or 1. In either case the contribution to entropy is zero, and the fermi system at $T = 0$ has zero entropy.

(e) Calculate the variance of the total number of particles $\langle N^2\rangle_c$, and comment on its zero-temperature behavior.

The total number of particles is given by $N = \sum_{\vec{k}} n_{\vec{k}}$. Since the occupation numbers are independent,

$$\langle N^2\rangle_c = \sum_{\vec{k}}\langle n_{\vec{k}}^2\rangle_c = \sum_{\vec{k}}\left(\langle n_{\vec{k}}^2\rangle_- - \langle n_{\vec{k}}\rangle_-^2\right) = \sum_{\vec{k}}\langle n_{\vec{k}}\rangle_-\left(1 - \langle n_{\vec{k}}\rangle_-\right),$$

since $\langle n_{\vec{k}}^2\rangle_- = \langle n_{\vec{k}}\rangle_-$. Again, since at $T = 0$, $\langle n_{\vec{k}}\rangle_- = 0$ or 1, the variance $\langle N^2\rangle_c$ vanishes.

(f) The number fluctuations of a gas are related to its compressibility κ_T, and number density $n = N/V$, by

$$\langle N^2\rangle_c = Nnk_B T\kappa_T.$$

Give a *numerical estimate* of the compressibility of the fermi gas in a metal at $T = 0$ in units of $Å^3\,eV^{-1}$.

To obtain the compressibility from $\langle N^2 \rangle_c = N n k_B T \kappa_T$, we need to examine the behavior at small but finite temperatures. At small but finite T, a small fraction of states around the fermi energy have occupation numbers around 1/2. The number of such states is roughly $N k_B T / \varepsilon_F$, and hence we can estimate the variance as

$$\langle N^2 \rangle_c = N n k_B T \kappa_T \approx \frac{1}{4} \times \frac{N k_B T}{\varepsilon_F}.$$

The compressibility is then approximated as

$$\kappa_T \approx \frac{1}{4 n \varepsilon_F},$$

where $n = N/V$ is the density. For electrons in a typical metal $n \approx 10^{29}\,m^{-3} \approx 0.1\,Å^3$, and $\varepsilon_F \approx 5\,eV \approx 5 \times 10^4\,K$, resulting in

$$\kappa_T \approx 0.5\,Å^3\,eV^{-1}.$$

****

6. *Stoner ferromagnetism*: the conduction electrons in a metal can be treated as a gas of fermions of spin 1/2 (with up/down degeneracy), and density $n = N/V$. The Coulomb repulsion favors wave functions that are anti-symmetric in position coordinates, thus keeping the electrons apart. Because of the full (position *and* spin) anti-symmetry of fermionic wave functions, this interaction may be approximated by an effective spin–spin coupling that favors states with parallel spins. In this simple approximation, the net effect is described by an interaction energy

$$U = \alpha \frac{N_+ N_-}{V},$$

where N_+ and $N_- = N - N_+$ are the numbers of electrons with up and down spins, and V is the volume. (The parameter α is related to the scattering length a by $\alpha = 4 \pi \hbar^2 a / m$.)

(a) The ground state has two fermi seas filled by the spin-up and spin-down electrons. Express the corresponding fermi wavevectors $k_{F\pm}$ in terms of the densities $n_\pm = N_\pm / V$.

In the ground state, all available wavevectors are filled up in a sphere. Using the appropriate density of states, the corresponding radii of $k_{F\pm}$ are calculated as

$$N_\pm = V \int_{k < k_{F\pm}} \frac{d^3 k}{(2\pi)^3} = V \int_0^{k_{F\pm}} \frac{4\pi}{(2\pi)^3} k^2 dk = \frac{V k_{F\pm}^3}{6\pi^2},$$

leading to

$$k_{F\pm} = \left(6\pi^2 n_\pm \right)^{1/3}.$$

(b) Calculate the kinetic energy density of the ground state as a function of the densities $n_\pm$, and fundamental constants.

For each fermi sea, the kinetic energy is given by

$$E_{\mathrm{kin}\pm} = V \int_{k<k_{F\pm}} \frac{\hbar^2 k^2}{2m} \frac{d^3 k}{(2\pi)^3} = V \frac{\hbar^2}{2m} \frac{4\pi}{(2\pi)^3} \frac{k_{F\pm}^5}{5},$$

where m is the mass of the electron. The total kinetic energy density is thus equal to

$$\frac{E_{\mathrm{kin}}}{V} = \frac{\hbar^2}{2m} \frac{1}{10\pi^2} \left(k_{F+}^5 + k_{F-}^5 \right) = \frac{\hbar^2}{2m} \frac{3}{5} \left(6\pi^2 \right)^{2/3} \left(n_+^{5/3} + n_-^{5/3} \right).$$

(c) Assuming small deviations $n_\pm = n/2 \pm \delta$ from the symmetric state, expand the kinetic energy to *fourth order in δ.*

Using the expansion

$$(x+\delta)^{5/3} = x^{5/3} + \frac{5}{3} x^{2/3}\delta + \frac{10}{9} x^{-1/3} \frac{\delta^2}{2} - \frac{10}{27} x^{-4/3} \frac{\delta^3}{6} + \frac{40}{81} x^{-7/3} \frac{\delta^4}{24} + \mathcal{O}(\delta^5),$$

the kinetic energy calculated in the previous part can be written as

$$\frac{E_{\mathrm{kin}}}{V} = \frac{\hbar^2}{2m} \frac{6}{5} \left(6\pi^2 \right)^{2/3} \left[\left(\frac{n}{2} \right)^{5/3} + \frac{5}{9} \left(\frac{n}{2} \right)^{-1/3} \delta^2 + \frac{5}{243} \left(\frac{n}{2} \right)^{-7/3} \delta^4 + \mathcal{O}(\delta^6) \right].$$

(d) Express the spin–spin interaction density U/V in terms of n and δ. Find the critical value of α_c, such that for $\alpha > \alpha_c$ the electron gas can lower its total energy by spontaneously developing a magnetization. (This is known as the *Stoner instability.*)

The interaction energy density is

$$\frac{U}{V} = \alpha n_+ n_- = \alpha \left(\frac{n}{2} + \delta \right) \left(\frac{n}{2} - \delta \right) = \alpha \frac{n^2}{4} - \alpha \delta^2.$$

The total energy density has the form

$$\frac{E}{V} = \frac{E_0 + \alpha n^2/4}{V} + \left[\frac{4}{3} \left(3\pi^2 \right)^{2/3} \frac{\hbar^2}{2m} n^{-1/3} - \alpha \right] \delta^2 + \mathcal{O}\left(\delta^4 \right).$$

When the second-order term in δ is negative, the electron gas has lower energy for finite δ, that is, it acquires a spontaneous magnetization. This occurs for

$$\alpha > \alpha_c = \frac{4}{3} \left(3\pi^2 \right)^{2/3} \frac{\hbar^2}{2m} n^{-1/3}.$$

(e) Explain qualitatively, and sketch the behavior of, the spontaneous magnetization as a function of α.

For $\alpha > \alpha_c$, the optimal value of δ is obtained by minimizing the energy density. Since the coefficient of the fourth-order term is positive, and the optimal δ goes to zero continuously as $\alpha \to \alpha_c$, the minimum energy is obtained for a value of $\delta^2 \propto (\alpha - \alpha_c)$. The magnetization is proportional to δ, and hence grows in the vicinity of α_c as $\sqrt{\alpha - \alpha_c}$, as sketched below.

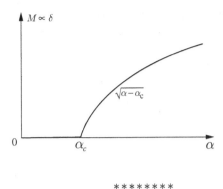

$$\ast\ast\ast\ast\ast\ast\ast\ast$$

7. *Boson magnetism*: consider a gas of non-interacting spin 1 bosons, each subject to a Hamiltonian

$$\mathcal{H}_1(\vec{p}, s_z) = \frac{\vec{p}^{\,2}}{2m} - \mu_0 s_z B,$$

where $\mu_0 = e\hbar/mc$, and s_z takes *three* possible values of $(-1, 0, +1)$. (The orbital effect, $\vec{p} \to \vec{p} - e\vec{A}$, has been ignored.)

(a) In a grand canonical ensemble of chemical potential μ, what are the average occupation numbers $\left\{ \langle n_+(\vec{k}) \rangle, \langle n_0(\vec{k}) \rangle, \langle n_-(\vec{k}) \rangle \right\}$ of one-particle states of wavenumber $\vec{k} = \vec{p}/\hbar$?

   Average occupation numbers of the one-particle states in the grand canonical ensemble of chemical potential μ are given by the Bose–Einstein distribution

$$n_s(\vec{k}) = \frac{1}{e^{\beta[\mathcal{H}(s) - \mu]} - 1} \qquad (\text{for } s = -1, 0, 1)$$

$$= \frac{1}{\exp\left[\beta\left(\frac{p^2}{2m} - \mu_0 sB\right) - \beta\mu\right] - 1}.$$

(b) Calculate the average total numbers $\{N_+, N_0, N_-\}$ of bosons with the three possible values of s_z in terms of the functions $f_m^+(z)$.

   The total numbers of particles with spin s are given by

$$N_s = \sum_{\{\vec{k}\}} n_s(\vec{k}), \quad \Longrightarrow \quad N_s = \frac{V}{(2\pi)^3} \int d^3k \frac{1}{\exp\left[\beta\left(\frac{p^2}{2m} - \mu_0 sB\right) - \beta\mu\right] - 1}.$$

After a change of variables, $k \equiv x^{1/2}\sqrt{2mk_BT}/h$, we get

$$N_s = \frac{V}{\lambda^3} f_{3/2}^+\left(z e^{\beta\mu_0 sB}\right),$$

where

$$f_m^+(z) \equiv \frac{1}{\Gamma(m)} \int_0^\infty \frac{dx\, x^{m-1}}{z^{-1}e^x - 1}, \qquad \lambda \equiv \frac{h}{\sqrt{2\pi mk_BT}}, \qquad z \equiv e^{\beta\mu}.$$

(c) Write down the expression for the magnetization $M(T, \mu) = \mu_0(N_+ - N_-)$, and by expanding the result for small B find the *zero-field susceptibility* $\chi(T, \mu) = \partial M / \partial B|_{B=0}$.

The magnetization is obtained from

$$M(T, \mu) = \mu_0 \left(N_+ - N_-\right)$$
$$= \mu_0 \frac{V}{\lambda^3} \left[f_{3/2}^+ \left(z e^{\beta \mu_0 B}\right) - f_{3/2}^+ \left(z e^{-\beta \mu_0 s B}\right)\right].$$

Expanding the result for small B gives

$$f_{3/2}^+ \left(z e^{\pm \beta \mu_0 B}\right) \approx f_{3/2}^+ \left(z[1 \pm \beta \mu_0 B]\right) \approx f_{3/2}^+(z) \pm z \cdot \beta \mu_0 B \frac{\partial}{\partial z} f_{3/2}^+(z).$$

Using $z \mathrm{d} f_m^+(z) / \mathrm{d}z = f_{m-1}^+(z)$, we obtain

$$M = \mu_0 \frac{V}{\lambda^3} (2 \beta \mu_0 B) \cdot f_{1/2}^+(z) = \frac{2 \mu_0^2}{k_B T} \frac{V}{\lambda^3} \cdot B \cdot f_{1/2}^+(z),$$

and

$$\chi \equiv \left. \frac{\partial M}{\partial B}\right|_{B=0} = \frac{2 \mu_0^2}{k_B T} \frac{V}{\lambda^3} \cdot f_{1/2}^+(z).$$

To find the behavior of $\chi(T, n)$, where $n = N/V$ is the total density, proceed as follows.

(d) For $B = 0$, find the high-temperature expansion for $z(\beta, n) = e^{\beta \mu}$, correct to second order in n. Hence obtain the first correction from quantum statistics to $\chi(T, n)$ at high temperatures.

In the high-temperature limit, z is small. Use the Taylor expansion for $f_m^+(z)$ to write the total density $n(B = 0)$, as

$$n(B = 0) = \left. \frac{N_+ + N_0 + N_-}{V}\right|_{B=0} = \frac{3}{\lambda^3} f_{3/2}^+(z)$$
$$\approx \frac{3}{\lambda^3} \left(z + \frac{z^2}{2^{3/2}} + \frac{z^3}{3^{3/2}} + \cdots\right).$$

Inverting the above equation gives

$$z = \left(\frac{n \lambda^3}{3}\right) - \frac{1}{2^{3/2}} \left(\frac{n \lambda^3}{3}\right)^2 + \cdots.$$

The susceptibility is then calculated as

$$\chi = \frac{2 \mu_0^2}{k_B T} \frac{V}{\lambda^3} \cdot f_{1/2}^+(z),$$
$$\chi/N = \frac{2 \mu_0^2}{k_B T} \frac{1}{n \lambda^3} \left(z + \frac{z^2}{2^{1/2}} + \cdots\right)$$
$$= \frac{2 \mu_0^2}{3 k_B T} \left[1 + \left(-\frac{1}{2^{3/2}} + \frac{1}{2^{1/2}}\right) \left(\frac{n \lambda^3}{3}\right) + O\left(n^2\right)\right].$$

(e) Find the temperature $T_c(n, B = 0)$ of Bose–Einstein condensation. What happens to $\chi(T, n)$ on approaching $T_c(n)$ from the high-temperature side?

Bose–Einstein condensation occurs when $z = 1$, at a density

$$n = \frac{3}{\lambda^3} f^+_{3/2}(1),$$

or a temperature

$$T_c(n) = \frac{h^2}{2\pi m k_B} \left(\frac{n}{3\zeta_{3/2}}\right)^{2/3},$$

where $\zeta_{3/2} \equiv f^+_{3/2}(1) \approx 2.61$. Since $\lim_{z\to 1} f^+_{1/2}(z) = \infty$, the susceptibility $\chi(T, n)$ diverges on approaching $T_c(n)$ from the high-temperature side.

(f) What is the chemical potential μ for $T < T_c(n)$, at a small but finite value of B? Which one-particle state has a macroscopic occupation number?

Since $n_s(\vec{k}, B) = \left[z^{-1} e^{\beta \mathcal{E}_s(\vec{k}, B)} - 1\right]^{-1}$ is a positive number for all $\vec{k}$ and s_z, μ is bounded above by the minimum possible energy, that is,

$$\text{for} \quad T < T_c, \quad \text{and} \quad B \text{ finite}, \quad z e^{\beta \mu_0 B} = 1, \quad \Longrightarrow \quad \mu = -\mu_0 B.$$

Hence the macroscopically occupied one-particle state has $\vec{k} = 0$, and $s_z = +1$.

(g) Using the result in (f), find the spontaneous magnetization,

$$\overline{M}(T, n) = \lim_{B \to 0} M(T, n, B).$$

The contribution of the excited states to the magnetization vanishes as $B \to 0$. Therefore the total magnetization for $T < T_c$ is due to the macroscopic occupation of the $(k = 0, s_z = +1)$ state, and

$$\overline{M}(T, n) = \mu_0 V n_+(k = 0)$$

$$= \mu_0 V \left(n - n_{\text{excited}}\right) = \mu_0 \left(N - \frac{3V}{\lambda^3} \zeta_{3/2}\right).$$

****

8. *Dirac fermions* are non-interacting particles of spin 1/2. The one-particle states come in pairs of positive and negative energies,

$$\mathcal{E}_\pm(\vec{k}) = \pm\sqrt{m^2 c^4 + \hbar^2 k^2 c^2},$$

independent of spin.

(a) For *any* fermionic system of chemical potential μ, show that the probability of finding an occupied state of energy $\mu + \delta$ is the same as that of finding an unoccupied state of energy $\mu - \delta$. (δ is any constant energy.)

According to fermi statistics, the probability of occupation of a state of energy $\mathcal{E}$ is

$$p[n(\mathcal{E})] = \frac{e^{\beta(\mu - \mathcal{E})n}}{1 + e^{\beta(\mu - \mathcal{E})}}, \quad \text{for} \quad n = 0, 1.$$

For a state of energy $\mu + \delta$,

$$p[n(\mu + \delta)] = \frac{e^{\beta \delta n}}{1 + e^{\beta \delta}}, \quad \Longrightarrow \quad p[n(\mu + \delta) = 1] = \frac{e^{\beta \delta}}{1 + e^{\beta \delta}} = \frac{1}{1 + e^{-\beta \delta}}.$$

Similarly, for a state of energy $\mu - \delta$,

$$p[n(\mu - \delta)] = \frac{e^{-\beta \delta n}}{1 + e^{-\beta \delta}}, \quad \Longrightarrow \quad p[n(\mu - \delta) = 0] = \frac{1}{1 + e^{-\beta \delta}} = p[n(\mu + \delta) = 1],$$

that is, the probability of finding an occupied state of energy $\mu + \delta$ is the same as that of finding an unoccupied state of energy $\mu - \delta$.

(b) At zero temperature all negative energy Dirac states are occupied and all positive energy ones are empty, that is, $\mu(T = 0) = 0$. Using the result in (a) find the chemical potential at finite temperature T.

The above result implies that for $\mu = 0$, $\langle n(\mathcal{E}) \rangle + \langle n(-\mathcal{E}) \rangle$ is unchanged for any temperature; any particle leaving an occupied negative energy state goes to the corresponding unoccupied positive energy state. Adding up all such energies, we conclude that the total particle number is unchanged if μ stays at zero. Thus, the particle–hole symmetry enforces $\mu(T) = 0$.

(c) Show that the mean excitation energy of this system at finite temperature satisfies

$$E(T) - E(0) = 4V \int \frac{d^3 \vec{k}}{(2\pi)^3} \frac{\mathcal{E}_+(\vec{k})}{\exp\left(\beta \mathcal{E}_+(\vec{k})\right) + 1}.$$

Using the label $+(-)$ for the positive (energy) states, the excitation energy is calculated as

$$E(T) - E(0) = \sum_{k, s_z} \left[\langle n_+(k) \rangle \, \mathcal{E}_+(k) + (1 - \langle n_-(k) \rangle) \, \mathcal{E}_-(k) \right]$$

$$= 2 \sum_k 2 \langle n_+(k) \rangle \, \mathcal{E}_+(k) = 4V \int \frac{d^3 \vec{k}}{(2\pi)^3} \frac{\mathcal{E}_+(\vec{k})}{\exp\left(\beta \mathcal{E}_+(\vec{k})\right) + 1}.$$

(d) Evaluate the integral in part (c) for *massless Dirac particles* (i.e., for $m = 0$).

For $m = 0$, $\mathcal{E}_+(k) = \hbar c |k|$, and

$$E(T) - E(0) = 4V \int_0^\infty \frac{4\pi k^2 dk}{8\pi^3} \frac{\hbar c k}{e^{\beta \hbar c k} + 1} \qquad (\text{set } \beta \hbar c k = x)$$

$$= \frac{2V}{\pi^2} k_B T \left(\frac{k_B T}{\hbar c} \right)^3 \int_0^\infty dx \frac{x^3}{e^x + 1}$$

$$= \frac{7\pi^2}{60} V k_B T \left(\frac{k_B T}{\hbar c} \right)^3.$$

For the final expression, we have noted that the needed integral is $3! f_4^-(1)$, and used the value of $f_4^-(1) = 7\pi^4/720$.

(e) Calculate the heat capacity, C_V, of such massless Dirac particles.

The heat capacity can now be evaluated as

$$C_V = \left. \frac{\partial E}{\partial T} \right|_V = \frac{7\pi^2}{15} V k_B \left(\frac{k_B T}{\hbar c} \right)^3.$$

(f) Describe the qualitative dependence of the heat capacity at low temperature if the particles are massive.

When $m \neq 0$, there is an energy gap between occupied and empty states, and we thus expect an exponentially activated energy, and hence heat capacity. For the low energy excitations,

$$\mathcal{E}_+(k) \approx mc^2 + \frac{\hbar^2 k^2}{2m} + \cdots ,$$

and thus

$$E(T) - E(0) \approx \frac{2V}{\pi^2} mc^2 e^{-\beta mc^2} \frac{4\pi\sqrt{\pi}}{\lambda^3} \int_0^\infty dx\, x^2 e^{-x}$$

$$= \frac{48}{\sqrt{\pi}} \frac{V}{\lambda^3} mc^2 e^{-\beta mc^2} .$$

The corresponding heat capacity, to leading order thus behaves as

$$C(T) \propto k_B \frac{V}{\lambda^3} \left(\beta mc^2\right)^2 e^{-\beta mc^2} .$$

$$* * * * * * * *$$

Index